Dead Meat

Elisabeth Abergel

Dead Meat

Competing Vitalities, Cultivated Meat Imaginaries and Anthropocene Diets

palgrave
macmillan

Elisabeth Abergel
University of Quebec at Montreal
Montreal, Canada

ISBN 978-981-97-9048-7 ISBN 978-981-97-9049-4 (eBook)
https://doi.org/10.1007/978-981-97-9049-4

Cover credit: AlonzoDesign

This Palgrave Macmillan imprint is published by the registered company Springer Nature Singapore Pte Ltd.
The registered company address is: 152 Beach Road, #21-01/04 Gateway East, Singapore 189721, Singapore

Acknowledgments

This book was made possible thanks to a fellowship at the Rachel Carson Center in Munich. While I was there, I benefitted from the generous input of many friends and colleagues, too many to name. Among them Jenia Murkherjee and Anna Pilz whose discipline and dedication to research were great sources of inspiration.

I would like to thank Céline Lafontaine for the many conversations and exchanges about writing and research and for her enthusiastic reception of what may have sometimes seemed like improbable ideas. To my students, new and old: Sophie, Claire, Julie, Rivellie, Nicolas, and many others who unknowingly participated in this work.

Thanks to my friends and accomplices at C-BAN for convincing me to include the Biosphere II story in the book: Lucy, Cathy, Devlin, Brewster, etc.

I am incredibly grateful to my family, Marcelle, Joseph, Cécile, for their support. To my mother Julie Abergel and my dearly departed father Jacques. To Albert and Dominique for providing me with a workspace with a view of a castle so I could write. To Joseph, Marie, Luz, Michèle, and all those who have generously welcomed me while I was doing research for this book. To Jacinthe and Jamie for their political insights. To Maggie, Oona, Samson, Niki, and Sparkie for giving me hope and bringing me joy.

This book is dedicated to Stephen Lisinski, who not only endured endless discussions about meat but also graciously prepared countless meals without it. His unwavering support and creativity have been a source of inspiration throughout this journey.

Contents

1

Introduction

Cell-Based Meat and Meat Imaginaries

> "A human being is primarily a bag for putting food into; the other functions and faculties may be more godlike, but in point of time they come afterwards. A man dies and is buried, and all his words and actions are forgotten, but the food he has eaten lives after him in the sound or rotten bones of his children. I think it could be plausibly argued that changes of diet are more important than changes of dynasty or even of religion…Yet it is curious how seldom the all-importance of food is recognized. You see statues everywhere to politicians, poets, bishops, but none to cooks or bacon-curers or market gardeners."
>
> —George Orwell, The Road to Wigan Pier

Throughout history we have adapted our food systems and transformed our environments in order to meet specific challenges (wars, famines, natural disasters, disease, etc.) and created the material conditions to develop and expand the edible resources around us. The environmental historian Donald Worster says that 'our most vital, constant and concrete' connection to nature has been growing, procuring, and eating food (Worster, 1990). The history of food and agriculture has been marked by our most intimate entanglements with nature on the one hand, as eating

E. Abergel, *Dead Meat*, https://doi.org/10.1007/978-981-97-9049-4_1

literally metabolizes organic matter, but also, on the other hand, by a desire to transform the life conditions and the beings that we consume, to keep us alive. The advances made in the life sciences in the last 50 years have led us to this point, where the creation and production of new forms of vitality, in the laboratory, have become possible for biomedical and agricultural research and now, we are told, for the purpose of producing lab-grown edible animal tissues as a replacement for traditional meat. Throughout the various chapters, I aim to explore the fundamental ideas that compelled me to delve into the world of cellular agriculture (CA) and its implications. At the heart of this inquiry is the examination of how these newly engineered forms of vitality challenge and redefine the processes by which food is understood as *purely biological.* This ontological shift has profound implications, not only for the future of food production but also for the relationship between consumers and what they consume. By analyzing these transformations, I seek to understand how these technological innovations are reshaping our conceptions of life and nourishment, ultimately altering the very fabric of what it means to eat today.

More recently, biotechnological innovations have had a major impact on agriculture and the consolidation of the agro-industrial complex on a global scale. Humans have always manipulated life and living organisms, but the techniques of recombinant DNA and genetic engineering have rendered possible the production of transgenic plants and animals expressing agronomic traits, moving and splicing DNA fragments of organisms from unrelated species and varieties. These techniques have made permeable interspecific boundaries. Whether it is herbicide-tolerant soya plants or the OncoMouse™, a transgenic mouse used in biomedical experiments to study cancer, these organisms represent the technological production of whole organisms, in the service of the life sciences all the while embodying the promise of biotechnology (Landecker, 2007; Myelnikov, 2020). As Landecker suggests in her work, these plants and animals have become technologized. Genetic engineering also offers the possibility to manufacture molecules and proteins of commercial importance, using fermentation techniques, by splicing genes and pasting them into carrier microorganisms and yeasts that are then cultured in industrial quantities in bioreactors. As we delve into the world of cellular meat,[1] it is important to consider it not as a point of historical rupture in how meat is produced but as the continuation of the logic of industrial agriculture.

While traditional meat is derived from living animals, synthetic meat, also known as in vitro meat, cell-based meat, cellular meat, cultivated meat, lab-grown meat, clean meat, or cultured meat, is produced from animal cell cultures in a laboratory (Datar & Betti, 2010; Post, 2012; Post & Hocquette, 2017). Cellular agriculture (CA), a term describing a series of scientific practices that use animal cells as the primary units of agricultural production, can be considered a new stage in our understanding of living materials. This approach aims to eliminate the whole animal from the production process, to replicate foods like meat, milk, or eggs, and other commodities such as leather, silk, or rhino horns. Inspired by tissue engineering techniques from regenerative medicine and fermentation techniques from biotechnology, cellular agriculture combines various cell culture processes (animal cells and tissues) and acellular culture processes (proteins, fats, flavors, and other molecules fermented in bacteria or yeast) (Cellular Agriculture Society, 2023). CA describes the different technical processes necessary for producing complex biological products, both living and non-living. More specifically, it involves a series of bioassembly techniques (Groll et al., 2016) and the construction of three-dimensional edible tissue.

CM consists of animal tissues cultivated from stem cells harvested from living animals, then multiplied and differentiated in the laboratory. These cells are used to develop food products such as meat, but also cellular cow's milk and eggs and human breast milk. By using cellular materials, researchers aim to replicate the organoleptic properties of meat, without the need for traditional livestock farming. The animals most considered for industrial meat production include cattle, poultry, pigs,

[1] Throughout the book, I will use the terms Cell-Based Meat, Cellular Meat, Cultivated Meat, Cultured Meat, Lab-Grown Meat, Synthetic Meat, and In Vitro Meat interchangeably, along with the abbreviation CM. Terminology varies among industry players; although many adopted "cultivated meat" in 2022, international organizations like the FAO and WHO use "cell-based food." The term "meat" is sometimes left out due to some countries' refusal to classify cultivated products as meat and because it excludes fish products.

and fish (including mollusks), which represent the most consumed species worldwide. However, other varieties of animals may enter this production process, in response to market demands and consumer choices. The meat produced primarily comprises muscle and fat and, in the case of meat produced by Aleph Farms, may include connective tissue and blood vessels (Niamh, 2018).

A significant turning point in popularizing and expanding the promises and imaginaries surrounding in vitro meat production occurred during a 2013 press conference in London, where the first lab-grown hamburger was unveiled. Professor Mark Post, an angiogenesis specialist initially working on the regeneration of cardiac cells and tissues, demonstrated with this first hamburger that it was possible to produce and replicate animal proteins, without resorting to livestock farming. As of 2024, there are over a hundred companies developing cultivated meat products, either meat or seafood as well as dairy or human breast milk, using the techniques of cellular agriculture. If one accounts for the companies producing blended products, that is, cultivated meat with plant-based protein, the Good Food Institute (GFI), a non-profit charitable organization dedicated to the development and commercialization of cellular agriculture, has a database which shows 171 entries, while the site outlines over 1500 companies involved in the alternative protein space, which includes plant-based meat analogues, plant-based dairy and eggs, precision fermentation, as well as the development of key ingredients for the industry (Good Food Institute, 2024). Companies such as Good Meat, Eat Just, Shiok Meats/UMAMI Bioworks, Upside Foods, and Cell Farm have been authorized to sell their products in several countries.

This book is about the forces shaping our food future, mostly the meaning, the value, and the representation we attribute to what we understand as food and, more specifically, what meat imaginaries reveal about the powers of technoscience,[2] in this particular moment in our planetary history. Meat imaginaries, one of the leading concepts developed throughout the analysis, refer to the speculative claims made about the engineering and industrial-scale production of meat using cultured

[2] The term technoscience reveals a complexity of meanings and actors, it is "a collection of institutions, normative structures, practices, ideologies, and vocations" (Sunder Rajan, 2012).

stem cells in order to create edible tissues in vitro as a replacement for conventional animal products as part of an Anthropocene diet. This diet emphasizes sustainability, reduced environmental footprint or hoofprint, and the ethical treatment of animals, aiming to minimize harm to the planet, while meeting nutritional needs. Central to this is the inclusion of cell-based meat—cultured in labs without the need for traditional livestock farming—alongside plant-based foods and sustainable agricultural practices.

Imagine eating "cultivated meat made from animal cells grown in bioreactors instead of living breathing animals" (Watson, 2022). Or as one writer puts it, imagine purchasing packaged meat with a label that reads: "no animals were harmed in the making of this product," because it was produced by growing skeletal muscle tissue outside of the organism (Mattick, 2018).

Now imagine the Ouroboros Steak concept, developed by a group of scientists and designers in 2020. This steak comes in a kit designed to literally 'grow-your-own-steak,' allowing users to grow a piece of meat from their own cells and blood. The term Ouroboros, derived from the ancient Egyptian symbol of a snake eating its own tail, reflects the cyclical nature of the concept. First showcased at the Philadelphia Museum of Art, these bite-sized pieces of meat were meant as a critique of the cell-based meat industry's ethics. Andrew Pelling, the scientist behind the Steak, used human serum from expired blood donations to grow the cells, instead of the animal serum typically used in tissue culture, leading him to assert that no animals were harmed in making the product. This design experiment aimed to "expose some of [the industry's] underlying constraints in order to see beyond the hype," considering that the industry still relies on Fetal Bovine Serum (FBS), collected from slaughterhouses, despite claims from some cell-based meat companies of having developed animal-free cell growth medium. The experiment was not about 'eating ourselves' to solve the meat problem, nor was it seen as an endorsement of cannibalism. Instead, the scientists-designer team is posing the questions, "what would be the sacrifice we need to make in order to be able to keep consuming meat at the pace that we are? In the future, who will be able to afford animal meat and who may have no other option than culturing meat from themselves?" (Hahn, 2020). The Ouroboros

Steak concept, with its eerie proposition of growing meat from our own cells, is more than just a provocative design experiment—it's a mirror held up to humanity's deepest anxieties and ethical quandaries. As we imagine a future where traditional meat is a luxury only the wealthy can afford, this concept forces us to grapple with unsettling questions: What sacrifices are we willing to make to sustain our carnivorous habits, even if it means blurring the lines between human and animal, nature and culture? The idea of "eating ourselves" isn't just a commentary on the sustainability of our diets; it's a stark reflection of our adaptability, our resilience, and the moral boundaries we might cross when faced with existential challenges. Yet, beneath this innovation lies a sobering truth about social inequality and the potential future, where the divide between those who can maintain their way of life and those who cannot grows ever wider. The Ouroboros Steak compels us to confront who we are as a species and how far we will go to preserve a lifestyle that may no longer be viable in the world we are creating.

For those contemplating the production of 'meat without animals,' this entails reimagining meat as a "future concept that must be liberated from its naturalized operative meaning," according to artist Adam Wolpa (Wolpa, 2016). This is where growing edible animal tissues becomes ontologically blurry and, to a certain extent, morally suspect. Such an exercise implies complex shifts that may dramatically alter our relationship to food, animals, and our own bodies (Van Der Weele & Driessen, 2013). Meat is one of those categories of foods that is woven into the cultural and social fabric. The rearing of animals used for meat production, the various species that count as food, and their status in the animal world are the stuff of civilizations. The language and categories we invent to classify animals and animal species and their relationship with humans are indicative of particular worldviews. Carnism creates a specific social order, by instituting strict rules about how animals are to be sacrificed, killed, and consumed. Which parts of the animal are edible and culturally and socially acceptable shed light on the nature of religious experience as well as the ethics of ritualistic killing. Anthropologists have studied how these beliefs organize specific rules of social and cultural conduct, while also prescribing dietary laws that provide basic rules of hygiene, such as the Kosher and Halal custom of forbidding pork.

Another religious restriction entails rules concerning the consumption of fresh meat: dead animals or animals that have recently died are strictly forbidden since dead animals may have carried diseases and old flesh might be rotten. Whether killed by predators, the victim of roadkill or old age, the manner in which the animal dies also determines its edibility. Hence meat-eating provides a unique way of understanding the often-blurry boundaries between religion and ordinary life. As Mary Douglas writes "there can be a marvelous correspondence between the avoidance of contagious disease and ritual avoidance" (Douglas, 2001, p. 30). Whether sacred or secular, the topic of meat confronts us with our own animality and bodily existence, as well as our desire for self-preservation. I therefore question the possibility of liberating meat from its corporeal origins.

Is cultured meat (CM) truly a sustainable and ethical alternative to traditional animal meat? It's important to note that CM isn't aimed at vegans or vegetarians but rather at meat-eaters, who are often seen as contributing to the problem. This raises important questions about whether CM is simply enabling the continuation of carnivory, which many consider problematic in itself. As one visitor to the Ouroboros project remarked, "Finally, we can have our steak and eat it too," highlighting the irony and complexity of this supposed solution.

Cultivated meat signals "a watershed moment in the history of food," according to Dr Uma Valeti, whose training is in cardiovascular medicine and who is the founder and CEO of UPSIDE Foods, one of the leading start-ups in this field (Watson, 2022). It's also been described as a game changer by others. As a demonstration of human ingenuity and technological possibilities, cultured meat proposes to resolve many of today's agricultural, ethical and environmental challenges, by eliminating whole animals out of the food production process, relying instead on the engineering, using tissue culture, of animal cells and the fermentation of their cellular components.

The topic of cell-based meat, as a potential solution to the global food crisis, has garnered significant attention in academic literature as well as in popular media. Some proponents of cell-based meat frame it as a potential solution to the food crisis through the lens of Malthusian theory, which suggests that population growth will eventually outstrip the

Earth's resources. Their main argument rests on the idea that industrial agriculture's diminishing returns, coupled with demographic projections and global climate upheavals, necessitate alternative and visionary approaches. Others have posited that cell-based meat represents a New Green Revolution or "ever-green revolution" comparable to the Green Revolution of the mid-twentieth century in terms of its magnitude (Coller, 2023). In addition, cellular meat has been hailed as an advance in what is known as the Fourth Industrial Revolution, a vast array of initiatives and discourses about technology-driven change which is merging "the physical, digital and biological worlds in ways that create both huge promise and potential peril" (WEF, 2024). In 2019, the World Economic Forum stated that "Producing meat in the laboratory without the involvement of living animals is a huge technical feat made possible by the Fourth Industrial Revolution" (World Economic Forum, 2019). From a philosophical perspective, assertions about cell meat being revolutionary highlight a fundamental shift in humanity's relationship with nature and food production.

As we navigate this new epoch, it becomes evident that we are confronted with difficult choices regarding how, in light of climate instability, biodiversity losses, and weather extremes, we envision sustainability and critically engage with food production and consumption, as activities that involve human and non-human relationships. The environmental and agricultural crises are consubstantial and cannot be viewed separately, despite technological promises about future foods to reduce our dependency on nature and animals for nourishment.

If meat is the problem, why is CM the solution? It is my aim to explore some facets and implications of such a question, using synthetic meat or cultivated meat as a way to get us out of environmental, health, economic, and ethical predicaments. Of course, this book is not an answer to these questions. It is a modest attempt to think through them, cell meat being the proverbial tip of the iceberg when it comes to achieving the promises and imaginaries of technoscience.

In the title of this book, the term "dead meat" might sound redundant because meat is what comes from a dead animal, but the industrial technologization of animals throughout history means that on some level,

> we confront the vegetal frontiers of meat as scientific engineering of animals renders them less sentient, less animal, and more like vegetables. (Subramaniam, 2021, p. 255)

The quotation refers to transgenic organisms that are engineered with genes that blur species boundaries, like animal genes in plants or plant genes in animals. However, the blurring between animals and vegetables is clearly expressed in the cellular construction of meat analogues. Animal-based and plant-based fake meats are now blended and sold as hybrid meats. Similarly, vegetable materials are often used as scaffolds or supports that allow animal cells to organize and form into tissues.

The notion of dead meat describes cultured edible tissues becoming 'meat' apart from live animals as sentient beings, but also the fact that as our planet reaches its biophysical limits, the survival of our species is at stake. The slang expression 'dead meat' conveys the idea that we are doomed or at least that we are in serious trouble or worse heading for a lifeless future. The expression 'competing vitalities' entails the idea that new possibilities, offered by the life sciences, encouraged by technoscientific advances and convergence, produce new life forms created in the laboratory, and which are now competing with existing beings in terms of their survival but also their place in the world, particularly in what has been called the bioeconomy. As these new forms of vitality, disembodied from their original source being, proliferate, they not only present us with ontological questions as to their true "nature" but they also compete in the marketplace by extracting and producing value from biological resources. This is an example of what Melinda Cooper calls in vivo economies, describing how "life science research became closely attuned to the dynamics of value-creation in financial markets, as large pharmaceutical firms and small-scale biotech start-ups adapted to the demands of institutional investors" (Cooper, 2015; Waldby & Cooper, 2014). The production of "meat" without animals via the in vitro culture of stem cells into edible tissues takes on a particular significance and validation through the concept and narratives of the Anthropocene. It plays into the idea of Anthropocene diets, which is implied by CM as a central component to a sustainable and secure food future, as exemplified by the EAT-Lancet report about sustainable diets. The report produced by the EAT

Foundation, a science-based global platform dedicated to food system transformation, and the medical journal *The Lancet* was published in 2019. Entitled *Food in the Anthropocene: The EAT-Lancet Commission on Healthy Diets from Sustainable Food Systems*, the report introduces the idea of a Great Food Transformation designed to promote human health and environmental sustainability. More importantly, noting that the Anthropocene is itself a narrative, it is interesting to situate CM within those stories about planetary pasts and futures.

I think it's important for me to state my position regarding the promise of CM. I do not support the industrial exploitation of farm animals or the suffering resulting from the inhumane conditions imposed upon them. My critique is not directed toward advocating for the status quo in industrial animal farming. Instead, I am concerned with the idea of focusing on fake meat as a way to solve the complex meat-environment-ethics problem. Although a way out of this conundrum is to stop eating animal meat altogether, I am not against carnivory. I subscribe to Val Plumwood's point of view, advocating for an ethic that doesn't completely avoid the killing and eating of animals but strives to limit the unnecessary suffering associated with meat consumption. She sums it up in the following statement:

> Meat is the result of a reductionist-instrumentalist framework, but the concept of animal food allows us to resist the reductions and denials of meat by honoring the edible life form as much more than food, and certainly much more than meat. (quoted in Pilgrim, 2013, p. 115)

My critique involves the creation of new forms of technologized life, what I termed competing vitalities, as a way to dissociate humans from the complexity of the actions involved in doing the moral/ethical groundwork that our circumstances necessitate. As such, my views concern the idea of meat and our need to see it as both problem and solution. My interest does not lie in cultivated meat itself, but rather in its utopian potential—the capacity to recreate material matter as a diversion from the social and moral dimensions of our current predicament (Douglas, 2001). This sociological phenomenon offers a lens through which we can examine our epoch, highlighting the inherent contradictions it encompasses.

Biosphere II and CM: Replicating Life in a Closed System

Some notes on the impetus for this work, my interest in cellular meat and where I speak from: moving from the natural sciences to the social sciences, I became interested in cell-based meat (CM) as it combines my background in biotechnology with the study of the social and historical transformation of agriculture, food systems, including their ecological impacts. This reflects my sustained interest in examining how ideology and utopia, as imaginaries, interact with technological practices and discourses, and how these dynamics, in turn, affect and are affected by material and social conditions.

I started my journey working in a bustling laboratory, fresh with a master's degree in plant molecular genetics working for a biotech start-up and moving on a few years later, to a large pharmaceutical vaccine company. I started working on agricultural biotechnology applications and as I changed laboratories and companies, the focus of my work varied as did the living organisms that were my research subjects. I first found myself in a small university laboratory setting, my days consumed by plant molecular genetics and the intricate manipulations of plant DNA, in pursuit of creating genetically modified organisms (GMOs) with enhanced traits. At the time, I was working on Sorghum, a plant I had never seen nor had any experience with. My project involved characterizing and extracting genetic material from *Sorghum bicolor* that could be used to *improve* or as they say to genetically *enhance* other plants, to make them more agronomically efficient and commercially profitable. Sorghum is an ancient grain and tropical plant, grown in sub-Saharan Africa and other parts of the world. It is a multipurpose plant that is extremely resilient, as it naturally grows and tolerates poor soils and arid conditions. It is a magical plant that can do so many things and have so many uses, such as food, fiber, energy, etc., now being hailed as a climate resilient plant. At one point early in my scientific career, I attended a conference in Arizona. One of the side activities was a visit to Biosphere II in the middle of the Arizona desert and a chance to meet and talk with biospherians, through a special glass enclosure, as they were still conducting their work inside the immense greenhouse designed to recreate several ecosystems and biomes. It was there, as we were dropped off on the site that housed the Biosphere project, that I saw sorghum growing in a field for the first time. I had only ever worked with its DNA and RNA, extracted from young shoots that a technician grew in our university greenhouse. I collected the weeks-old, grass-like seedlings and proceeded to isolate their genetic material. During that time, a friend of mine, who had no scientific training, came to the lab and commented on the fact that it didn't seem to matter what organisms my colleagues and I worked on, there were no signs of their physical presence. Only small plastic test-tubes filled with liquid that contained DNA, at various processing stages, or stacks of petri dishes alongside the various pieces of plasticware and equipment on our benches. I recall that she said that what struck her about molecular biologists is that we could be working on elephants or ants, there were no physical signs of the life form we were handling. She also commented on the fact that we looked like we were doing kitchen work, as we followed experimental protocols and adapted instructions through our different experiments and noted everything in our lab books. After working in a small lab, I then worked for several biotech start-ups, characterizing genes for pharmaceutical research, working on a variety of living organisms, depending on the research project. At that time, I hadn't realized that I would be switching from molecular genetics to the social sciences and devoting much of my academic career to researching food and agriculture. My friend's remarks resonated with me.

The whole point of Biosphere II was to recreate the Earth's climates in a self-enclosed space that could generate its own oxygen, power, and food. The intent was to recreate a series of ecosystems found on Earth, in order to understand our planet better. Several key ecosystems—an ocean, a tropical forest, a desert, etc.—were artificially constructed and redesigned and populated with what were considered key species of plants, animals, insects, birds, and fish, in the hopes that, at minimum, ecological cycles would be recreated and provide 'natural' conditions for survival and producing food. The vast structure, a gigantic greenhouse the size of several football fields, provided enough surface area for a waterfall and a small ocean-like environment, a mangrove, a desert, as well as other biomes for growing food and raising animals. Designed like a space mission, the eight biospherians wore jumpsuits with the biosphere logo emblazoned on their chest. We were told that the results of this human experiment might enable long-haul space travel and colonization because it was becoming obvious to the creators and backers of the project that the Earth might one day become uninhabitable.

By the time we got to the site in the Arizona desert, the biospherians were nearing the end of their two-year experiment within the structure. They were running out of food, had to deal with a cockroach infestation and an out-of-control pig population that threatened their food supply, and oxygen had to be pumped into the vast structure so they could breathe and keep the whole thing alive. As we asked them questions through the glass structure, I got the sense that things were not running very smoothly. At that point, I started thinking about them living inside this large bubble for such a long time and as I read about their struggles at the end of their mission, I realized that this vision of the future of Earth was a complete fiction, a product of imaginary thinking about the powers of science and technology and childish science fiction accounts of space colonization. This was far from a wholistic Earth-centered, sustainable, nature-affirming, vision of the here and now.

As my career in the biotech industry evolved, I continued my research in industrial settings (biotech start-ups and in the end a large multinational pharmaceutical company), and my research was driven by results, the ability to produce the desired outcome without even trying to understand the organisms we were manipulating, altering, and/or extracting genetic material from. I had read and admired Evelyn Fox Keller's excellent book *A Feeling for the Organism*, about the life and work of the Nobel Prize winner Barbara McClintock (1902–1992), who pioneered the mechanisms surrounding transposable genes in Maize plants. The title of the book and the way she claimed to know the plants she worked with were a contrast to the research I was doing. There was not enough time, and my peers didn't think that there was any value in trying to understand the stories of the plants, animals, bacteria, or fungi we handled every day.

One day, while working on the genetic makeup of a soil bacterium that was isolated because it contained some potentially useful properties for altering antibiotic resistance, it struck me that biotech was exactly

Biosphere II. Behind this idea of making the world better, it carried with it all kinds of ideologies and premises about social organization and control over nature and life and about enclosing life and living systems into a kind of technological vision of the future. It was the material realization of the promise and hope built on a future designed by science and technology. Biosphere II failed. Yet here we are today with new technological tools and the full engagement of governments and industry to fulfill and expand this failed vision.

So, I left the lab behind, and all my experiments and my experimental subjects, and embarked on a new phase of my journey, enrolling in a PhD program in environmental studies that didn't confine me to test-tubes, pipettes, gels, and DNA sequences, but opened the doors to exploring policy issues and the regulatory landscapes surrounding GMOs in Canada. It brought me outside of the laboratory into the places where the organisms I researched lived and thrived. My work transformed from decoding genetic sequences to deciphering the intricate web of regulations designed to manage our environment and public health. Using my scientific background enabled me to study the details of science-based policies and public discourse about technology—in this case agricultural biotechnologies.

Outside the confines of the lab, in the farmers' fields, in the broader context of society, questions were being raised about the very work I was immersed in. The dialogue had shifted from genes and their potential to revolutionize our understanding of life and living systems and, more broadly, society, to genuine concerns—such as the environmental and public health safety of GMOs and our instrumental view of living systems. And, most of all, the lack of public information about what was happening. As I delved into the policy realm, I felt it was necessary to communicate scientific knowledge to a broader audience. The transition from lab coats to advocacy, from pipettes to policy briefs, wasn't seamless. However, navigating this uncharted territory introduced me to social sciences, science and technology studies, and environmental thought and activism.

Meeting with farmers, environmentalists, and policymakers, I witnessed diverse perspectives converging and diverging on the development of genetic technologies, and on the power for people to say 'NO' when these technologies create more problems than they pretend to fix, or when industrial monopolies threaten to undermine our capacity to produce food or feed ourselves, or to search for sustainable and socially just solutions to the difficulties we are facing. My scientific lens widened to include the human and social dimension, prompting me to advocate for serious critical approaches to so-called technological progress, but also for an inclusive dialogue that acknowledged societal values and concerns.

The journey from the controlled environment of the lab to the unpredictable landscape of activism and social sciences was challenging, but it brought to light the ways in which technologies starting with GMOs have evolved and expanded in profound ways. My work today focuses on cellular agriculture, and more specifically, cellular meat, as a strategy to resolve a series of issues like climate change, animal cruelty, and health issues among others. My premise is that this cell meat, like Biosphere II, lives in the imaginary world of the technological fix. Just like Biosphere II, it started with a dream of wanting to 'repair' the planet and provide an exit strategy for humanity. Early on the impetus for growing tissues outside animal bodies came from a project funded by NASA with the intent to produce food for long-haul space travel and with the goal of space colonization. More concretely, the cellular meat project is an interesting portal through which to examine the past, present, and future in these ecologically challenging times.

What cellular agriculture represents is an opportunity for technological convergence and the promise of a better world, through hypothetical, better, human-produced meat. It is a perfect case-study to understand how sociotechnical imaginaries, possible futures, and promissory economies get articulated. Ultimately, my concern is not so much that this industry may or may not exist or whether it fulfills its promises, but my interest lies in what cellular meat says about this moment in our planetary history as some materialization of modernist thought. The underlying message it carries represents a kind of escape from nature, and possibly from ourselves, forging new life in the laboratory as a form of storytelling about the Anthropocene.

Today, as I navigate the space between activism and academia, advocacy and academic research, always looking to make sense of these profound developments, in the role played by agriculture in warning us about biomedical and more ominous genetic interventions, I realize it all starts and ends with food, doesn't it? Remembering my research work in the lab, I never imagined that the animal cell cultures and the tissues that I worked on would one day become food.

Thinking Through/with Meat

The structure of the book follows a set of chapters that involve stories and explanations about the modern food system in connection with progress in science and technology and the implications of these developments on how we deal with the environmental crisis. The focus is on meat but as it turns out, this concept is not as straightforward as it seems. At a structural level, this means bringing together diverse sets of literature about industrial agriculture, food tech, animals, the making of edible tissues in the lab, diets, and eating and how these link with the environmental crisis, named as the Anthropocene. My analysis aims to connect these related domains systemically, not in linear causation.

Throughout the book, I refer to many theorists and build upon a vast literature in the social sciences, encountered over the course of my academic career. I also refer to a diversity of authors who have inspired me to think with and through the various questions I raise in this book. Much of my analysis is drawn from the thinkers of the Frankfurt School, Adorno and later Marcuse among them, who provide a complex critique of modernity and capitalism. These scholars sought to establish a critical theory of society, mostly through their analysis of the dialectic tension between reason and freedom and their critique of the Enlightenment. Much of my research draws upon a diverse body of literature, encompassing Marxian frameworks within rural sociology, as well as sociological, feminist, and cultural perspectives from food studies. In this work, I have attempted to create a dialogue between various scholars in the sociology of science and technology (STS), environmental humanities (EH), eco-feminist literature, and authors working on the political economy and political ecology of agriculture and food systems. Because of the technical nature of CM, I also refer to the scientific literature, in order to highlight the material aspects of what cell meat represents and what it does or doesn't do. Crossing disciplinary boundaries is fraught with danger, and my attempt at engaging various types of knowledges to think through this sociological phenomenon does not follow a systematic path and may leave open some questions or deeper philosophical issues. Additionally, many authors I cite would probably not agree with each other, some of the premises of their work, their disciplinary frames and political ideas may not always be compatible. I hope to work in the interstices and the liminal spaces that their collective works and visions offer. Environmental thought invites us to think wholistically. The complexity of the problems we are facing today demands that we move away from compartmentalized approaches.

Meat has been and continues to be the subject of various debates couched in what Chatterjee and Subramaniam call a "diverse set of politics," suggesting that carnivory cannot be understood outside of the historical, political, economic, cultural, geographical contexts that produce it (Chatterjee & Subramaniam, 2021). Their approach provides a framework for understanding the transient nature and complex meanings of 'meat.' It also provides us with an interesting way to study the impacts of cellular meat, because rather than defining what meat is, they view it as "politics, a site of transnational flows, colonial circuits and varied media significations of gender, race, and class" (p. 2) that include human and non-human entanglements, as well as a source of contrasting ideological forces. Moreover, meat is a constantly evolving and shifting category, as industrialized meat is not a single substance, it is highly technologized and can be animal but also vegetal, microbial, or fungal or a mix of these.

Chapter 2 provides an analysis of the connection between animal agriculture and the Anthropocene concept. Today, most of the meat we consume, unless it is bought directly from the farm, is already divorced from its animal. Meat is the commodity form, it is a standalone product that exists far from the animal bodies, the people, the sites and chains of production that made it consumable (Hannan, 2020). This chapter also introduces readers to cellular agriculture or cell ag (CA) and its focus on the production of alt-protein, whether cell-based animal or plant-based meat. The types of cellular products are surveyed to provide a context to the notion of cellular agriculture and the idea of food tech. It also provides some context into how scholars have looked at the relationship between the industrialization of agriculture and the environmental crisis, mostly from a political economy perspective.

As we navigate this new epoch, it becomes evident that we are confronted with difficult choices regarding how, in light of climate instability, biodiversity losses and weather extremes, we envision sustainability and critically engage with food production and consumption, as activities that involve human and non-human relationships. The environmental and agricultural crises are consubstantial and cannot be viewed separately, despite technological promises about future foods to reduce our dependency on nature and animals for nourishment. It is my impression that calculations of meat's impact on the environment tend to be based on an aggregate global meat industry, leaving out the farms and ranching operations that raise animals on grass and practice regenerative land management (Niman, 2021). Granted, the footprint or hoofprint of the "global meat complex" (GMC) is considerable. The Institute for Agricultural Trade and Policy (IATP) defines the GMC as a concentrated—vertically and horizontally integrated—"web of transnational corporations that controls the inputs, production and processing of mass quantities of food animals" (Sharma, 2018). However, despite the enormous power of the largest meat-producing companies like JBS, Tyson Foods, Cargill and Smithfield, in 2018, only a small portion of the meat produced in the world was actually traded, since much of it is grown by small producers and tends to supply regional markets (Ibid). Besides, these same companies are also investing in the alternative protein business, including in vitro meat or cultivated animal meat.

Chapters 3 and 4 follow a logical sequence. Chapter 3 provides some basic technical information about how CM is produced and some of the technical challenges the industry faces. This chapter gives readers a sense of the materiality of making CM, what it entails, and how various technologies converge to make it happen. The pattern that emerges from this chapter is that making CM is not straightforward, it is highly technologized, and it is an expensive proposition. It requires specialized knowledge and a developed scientific infrastructure, as well as a trained workforce. As the process of making the meat is described, my aim is to show the ways in which edible tissues materialize, as the result of converging technologies. The descriptions are also meant to convey how CM is radically transforming the relationship between farming and the environment, as meat now becomes biomanufactured and reconfigured in vats or bioreactors. However, numerous challenges, such as reducing production costs, improving quality, creating adequate regulations, and increasing production capacity, must be addressed before large-scale commercialization of lab-grown meat can occur (GFI, 2022; Stephens et al., 2018). Additionally, it remains uncertain whether the public will accept this meat or whether it will contribute to the security and sustainability of the food system. The energy consumption and ecological impacts associated with these products remain unknown, as no cellular meat production facilities have yet reached their maximum industrial yield, and several contradictory studies exist on the subject (Lynch & Pierrehumbert, 2019; Rodríguez Escobar et al., 2021; Tuomisto & de Mattos, 2011). Furthermore, the industry faces challenges, such as the need to eliminate animal-derived materials in culture media, technical barriers and limitations in terms of scalability, contamination risks, high production costs, inefficient resource mobilization, non-compliance with conventional meat standards, and concerns about food safety and human health.

Chapter 4 outlines the paradigm that drives cellular agriculture. It starts from cells as the unit of production of meat so that animal proteins can be made in reverse (i.e. starting from the cells rather than the animal). I use the concepts of biocapital (Rajan, 2006) and tissue economies (Cooper, 2008, 2015; Waldby & Mitchell, 2006) in order to make sense of recent developments in the life sciences and the ways in which they converge with politics to offer new speculative futures. In this sense, developments in stem cell science and biotechnology more broadly offer a glimpse of the ways living systems have become an important feature of how to solve the problems of industrial capitalism. This is what Cooper calls the broader societal project of "reinventing life beyond the limit," which includes developing life on Mars, but also as a strategy to perpetuate economic growth and as a renewed form of capital accumulation. The concept of lively capital refers to the ways in which the processes of life itself—biological, social, and political—become a form of capital that can be manipulated, commodified, traded, and exploited. This idea builds on the concept of biocapital, where biological materials and processes such as genetic information, bodily tissues, and even human experience become increasingly harnessed for economic gain. Lively capital emphasizes that life is not just something to be sustained or preserved, but also something that can be extracted, transformed, and utilized as a resource.

This chapter is about meat as technology. In vitro meat, Galusky suggests, must be understood as:

> the product of active (and increasing) human intervention. [...] It is also an object—something not simply reducible to the animal. Animals, of course, are a necessary part of the process (at least for now, pre-in vitro), but so are ecosystems and, in most contemporary encounters, distribution systems. We can claim that meat is a technology without expanding that claim to animals themselves. (Galusky, 2014, p. 938)

Understood as technology deeply embedded within a sociotechnical system, meat can be produced and designed according to human control and desires, it becomes a means of accomplishing something that reflects specific values and interests (Galusky, 2014).

To understand meat as technology, he suggests we must also consider the ways in which technological systems incorporate nature and question the type of assumptions made upon what kind of nature is being reproduced and whose vision prevails. Clearly, the process of making meat outside the animal body involves a series of procedures and interventions which try to mimic bodily functions, away from the kinds of interactions that exist between bodies in/and nature. We ask what kind of nature is reproduced in cellular agriculture? In what kind of world does cultivated meat or human cellular breast milk become necessary?

Chapter 5 traces the trajectory of CM, highlighting the various phases of development of CM first as a figment of the imagination until the establishment of its potential realization today. I briefly retrace the stories about synthetic meat that have existed as part of science fiction since the late nineteenth century and then show how CM developers have expanded the idea into a full-blown enterprise. Other authors have provided similar timelines, albeit in a slightly different format. As I was researching the book, I came across an article that roughly details a similar evolution which complemented my analysis (Stephens et al., 2019). As the story of CM gets told and retold in different ways, new and unexpected knowledge will emerge. A lot of the content of this chapter was drawn from my impressions and observations during cultivated meat events and conferences. The language used, the links to philanthropy, and the recurrent references to moonshot capitalism were issues I needed to delve into as they bring to light certain aspects of the CM industry which tie the stories together.

Similarly, we must examine the ways non-animal-based 'meat,' so-called fake meats or 'meatless meat,' gives rise to anti-meat politics (Chatterjee & Subramaniam, 2021) and associated eating cultures which is the topic of Chap. 6. Lab-grown meat, in particular, has been the cause of 'culture wars,' in terms of whether the term 'meat' or 'dairy' or 'milk' or 'cheese' should apply to cellular agricultural products. Beyond issues of naming and labeling, or even defending the economic interests of ranchers and cattle farmers, recently, conservative politicians in the U.S. and in Italy have been unsympathetic to the cellular agriculture industry. In the U.S., this hostility toward animal cell-based meat has culminated in several states enacting laws banning the products. This raises questions about the problematic ideological positioning of extreme right politics and of certain political and theoretical currents in ecological circles, referred to as technocritique. Chapter 6 also focuses on the links between eating and the construction of identity and the implications of CM in terms of biosocialities and geosocialities.

Chapter 7 delves into the significance of meat in Western diets and what it represents. It would be an error to consider it simply as fleshy matter or any other food item. Instead, when viewed as a social marker, it becomes a shifting signifier for deconstructing the boundaries of food production, consumption, human and non-human bodies, and environmental futures. Many authors, such as Roland Barthes, Sidney Mintz, and Claude Lévi-Strauss among others, have studied food as material culture, as a way to illuminate the social and symbolic significance of food beyond its functionality. As Roland Barthes suggests "food transmits a situation; it constitutes an information; it signifies," similarly, meat and what it represents convey ideological expectations about eaters and the eaten, and about life and the technoscientific achievements, as we negotiate our future on a warming planet. These concepts are key. LeBesco and Naccarato's book, *Edible Ideologies*, adopts a cultural perspective to examine how food and food-related practices are portrayed and the underlying myths that influence them (LeBesco & Naccarato, 2008). The book delves into the ways in which these representations contribute to the formation of both oppressive and oppositional ideologies and behaviors. The authors argue that cultural representations of food, and actual food practices, contain social, cultural, ethical, and political assumptions about negotiating social change and protecting the dominant ideology. Extending this idea further, it appears that epistemologies of meat-eating might be explored, in terms of how they shape worldviews or how they enable thoughts and actions around how we inhabit the Earth in times of planetary crisis. Cultural alimentary shifts such as meat avoidance also echo this statement. Cell-based meats appear to seamlessly navigate between carnivory and veganism, as they satisfy a desire for meat-eating, without the ethical and ecological burdens of industrial animal agriculture.

In the context of cellular meat (CM), the notion of authenticity is profoundly destabilized by the intricate processes of bioassembly, transformation, and substitution that constitute its production. The claim of CM as "real" meat invites a deeper interrogation about the markers of authenticity used to determine the boundaries between authentic and inauthentic, and the original versus imitations, in terms of a technologically mediated food system. In this context, authenticity is not simply a matter of mimicking the original, but it becomes a complex negotiation of value, technology, and the ideological constructs surrounding what we consume as food. This is explored in detail, by juxtaposing a bioart installation entitled *Disembodied Cuisine* and the public tasting of the first lab-grown hamburger, as a way to understand the social construction of taste but also on some level the relationship between art and science. Does art imitate life or does life imitate art?

Chapter 8 describes the process of making animals and their bodies invisible. Carol J. Adams, using the concept of the absent referent, urges us to not simply think about *what* we are eating when eating animal flesh but to consider *who* is consumed. It is interesting to ponder the sociological and cultural significance of the absence of bodies when consuming 'post-animal' flesh. Ultimately, the politics of "what and who becomes meat," according to Chatterjee and Subramaniam (2021), and I would add of 'how meat becomes,' are of critical importance for this book because synthetic meat embodies a series of binary opposites such as synthetic and natural, reality and fiction, authentic and fake, tradition and innovation, urban and rural, etc. Furthermore, the emergence of laboratory-grown meat provides valuable insights about the dematerialization of life and living matter, illuminating the intricate dynamics between the material and immaterial aspects of technologically mediated food production and the emergence of new eating cultures. As discussed in Chatterjee and Subramaniam (2021), meat serves as a catalyst for examining power dynamics and the formation of individual and national identities. However, the recent development of lab-grown meat introduces new dimensions to this discourse, necessitating an exploration of the interplay between the construction of new social worlds and the technoscientific imaginaries that shape and are shaped by them.

Chapter 9 examines techno-utopia, in relation to Anthropocene narratives. Meat and meat discourse are ambiguous to say the least because meat is political. It is an episteme, in the sense that meat embodies an underlying set of knowledge structures and discursive formations surrounding its production, consumption, and regulation. These, in turn, shape how various subjects, within food and agriculture, are understood and managed within a historical period or cultural context. As Foucault argues in *The Archaelogy of Knowledge* (1969), writing about the concept of meat involves an exploration of its constantly evolving multifaceted and complex meanings that consider species entanglements and consider meat beyond its material immanence and fleshy existence, but also as an episteme, including its form as process and product, embodying a series of societal dysfunctions and dystopian visions of the future (Coole & Frost, 2013). The discussion in this chapter centers around the various themes that emerge within CM discourse, as well as visual representations of what the future of eating CM might look like. Because the Anthropocene has served as a basis for legitimating the development of Alt-proteins and "the shift towards the biotechnification of food production" (Sexton, 2018), it is important to understand how various discourses surrounding the environmental crisis intersect with CM as a potential solution to managing and possibly overcoming troubled times. Anthropocene narratives such as eco-catastrophism, ecomodernism, and planetary realism identify different end-of-world imaginaries that position humans in various social and political roles within each scenario (Rothe, 2020). The first of these narratives includes managing catastrophic events through Earth system stewardship where humans, through a Great Food Transformation, will set up the conditions to maintain safe operating limits to manage climate change. The second narrative, ecomodernism, relies on a worldview that places humans at the architects of planetary change through science and technology. By harnessing the powers of technoscience, natural biological evolution can be superseded by human-led evolution, viewing climate change as reversible. In this case, hope and prosperity are attained through technological innovation. Finally, the third Anthropocene narrative, planetary realism, focuses on strategies that enable humans to adapt to the uncertainty brought about by climate change, recognizing its unknowability and unpredictability. This narrative relies on the concept of resilience as essential for surviving catastrophe, thereby challenging anthropocentric accounts that view humans as the central dominating force in nature. In doing so, it critiques modernist notions of control and domination. Planetary realism interprets human survival as intrinsically linked to the health and well-being of other species, emphasizing the importance of learning from the survival strategies and specialized knowledge of vulnerable populations—particularly Black and Indigenous peoples—in the face of catastrophic events. Specific imaginaries and visions surrounding cellular meat (CM) examine how these narratives both construct and are shaped by technoscientific culture. Through these imagined futures, the chapter provides interpretations of what CM signifies in the context of managing climate change and proposing potential solutions. This analysis highlights how technoscientific culture not only creates these visions but also influences our understanding of CM's role in addressing global challenges.

Starting from the metaphor of tissue repair, tissue engineering works by restoring the architecture and function of tissues, Chap. 10 discusses the idea of environmental repair as a paradigm for both CM and managing the effects of climate change. Drawing a parallel between repair at the cellular and ecological level enables us to offer insight into the enactment of repair as materially configured through science and technology. The belief that life, ecosystems, agriculture, animals, and the planet can be fixed demonstrates an understanding of the world that is based on instrumental rationality. The same instrumental reality that is responsible for climate change and planetary damage. Repair demands both material and immaterial practices and work, starting with the premise that things and beings are fixable and that the knowledge to repair them either exists or needs to be uncovered as innovation. In this particular case, rather than focusing on studies that interpret the notion of repair as a form of 'care,' the approach taken is that repair represents a failed attempt to heal our relationship to nature through technology. The main argument is that repair work speaks of a particular way of negotiating the social order. The social breakdown that is announced by climate change serves to reinforce the need for technological solutions and technofixes. As a result, the repair paradigm provides the impetus for "patching up, reconfiguring, interpolating, and reassembling settings from previous forms of existence" (Denis et al., 2016). The phenomenology of repair is about human and non-human relationships to damage, decay, and efforts to mend what has been broken. In the context of environmental or technological repair, it uncovers how individuals and societies experience and conceptualize the act of repairing ecosystems, machinery, or even social relations, considering not just the technical aspects but also the emotional, moral, and existential implications of these processes.

Chapter 11 takes up the idea of competing vitalities by describing two specific framings of how various forms of life—natural and synthetic—vie for dominance, legitimacy, or even survival within the food system. A dynamic tension exists between traditional forms of vitality, such as the bodies of living animals and farming communities, and the engineered vitality of lab-grown tissues cultivated to mimic or replace those animals and, eventually, forms of agriculture. The chapter also discusses questions raised about the values and authenticity of different life forms, illustrated in the tension between real and fake meat. This tension also reflects deeper philosophical debates about what constitutes life and vitality, challenging our understanding of nature, technology, and the boundaries between them.

Dead Meat: Competing Vitalities, Cultivated Meat Imaginaries and Anthropocene Diets is by necessity concerned with the material and imaginary configurations of cellular agriculture, mainly because cultivated meat has yet to become a commercial reality. As a result, much of the discourse surrounding CM, since the idea of artificial flesh came about, focuses on the technological possibilities of its realization and the future it portends. As STS scholars have shown, scientific promises have a long history and have become the primary means of communicating scientific information to the public. The rising number of emerging technologies and the pace of technological development, in the last decades, has resulted in a hyperinflation of scientific promises (Audétat et al., 2015). The use of terms like techno-optimism, techno-utopia, and sociotechnical imaginaries all speak to this phenomenon. However, what distinguishes promises from these other notions is the tangible existence of what scholars call a promissory economy, within which expectations and hopes play an important role. This includes expectations about investments in new technologies, as well as in research and development (R&D), signaling what sociologists Brown and Michael have termed a sociology of expectations (Brown & Michael, 2003). Their work and subsequent STS scholarship have focused on complex sociotechnical change, using such concepts as anticipation, visions, expectations, foresight, and other forward-looking concepts, providing an understanding of the mechanisms that enact the "past and present futures," as the sociologist of time Barbara Adam would say (Adam, 2010). Promises mobilize funds and political interest for current innovation; they are speculative; they are firmly anchored in the present and its socioeconomic power structures. In this sense, promises work by "freezing the present," as a means of maintaining the politics of emerging technologies in the space of regulations, thereby depoliticizing the technology question (Bensaude-Vincent, 2015). While emerging technologies like artificial intelligence, synthetic biology, or nanotechnologies have played a key role in influencing visions of the future, so too have imagined futures impacted new technologies. Values and norms are not unidirectionally shaped when it comes to scientific knowledge:

> [It]…is not a transcendent mirror of reality. It both embeds and is embedded in social practices, identities, norms, conventions and discourses, instruments and institutions—in short, in all the building blocks of what we term the social. The same can be said more forcefully of technology. (Jasanoff & Kim, 2015, p. 3)

There is a constant interplay between the legitimation of technologies and stories about the future, which means delving into the stories that shape and render some technologies more performative. As new technologies take hold, the lack of social systems to guide them leads to a kind of technological wild west (Selin, 2008). This is because:

> This ability to not only imagine technologies but also produce technological futures is not matched by our scholarly – or practical – understanding of what kind of socio-technical systems we are creating. Grasping complexity over the long term and accounting for the ongoing myriad of interactions between values, machines, and regimes has proven daunting for the social sciences. Technological knowledge itself is often shrouded in uncertainties and at the same time supported by a hitherto unprecedented speed of development. Quite simply, processes of discovery and invention in the natural sciences and the translation of such knowledge into commodities are faster than our ability to safely regulate or make sense of the new social dynamics engendered by such processes. (Selin, 2008, p. 1879)

Thus, interest in studying the future in STS scholarship, with the work of Jasanoff and Kim (2015) and others before them, provides a renewed understanding of sociotechnical imaginaries and brings a complex analysis about the ways in which shared assumptions about social life and desirable futures articulate visions "of what ought or ought not, to be lived." Therefore, the concept of imaginaries as interpreted as social practice is at the core of this book. It consists of a "constant interplay between utopia and dystopia" (Ibid).

Imaginaries define how the social actors involved in cellular agriculture produce and perform different versions of the collective good (Jasanoff & Kim, 2015). These visions are woven within a series of narratives about the present such as climate change or the Anthropocene. The type of storytelling, characteristic of CM, involves, as mentioned earlier, constant interplays between utopian and dystopian visions of the world that include ruptures and pivots between old and new possible worlds. CM focuses on the productive possibilities of the present, with a sense of urgency dictated by both the pressing global problems we are facing and the need for investments, but advocates are envisioning an optimistic future where agriculture, humans, and the world will be made better. In terms of historical change, to a certain extent this future vision is predicated on eliminating the past. Science fiction and technological innovation are often entangled and mutually influence each other. These two spheres imagine possible futures and this is also true in the CM space. Visions of the future are both discursively implied and visually depicted, in relations to the way meat will be produced, in factories rather than farms. Focusing on narratives is important because stories and narratives articulate and support political ideas and also reveal social and economic structures. They participate in what humanities scholars would call world-making.

Chapter 10 examines the implications of CM in terms of planetary repair. The importance of questioning technofixes, when faced with a climate emergency, involves setting up a critique of technology as well as industrial agriculture and capitalism more broadly, while analyzing how the life industries and the bioeconomy interact, within the broader context of the Anthropocene. As Banu Subramaniam says, in her piece about the politics of beef-eating in India, meat has become a central node for a politics of purity and sets up a negotiation around boundaries and biological categories—animal, vegetable, cow, chicken, pork, etc. Technological meat or lab-grown meat brings up notions about degrees of what or who is "killable" (Chatterjee & Subramaniam, 2021). As we shall examine later, the production of cellular beef or chicken involves the sacrifice of other life forms. What I am arguing, in a broader context, is that while it is an object of politics, cell meat represents attempts to depoliticize meat, relegating its emergence as something antipolitical in the sense that it remains trapped within the technosphere and technonature, eluding the larger issues surrounding economic growth, social and environmental inequalities and instead, focusing attention on the idea of repair, the capacity to fix the damage caused by human actions and restore ecosystems while ignoring the problematics of technosolutionism.

Bibliography

Adam, B. (2010). History of the Future: Paradoxes and Challenges. *Rethinking History, 14*(3), 361–378. https://doi.org/10.1080/13642529.2010.482790

Anthis, J. R., Berenstein, P., Beurrier, Biehl, C., Birdie, P., Burley, J., Chantre, C., Davis, J., Dohadwala, S., d'Origny, G., Espirito Santo, V., Espitia, F., Gonzalez, G., Grandic, I., Green, C., Groenewegen, L., Guvenilir, A., Hanyu, Y., Kane, M. H., & Zimmerman, J. (2023). In K. Spiro, S. Hall, J. Darling, & M. Benami (Eds.), *Modern Meat: The Next Generation of Meat from Cells* (1st ed.). Cellular Agriculture Society.

Audétat, M., et al. (2015). *Sciences et technologies émergentes: Pourquoi tant de promesses?* Hermann.

Bensaude-Vincent, B. (2015). Promesses et régime d'historicité en technosciences. In M. Audétat (Ed.), *Pourquoi tant de promesses?* (pp. 49–68). Éditions Herman.

Brown, N., & Michael, M. (2003). A Sociology of Expectations: Retrospecting Prospects and Prospecting Retrospects. *Technology Analysis & Strategic Management, 15*(1), 3–18. https://doi.org/10.1080/0953732032000046024

Chatterjee, S., & Subramaniam, B. (Eds.). (2021). *Meat! A Transnational Analysis*. Duke University Press.

Coller, J. (2023). *Council Post: How Business Innovation Can Turn The Food Crisis Into A New Green Revolution*. Forbes. Retrieved from https://www.forbes.com/sites/forbesbusinesscouncil/2023/07/10/how-business-innovation-can-turn-the-food-crisis-into-a-new-green-revolution/

Coole, D. H., & Frost, S. (2013). *New Materialisms: Ontology, Agency, and Politics*. Duke University Press.

Cooper, M. (2008). *Life as Surplus: Biotechnology and Capitalism in the Neoliberal Era*. University of Washington Press.

Cooper, M. (2015). In Vivo Economies: Temporalities of Life and Value. *Journal of Cultural Economy, 8*(3), 257–259. https://doi.org/10.1080/17530350.2015.1039461

Datar, I., & Betti, M. (2010). Possibilities for an In Vitro Meat Production System. *Innovative Food Science & Emerging Technologies, 11*(1), 13–22. https://doi.org/10.1016/j.ifset.2009.10.007

Denis, J., Mongili, A., & Pontille, D. (2016). Maintenance & Repair in Science and Technology Studies. *Tecnoscienza: Italian Journal of Science & Technology Studies, 6*, 5–16.

Douglas, M. (2001). *Purity and Danger: An Analysis of the Concepts of Pollution and Taboo* (Repr). Routledge.

Galusky, W. (2014). Technology as Responsibility: Failure, Food Animals, and Lab-grown Meat. *Journal of Agricultural & Environmental Ethics, 27*(6), 931–948. https://doi.org/10.1007/s10806-014-9508-9

GFI, G. F. I. (2022). *2022 Cultivated Meat State of the Industry Report.* Retrieved from https://gfi.org/wp-content/uploads/2023/01/2022-Cultivated-Meat-State-of-the-Industry-Report-2-1.pdf

Good Food Institute. (2024). *Alternative Protein Company Database* [Dataset]. Retrieved from https://gfi.org/resource/alternative-protein-company-database/

Groll, J., Boland, T., Blunk, T., Burdick, J. A., Cho, D.-W., Dalton, P. D., Derby, B., Forgacs, G., Li, Q., Mironov, V. A., Moroni, L., Nakamura, M., Shu, W., Takeuchi, S., Vozzi, G., Woodfield, T. B. F., Xu, T., Yoo, J. J., & Malda, J. (2016). Biofabrication: Reappraising the Definition of an Evolving Field. *Biofabrication, 8*(1), 013001. https://doi.org/10.1088/1758-5090/8/1/013001

Hahn, J. (2020, November 13). *Grow-your-own Human Meat Kit is "Technically" not Cannibalism.* Dezeen. Retrieved from https://www.dezeen.com/2020/11/13/ouroboros-steak-meal-kit-andrew-pelling-grace-knight-orkan-telhan/

Hannan, J. (2020). *Meatsplaining: The Animal Agriculture Industry and the Rhetoric of Denial.* Sydney University Press.

Jasanoff, S., & Kim, S. H. (2015). *Dreamscapes of Modernity: Sociotechnical Imaginaries and the Fabrication of Power.* University of Chicago Press. Retrieved from https://www.press.uchicago.edu/ucp/books/book/chicago/D/bo20836025.html

Landecker, H. (2007). *Culturing Life: How Cells Became Technologies* (First paperback ed). Harvard Univ. Press.

LeBesco, K., & Naccarato, P. (Eds.). (2008). *Edible Ideologies: Representing Food and Meaning.* State Univ. of New York Press.

Lynch, J., & Pierrehumbert, R. (2019). Climate Impacts of Cultured Meat and Beef Cattle. *Frontiers in Sustainable Food Systems, 3.* https://doi.org/10.3389/fsufs.2019.00005

Mattick, C. S. (2018). Cellular Agriculture: The Coming Revolution in Food Production. *Bulletin of the Atomic Scientists, 74*(1), 32–35. https://doi.org/10.1080/00963402.2017.1413059

Myelnikov, D. (2020). Tinkering with Genes and Embryos: The Multiple Invention of Transgenic Mice c. 1980. *History and Technology, 35*(4), 425–452. https://doi.org/10.1080/07341512.2019.1694126

Niamh, M. (2018). *Aleph Farms CEO on its 3D Cultured Beef: "Unlike Other Companies, Our Meat Grows Together Like Real Meat."* Foodnavigator.Com. Retrieved from https://www.foodnavigator.com/Article/2018/05/02/Aleph-Farms-CEO-on-its-3D-cultured-beef-Unlike-other-companies-our-meat-grows-together-like-real-meat

Niman, N. H. (2021). *Defending Beef: The Case for Sustainable Meat Production.* Chelsea Green Publishing.

Pilgrim, K. (2013). 'Happy Cows,' 'Happy Beef': A Critique of the Rationales for Ethical Meat. *Environmental Humanities, 3*(1), 111–127. https://doi.org/10.1215/22011919-3611257

Post, M. J. (2012). Cultured Meat from Stem Cells: Challenges and Prospects. *Meat Science, 92*(3), 297–301.

Post, M. J., & Hocquette, J.-F. (2017). Chapter 16—New Sources of Animal Proteins: Cultured Meat. In P. P. Purslow (Ed.), *New Aspects of Meat Quality* (pp. 425–441). Woodhead Publishing. https://doi.org/10.1016/B978-0-08-100593-4.00017-5

Rajan, K. S. (2006). *Biocapital: The Constitution of Postgenomic Life.* Duke University Press.

Rodríguez Escobar, M. I., Cadena, E., Nhu, T. T., Cooreman-Algoed, M., De Smet, S., & Dewulf, J. (2021). Analysis of the Cultured Meat Production System in Function of Its Environmental Footprint: Current Status, Gaps and Recommendations. *Foods, 10*(12), 12. https://doi.org/10.3390/foods10122941

Rothe, D. (2020). Governing the End Times? Planet Politics and the Secular Eschatology of the Anthropocene. *Millennium, 48*(2), 143–164. https://doi.org/10.1177/0305829819889138

Selin, C. (2008). The Sociology of the Future: Tracing Stories of Technology and Time. *Sociology Compass, 2*(6), 1878–1895. https://doi.org/10.1111/j.1751-9020.2008.00147.x

Sexton, A. E. (2018). Eating for the Post-Anthropocene: Alternative Proteins and the Biopolitics of Edibility. *Transactions of the Institute of British Geographers, 43*(4), 586–600. https://doi.org/10.1111/tran.12253

Sharma, S. (2018). *Mighty Giants: Leaders of the Global Meat Complex.* Retrieved from https://www.iatp.org/blog/leaders-global-meat-complex

Stephens, N., King, E., & Lyall, C. (2018). Blood, Meat, and Upscaling Tissue Engineering: Promises, Anticipated Markets, and Performativity in the Biomedical and Agri-food Sectors. *BioSocieties, 13*(2), 368–388. https://doi.org/10.1057/s41292-017-0072-1

Stephens, N., Sexton, A. E., & Driessen, C. (2019). Making Sense of Making Meat: Key Moments in the First 20 Years of Tissue Engineering Muscle to Make Food. *Frontiers in Sustainable Food Systems, 3*. https://doi.org/10.3389/fsufs.2019.00045

Subramaniam, R. (2021). The Ethical Impurative: Elemental Frontiers of Technologized Meat. In S. Chatterjee & R. Subramaniam (Eds.), *Meat! A Transnational Analysis* (pp. 254–278). Duke University Press.

Sunder Rajan, K. (2012). *Lively Capital: Biotechnologies, Ethics, and Governance in Global Markets*. Duke University Press.

Tuomisto, H. L., & de Mattos, M. J. T. (2011). Environmental Impacts of Cultured Meat Production. *Environmental Science & Technology, 45*(14), 6117–6123. https://doi.org/10.1021/es200130u

Van Der Weele, C., & Driessen, C. (2013). Emerging Profiles for Cultured Meat; Ethics through and as Design. *Animals, 3*(3), 647–662. https://doi.org/10.3390/ani3030647

Waldby, C., & Cooper, M. (2014). *Clinical Labor: Tissue Donors and Research Subjects in the Global Bioeconomy*. Duke University Press.

Waldby, C., & Mitchell, R. (2006). *Tissue Economies: Blood, Organs, and Cell Lines in Late Capitalism*. Duke University Press.

Watson. (2022, November 16). *'A Watershed Moment in the History of Food…' Cultivated Meat Edges Closer to Commercialization in the US as UPSIDE Foods gets FDA Green Light*. Foodnavigator-Usa.Com. Retrieved from https://www.foodnavigator-usa.com/Article/2022/11/17/A-watershed-moment-in-the-history-of-food-Cultivated-meat-edges-closer-to-commercialization-in-the-US-as-UPSIDE-Foods-gets-FDA-green-light

WEF, W. E. F. (2024). *Fourth Industrial Revolution*. World Economic Forum. Retrieved from https://www.weforum.org/focus/fourth-industrial-revolution/

Wolpa, A. (2016). Seeing Meat without Animals: Attitudes for the Future. In B. Donaldson & C. Carter (Eds.), *The Future of Meat without Animals* (pp. 87–96). Roman and Littlefield International.

World Economic Forum. (2019). *Meat: The Future Series. Alternative Proteins* [White Paper]. World Economic Forum. Retrieved from https://www3.weforum.org/docs/WEF_White_Paper_Alternative_Proteins.pdf

Worster, D. (1990). Transformations of the Earth: Toward an Agroecological Perspective in History. *Journal of American History, 76*(4), 1087–1106. https://doi.org/10.2307/2936586

2

Animal Agriculture and the Anthropocene

Post-Animalities and the Industrialization of Agriculture

Before situating CM within the logic of productivist agriculture, it is important to understand the unique nature of agriculture within capitalist production. Accomplishing this requires that we delve into the many ways that living organisms involved in agriculture become gradually transformed as "living factories" (Fish, 2013) as both forms of capital and modes of production. This evolution is premised on an understanding of the role of technology, not simply as a set of biological, chemical, and mechanical tools and knowledges that transform natural processes and living beings into commodities, but, specifically, as providing the ground for the ontological alteration of nature in the image of capital (Fish, 2013, p. 13). Cellular agriculture stands as the culmination or pinnacle of industrial capitalism, epitomizing the intentional displacement and replacement of animals into post-animals. This process signifies not merely a substitution of traditional livestock but an ontological shift, where the essence of animality is redefined within the confines of techno-scientific production. Even though it uses all of the techniques of genetic engineering and biotechnology, CM is a more advanced state of control

E. Abergel, *Dead Meat*, https://doi.org/10.1007/978-981-97-9049-4_2

over living processes, as it encompasses the processes of life creation not simply at the genetic level. Cellular agriculture aims to synthesize artificial flesh within the controlled environment of the laboratory or bioreactor, which simulates a partially realized vision of nature.

A Marxian analysis of the industrialization of agriculture reveals that animals introduce unique challenges to both the production and labor processes, significantly impacting corporate profitability. This perspective underscores how the integration of living beings into mechanized systems of production not only complicates efficiency and control but also exposes inherent contradictions, in the capitalist drive for maximum exploitation and profit. By examining these dynamics, we can better understand the tensions and inefficiencies that arise when capitalist modes of production intersect with the complexities of biological life, ultimately influencing the economic outcomes for agribusinesses. In light of this political economy analysis, it becomes evident that the transformation of biological life into mechanized forms aligns seamlessly with the imperatives of agribusiness. The logic of the industrialization of the food system includes the relocation of on-farm activities toward more integrated parts of the food system. This process reflects a broader capitalist strategy to subordinate natural processes to industrial efficiency and profit maximization, further entrenching the commodification of living beings within the agricultural sector. The capitalist imperative prioritizes efficiency, scalability, and profit maximization, often at the expense of ethical considerations and ecological sustainability. Consequently, this drive leads to the commodification and instrumentalization of animal life, further entrenching the exploitative nature of industrial agriculture and exacerbating its socio-environmental impacts. My contention is that it is precisely the collateral damage of industrializing life that now necessitates the imperative of "saving it" through the replacement of animals in the production and labor process using cellularized meat products as the means to sustain the capitalist model of agriculture. This shift represents a continuation of the same logic, repackaged under the guise of ethical and sustainable innovation. Moreover, it underscores a paradox within capitalist logic: the very systems that have led to ecological degradation and ethical concerns are now promoting biotechnological innovations as solutions. By turning to cellular agriculture, agribusinesses seek to mitigate the adverse effects of conventional practices, yet this move simultaneously reaffirms the capitalist commitment to technological control and market expansion, rather than addressing the root causes of exploitation and environmental harm.

The desire to circumvent the economic, ethical, and environmental pitfalls inherent to traditional livestock farming fails to disrupt the capitalist logic that seeks to address crises through frameworks like eco-modernization, the "Good Anthropocene," or green capitalism. Instead, these solutions perpetuate the same commodification and technological interventions, reinforcing the underlying system rather than challenging its foundational premises. As we will see in Chap. 8, narratives of animal liberation and the Anthropocene are used to justify the 'disruption' of traditional industrial agriculture by cellular agriculture. However, as one observes the expansion of cellular agriculture in practice and through discourse—especially regarding meat, where the promises of scalability, efficiency, and profitability have yet to be realized—it becomes clear that these aspirations remain largely speculative. This gap between promise and practice highlights the persistent challenges and contradictions within the capitalist framework, calling into question the feasibility of truly sustainable and ethical alternatives within an economic system driven by commodification and technological intervention. Can capitalism and the logic of industrialization of 'animal products' actually save animals and the planet? Or is 'sustainable meat' or 'green meat,' whether it is animal, plant, or post-animal, simply a myth—an imaginary?

Agriculture is unique, as Goodman et al. (1987) suggest in their landmark book *From Farming to Biotechnology* because it "confronts capitalism with a natural production process," which means that its industrialization has been determined and shaped by overcoming or removing a series of natural constraints, such as biological time, in the various phases of plant and animal gestation. Historically, this has entailed the transformation and the breaking down of natural biological and temporal processes into specific sectors of agricultural industrial activity, with the purpose of reducing the importance of nature in rural production as a means to exert technological control over it (Ibid). The process of industrializing agriculture involves a series of appropriations of rural labor and biological production, as well as the development of industrial substitutes for farm products. Machines, fertilizers, hybrid seeds, genetically modified organisms (GMOs), as well as pesticides and other inputs incorporate forms of appropriation of the labor process of farmers because these have been transformed into commodity forms and no longer originate on the farm. Goodman et al. describe the historical evolution of the agro-industrial complex, seen as the convergence of mechanical, chemical, and genetic innovations, leading to the increased capitalization of rural production. As rural production became more discontinuous, it involved massive structural, technological, and geographical changes in the area of crop and livestock[1] production. The industrialization of agriculture has meant reorganizing biological time and rural space in order for capital interests to:

> reduce the importance of nature in rural production, and specifically as a force beyond their direction and control. This was achieved initially by relaxing the constraint of land as space via mechanization, and subsequently, by the continuing struggle to transform the secrets of biological production into scientific knowledge and industrial property. In effect, the process of natural reproduction of plants and animals is being internalized via science in the reproduction of industrial capitals. (Goodman et al., 1987, p. 3)

Effectively, the authors use Marxian analysis to develop the concept of appropriationism to describe the sustained restructuring of agricultural production to generate opportunities for capital accumulation. This entails intensifying the industrial penetration of farm-level activities in order to reduce the importance of land as the essential material element of rural production (p. 48). The authors refer to the elimination of the rural product, such as the replacement of butter by margarine, for instance, as forms of substitutionism, a process that gradually erodes and eradicates the rural base of agriculture, signaling the final phase of agricultural industrialization and agrarian livelihoods. These two concepts, appropriationism and substitutionism, shed light on what is happening more broadly in the agricultural world today, particularly in the

---

[1] I use the term livestock purposely to denote the fact that, as Kathryn Gillespie suggests, it literally describes "live stock," defining animals as live property (Gillespie, 2018). However, authors like Hribal (2003) use the term "living stock" to emphasize the commodity form of farm animals.

production of animal products without animals. The farming sector's decreasing reliance on nature in the rural process can be viewed through the dual processes of appropriation and substitution of time, space, products, and processes. Cultivated meat might be an extreme expression of this phenomenon; it is a process that enables living entities or specific forms of vitalities to survive outside the body. As Landecker demonstrates in her work on the history of cell culture (see Chap. 4), technoscience has facilitated the possibility for life to transcend the organic forms that originally enabled its existence (Landecker, 2007). It is not a case of life imitating nature but rather one where experimenting outside the boundaries of untethered life might fabricate a different kind of being, "no longer dependent for meaning on a dialectical relationship with a greater, enduring body" (Stafford, 2007, p. 392). The existent no longer determines the limits of biology, creating the possibility of a post-animal bioeconomy. Here, post-animalities further abstract from the animal as living commodity, fragmenting and commodifying animal bodies at the cellular level. The animal, no longer a whole organism, is now reduced to its most basic biological units—cells. By decoupling the production of biological materials from the existence of living organisms, CA pushes the boundaries of biology beyond its traditional constraints. This means that the production of meat can occur independently of traditional ecosystems and animal farming practices. On a more philosophical level, this shift suggests a fundamental redefinition of biology. If biology is no longer confined to the existence of living beings but is instead defined by the potential to engineer life at the cellular level, what implications does this have for our understanding of life itself? This development may signal the advent of a post-biological age, as envisioned by synthetic biology, where life is no longer shaped by evolutionary processes but is instead created through human desires and technological intervention. This prompts us to think about the implications of synthetic biology in shaping future forms of existence and their connections with nature within the realm of competing vitalities.

More broadly, these processes are in continuity with the concepts of substitutionism and appropriationism, relevant for understanding the impact of biotechnological transformations and the modernist goal to control time and space in agriculture. The spatiotemporal control of life is at the heart of the Cellular Agriculture project.

In terms of animal agriculture and its industrialization, it is clear that, historically, grain and livestock production have contributed to mass food consumption patterns based on animal protein. However, industrial animal agriculture, unlike grain or cereal production, is characterized by two factors: first, space appropriation via land is not a major biological requirement for animal rearing and, second, livestock is mobile, facilitating the "physical concentration of production" necessitating less human labor through the mechanization and robotization of milk production, automation in feeding systems, and increasing use of so-called smart livestock management techniques (Attard, 2022; Goodman et al., 1987). The appropriationism of animal production is achieved mostly by capturing the biological processes of selective breeding of cattle and reproductive techniques, such as the artificial insemination of dairy cows. Historically, through increased biological intervention and industrial intensification, animals have essentially become living factories (Fish, 2013) or imperfect protein machines (Galusky, 2022).

The industrialization of agriculture highlights the complex and difficult relationship between capitalism and nature. Goodman et al. refer to this as a process by which plant and animal life is internalized by science, in line with Marx's idea of subsumption.[2] Marx distinguished between formal and real subsumption in his analysis of the relationship between capital and labor.[3] Formal subsumption occurs when existing processes and relations are appropriated for profit without significant alteration to their fundamental nature. In this case, capital incorporates labor practices that originate outside its domain. In contrast, real subsumption involves a profound transformation of these processes and relations to align with the imperatives of capitalist production. It is the direct result of the growth of the capitalist economy. As capitalist expansion reaches its

[2] Marx introduces the concepts of formal and real subsumption in *Capital, Volume I* (1867) and in his *Unpublished Chapter VI*.

[3] A traditional Marxist interpretation of the concept of labor assumes a transhistorical dimension of humans' metabolic interaction with nature which is rejected by authors such as Moishe Postone, who interprets labor as existing within capitalism. Hence, the production process is understood as a labor process (Devyver, 2022; Postone, 2020).

limits, the processes of formal subsumption can no longer play a central role. Real subsumption, in this case, means that life itself becomes a mode of production but is also the site of mediation of social reproduction (human and non-human) whereby "the capitalist mode of production concretely (re)produces the natural environmental from within" turning farming into factory or industrial farming, for instance (Haug, 2018). Applying this distinction to the realm of agricultural biotechnology, we observe that real subsumption transforms life processes themselves, to meet the demands of capital. In the context of agricultural biotechnology, this transformation is particularly evident. The introduction of biotechnological innovations fundamentally alters growth yields, reproductive cycles, growth rates, pest resistance, and disease resistance in plants and animals. These alterations do not merely enhance existing agricultural practices; they redefine the organisms themselves, creating new forms of life specifically tailored to the requirements of capitalist production. This process not only changes the biological attributes of these organisms but also reshapes their relationship to the environment and to human societies and vice versa. As Knorr-Cetina's work demonstrates, molecular biology—the tools of biotechnology—has become a science of life without nature, leading to the transformation of living organisms into production sites and into what she calls *molecular machines* (Knorr-Cetina, 1999).

Furthermore, agricultural intensification driven by advancements in biotechnology represents a profound shift in the political economy of food production. This shift is characterized by the increasing commodification and control of life processes where nature itself becomes a site of technological mediation for serving human purposes. Agricultural biotechnology enables the creation of genetically modified organisms (GMOs) that are optimized for specific economic purposes, thereby transforming the fundamental nature of agricultural production. This transformation is a clear manifestation of real subsumption as it involves the reengineering of life processes to enhance productivity and profitability within the capitalist framework. Moreover, the redefinition of the relationship between organisms, the environment, and humans through biotechnology has significant implications for political economy because it reinforces the dominance of capital over nature. The fact that animals are considered private property is a central feature of capitalist relations, this carries implications for how animal bodies are exploited as "superexploited commodities" (Hannan, 2020). The distinction between animal and machine is erased when animals are treated as property:

> Under capitalism, animals become machines. They are treated like mechanical systems, into which farmers adds inputs and from which he collects outputs. If the goal is money, it makes sense to manipulate the biology of animals and create mutant beings that produce more flesh, milk, eggs, leather, wool and fur. The only limitation to this mad drive to extract every last drop of profit from animal bodies are those imposed by their biology. At some point, animals will break. They can no longer give birth. They can no longer produce milk. They can no longer lay eggs. At that point, they either get discarded, slaughtered or turned into some new product, like animal feed, in a system eager to find value where value was never before thought to exist. (Hannan, 2020, p. 29)

When industrial agriculture has pushed animal bodies to their physiological limits, cell meat emerges as a technological solution to address the "body problem" in agriculture. Not only does CM perpetuate further alienation from animals, but it does so in order to alleviate suffering. In addition, animals become alienated from their bodies when cellular agriculture enables their cells to be commodified, long after the animals themselves have died. In this process, animals are disconnected from their physical existence, as their cells are harvested, preserved, and utilized in ways that strip them of their original identity and vitality. Cell-based meat maintains the commodification of biological entities and the ethical and environmental challenges associated with such practices, rather than fundamentally resolving the issues inherent in the exploitation of living systems. Once again, CM epitomizes real subsumption because it fundamentally transforms, in addition to appropriating and substituting biological processes, the production of meat by creating value via cellular extraction. In traditional animal agriculture, animals are bred, raised, and slaughtered for meat. This can be seen as formal subsumption, where existing biological processes are harnessed for capitalist purposes. Cellular meat, however, bypasses the living animal entirely (except as a source of stem cells in some instances), growing meat directly from cultured cells in bioreactors. This includes yet another form of subsumption which includes the datafication of life, meaning the transformation of biological systems transformed into quantifiable data, which become deeply intricated with machine processes. This is clearly seen in bioprinting, where digital models or meat blueprints are used to guide cell growth and tissue formation. These digital designs, controlled by algorithms, allow cells to be arranged in specific structures. In cultured meat production, for example, automated systems use real-time data to regulate bioreactors and optimize conditions for cell growth. These systems subsume biological processes into a digital framework, making them part of an algorithm-driven environment. More broadly, datafication involves mining large biological datasets to identify key factors in biological processes, often using artificial intelligence. This means extracting and analyzing data, such as gene expression, protein folding, and other cellular mechanisms, to uncover complex mechanisms within living organisms as a way to design and improve productive processes. Through subsumption, biological life is increasingly managed and understood through digital and computational means. This means that biological processes become information technologies and that the manipulation of information informs the manipulation of material substances. However, this obscures the active role of cells in terms of interpreting information and transforming their environment (Fish, 2013). This shift represents a profound transformation of the relationship between humans, animals, and food production, as it aims to redefine the very process of what constitutes 'meat' within the capitalist system and existing power structures.

Dead Meat and the Anthropocene

The climate question and the Anthropocene as they relate to agriculture and food systems represents a double movement, on the one hand, a synthesis of productivism and its exacerbation of agro-industrial development anchored in postnatural and techno-utopian logics and, on the other, what is known as alternative agriculture that promotes ecological, local, and territorial approaches, with a tendency toward repeasantization. The agrarian question, as it relates to continuous changes in the global food and agricultural system, can roughly be summarized as the resulting tensions between the reproduction of capital and social reproduction (McMichael, 2013). McMichael emphasizes the role played by rural dispossession and the fact that an analysis of capital, as the sole driving force for the industrialization of agriculture and its impact on ecological degradation, obscures the importance of rural inhabitants and their potential to resist dispossession, where resistance stands as a reproductive necessity, in order to sustain local resources and ensure food security and the future of farming communities. This exemplifies Polanyi's idea of the double movement in *The Great Transformation*, which describes the tensions between the negative impacts of unchecked market expansion and the resulting need for social protection in capitalist societies (Polanyi, 1944). McMichael's analysis relates to the push and pull between the reproduction of capital versus social reproduction—understood as the processes through which societies sustain and perpetuate themselves across generations. In the context of global agriculture, social reproduction involves maintaining the basic conditions for small-scale subsistence farming, which is important for food security and often persists alongside market-oriented commercial farms, in transitional as well as in advanced economies. McMichael views social movements such as the food sovereignty movements as an alternative to the historical corporatization of food and describes the many self-organized peasant groups and communities across the world resisting global markets.

McMichael's narrative of peasant resistance amid industrial giants runs counter to the kind of scientific agriculture or hyper-modernized rural visions often promoted in popular culture, which promote various high-tech scenarios for feeding the world population in the future. Scenarios such as: ensuring food security in light of a global population expected to reach 9.1 billion by 2050; the negative impacts of climate change; overgrazing; soil degradation; post-harvest losses, to name a few. Some key international institutions have called for an alignment of diets with the imperatives of curbing climate change. In particular, the report entitled *Food in the Anthropocene: The EAT-Lancet Commission on Healthy Diets from Sustainable Food Systems* aims to address the needs of a growing population while also defining sustainable food systems (Willett et al.,

2019). The Commission lists the environmental and health-related risks posed by animal agriculture, supported by an impressive number of statistics, such as the 263 million tons of meat produced globally (in 2019) which is expected to double to 445 million tons by 2050, during which time the world's population is expected to triple from 3 to 9 billion people (p. 5). Greenhouse gas (GHG) emissions have been rising steadily, as more countries shift their agriculture to industrialized practices; this is accompanied by shifts in diets that rely increasingly on meat consumption. The Food and Agriculture Organization of the UN (FAO) estimates that livestock accounted for between 11.1% and 19.6% of global greenhouse emissions, coming from their respiratory process (CO_2), digestive process (methane), and manure which emits methane, nitrous oxides, ammonia, and CO_2 (Blaustein-Rejto, 2022; Steinfeld, 2006). Other factors, such as emissions from grazing land or land-use change caused by animal agriculture, are also factored in. These numbers include the impact of animal feed production. The range in values corresponds to the calculations and data sources used, as well as the method of conversion of emissions of different greenhouse gases. However, what is important are the solutions promoted to cut these emissions, one of which is to intensify animal production toward landless systems because these have lower land and feed requirements, therefore contributing to fewer GHG emissions (Hayek & Miller, 2021). These include the concentrated animal feeding operations or CAFOs mentioned earlier, where livestock management is oriented toward rapid productivity and relies on confinement. Advocates of 'sustainable intensification' contend that adoption of this system might work well in low- to middle-income countries where farm size is relatively small and where the demand for meat is increasing (Ibid). The FAO predicts that by 2050, the demand for meat and animal products, including dairy and eggs, in developing countries will double, driving significant growth in livestock production. However, an important way to offset climate change remains switching diets, moving away from a meat-based diet to alternative proteins such as plant-based and insect-based foods, including cell-based products like meat and dairy, as advocated by many international organizations, such as the FAO and others.

These numbers have made the issue of eating meat very complicated for climate-concerned producers and eaters. Chicken—the term describes both the animal and the meat—is the world's most popular meat, because it is the cheapest to produce per pound and it has the greatest feed efficiency, compared to other farmed animal species. In his book about cellular meat, Wyatt Galusky traces the evolution of chickens and chicken in the American diet referring to the "chicken of tomorrow" contest, as a product of technology leading up to the reconfiguration of chicken tissues as CM (Galusky, 2022). I want to discuss domesticated chicken(s) as a marker of a "human reconfigured biosphere" (Bennett et al., 2018). There are more chickens in the world than any other species of either wild birds or other farmed birds and either farmed pigs or cattle. Bennett et al. consider that chickens (*Gallus gallus*), starting with their wild ancestors, are the most common bird in the history of the world. Their claim is that chickens serve as strong candidates as markers of the beginning of the Anthropocene epoch, due to their rapid population explosion and geographical range, as well as their morphological changes over time. For the authors, the proliferation of chickens, in particular broiler chickens, and their standing biomass means that they constitute an important environmental burden and an unprecedented accumulation of chicken carcasses that are indicative of human-induced changes to the biosphere. Their meat fueled slave labor on the plantations and their eggs fed factory workers during the Industrial Revolution. Bennett et al.'s study suggests that chickens could act as a biostratigraphic marker species for the Anthropocene, just as other fossil species indicate past geological epochs, revealing the transformation of ecosystems to meet human consumption patterns (Ibid). Clearly, the advent of the broiler chicken came about as a result of advances in science and technology, contributing to its efficiency, rapid growth, and morphological changes. So, while the broiler chicken might be a marker of the Anthropocene, CM symbolizes a response to the Anthropocene potentially guiding us toward a post-Anthropocene future where technological innovation, it is hoped, will reduce anthropogenic pressure on the biosphere.

Does the intentional phasing out of animal farming through cellular agriculture, largely driven by appropriationism and substitutionism, signal a broader shift in how humans interact with and exploit biological systems, while also addressing climate change? Cellular agriculture takes place within a specific period in human history, identified as the

Anthropocene, also known as the 'age of Man' whereby human activities over the centuries have altered the planet's climactic and geological systems in profound and devastating ways (Lewis & Maslin, 2015).

The Anthropocene, following the Holocene, has been defined as the way in which humanity and human activity have geologically altered the planet and our relationship to Nature. Essentially, humanity had created its own geological epoch, reaching its pinnacle by the middle of the twentieth century (Lynch & Pierrehumbert, 2019). However, a more complex understanding of the Anthropocene includes critical ways of analyzing environmental destruction and climate change, as a product of colonial violence, informed not by what might transpire in the future but the ways in which genocides and ecocides are historically interconnected. This includes understanding how the exploitation of animals mirrors broader patterns of exploitation and domination of humans, bringing a more inclusive perspective of the relationships between humans and non-human animals into the context of the Anthropocene.

Termed as the Capitalocene (Malm, 2015; Malm & Hornborg, 2014; Moore, 2016) marked by the Great Acceleration (Steffen et al., 2015), the Plantationocene (Haraway, 2015), etc., these different interpretations of the Anthropocene provide a temporal and spatial context to the emergence of this phenomenon. More information about the Anthropocene is provided throughout various sections of the book, but for now suffice it to say that, as we've shown, CM has been promoted as a solution to the ecological devastation caused by industrial meat production.

In the Anthropocene epoch, agriculture and food production have been identified as significant contributors to climate change while also being highly vulnerable to its impacts. Despite an extensive and expanding body of research highlighting the critical role of agriculture in global environmental change—from soil and water to atmospheric carbon—there has often been a lack of engagement with critiques of the Anthropocene concept (Reisman & Fairbairn, 2021). A notable example of addressing agri-food systems within this new epoch is the EAT-Lancet Commission report on Food in the Anthropocene. This report uncritically endorses the concept of the Anthropocene and suggests a "universal healthy diet" when calling for "The Great Food Transformation" (Willett et al., 2019) to reduce the impact, during a time when humanity has become the "dominating driver of change" (Reisman & Fairbairn, 2021, p. 688). The EAT-Lancet report recommends a predominantly plant-based diet, as well as the consumption of meat or fish with low environmental impact, thus urging less global reliance on livestock as a source of animal protein, because of its role in producing and emitting greenhouse gases. Critics of the report have noted that, despite its reference to the Anthropocene and planetary boundaries, the report fails to shift eating away from chemically intensive agriculture and an industrial food paradigm, pointing to the big agricultural corporations that sponsored the Lancet study (Shiva, 2019).

The EAT-Lancet Commission report presents a reference diet, a "healthy diet" suited for an estimated population of 10 billion people, while remaining within a safe operating space (Ibid). The authors claim that their universal framework can be applied across eating cultures and production systems, "with a high potential of local adaptation and scalability" (p. 1). However, attaining these goals hinges on maintaining or lowering red meat and dairy consumption levels. The reference diet, or Anthropocene diet, is part of what they call the Great Food Transformation which calls for a "transformation of the global food system."

This healthy reference diet largely consists of vegetables, fruits, whole grains, legumes, nuts, and unsaturated oils and includes a low to moderate amount of seafood and poultry and a low quantity of red meat, processed meat, added sugar, refined grains, and starchy vegetables (Willett et al., 2019, p. 1). The EAT-Lancet report also places emphasis on climate mitigation technologies for curbing the excesses of the current food system. While they refer to existing technologies, the series of reports by the World Economic Forum delves into the special challenge that meat poses for the future. They encourage investigating and investing in "new protein possibilities" linked to innovation and the Fourth Industrial Revolution (discussed later) which includes insects, mycoproteins, algae, plants, etc. One section in particular dedicated to cell meat offers a glimpse into the future of meat production. More importantly, it emphasizes how cultured meat aligns with environmental objectives, including the UN Sustainable Development Goals (SDGs) and the Paris Agreement, which encourages member states to strive to limit global temperature increases to 1.5 °C above pre-industrial levels.

Companies and innovators are also racing to develop "cultured" meat. These are meat products grown from starter cells taken from live animals. This novel protein source could be produced in urban "breweries," requiring just 2% of the land that the global livestock industry uses today. This would relieve 25% of the pressure on global agricultural water demands by 2030 and produce just 4% of the GHG emissions of today's global industrialized cattle herd, according to Professor Marc Van der Post of Maastricht University, a leading innovator in the cultured meat field and Chief Scientific Officer of Mosa Meat. With a much lower land and resource footprint and no need for antibiotics, this meat could also be healthier for people and the planet (World Economic Forum, 2018, p. 10).

Thanks to technological advances such as cell meat, it becomes possible to mimic natural products and possibly replace them, while their ecological value also lies in their ability to "deliver" on curbing the negative impacts of traditional agriculture.

When it comes to sustainable intensification of animal agriculture, the Breakthrough Institute, a research organization that promotes technological solutions and that is the home of the ecomodernist manifesto, is a fierce advocate of land-based intensity as well, which means applying the latest crop biotechnologies. The Institute supports the government funding of cultivated meat, as one technology among many, to curb GHG from agriculture. We will return to ecomodernism in the chapter dedicated to Anthropocene narratives. Such technologically deterministic perspectives tend to remove the possibility for alternative value systems in agriculture and food to emerge.

While today the space constraints associated with intensive animal farming are addressed by extreme forms of animal confinement, it remains true that one of the main causes of deforestation and habitat destruction is cattle ranching. We've all seen images of the vast expanses of CAFOs (Concentrated Animal Feeding Operations), large agricultural facilities where animals are confined in small spaces, sheds, or feedlots, for the purpose of maximizing meat, egg, or dairy production, while keeping costs at a minimum. These industrial operations are highly intensive and are used in livestock management—mostly for poultry, swine, or dairy production—for high productivity, often using concentrated feed (using chemically intensive crops like genetically modified corn or soy) to fatten the animals before slaughter.

In industrial terms, animals are considered inefficient relative to the energy and inputs required to convert their flesh into protein. It takes approximately 15,000 liters of water to produce a kilo of beef versus 1600 liters to grow a kilo of wheat. According to a report by the think tank RethinkX[4] the new protein disruption brought about by cellular agriculture will usher a new model of food production that will be more efficient:

> This rapid improvement is in stark contrast to the industrial livestock production model, which has all but reached its limits in terms of scale, reach, and efficiency. As the most inefficient and economically vulnerable part of this system, cow products will be the first to feel the full force of modern food's disruptive power. Modern alternatives will be up to 100 times more land efficient, 10–25 times more feedstock efficient, 20 times more time efficient, and 10 times more water efficient. They will also produce an order of magnitude less waste. (Tubb & Seba, 2020, p. 6)

As mentioned before, farm animals represent the final challenge to achieve the ultimate stage of industrialization of agriculture, part of what I have called the "body problem." By harnessing and culturing animal cells, as the basic living units of life, the production of edible meat tissues can now be imagined without resorting to raising and slaughtering billions of animals. While this may seem like a good idea, it raises questions about the significance, challenges, and prospects of a post-animal agricultural system. As environmental conditions are increasingly challenged, the process of agricultural industrialization has shifted toward a focus on the capacities and potentialities of living materials, namely, stem cells and biofabricated tissues and organs, to be used in new and novel ways, as foodstuff. The relevance of this, as it relates to the engineering of animal

[4] The RethinkX Project and this report about Rethinking Food and Agriculture analyzes the impact of technological disruption using a methodology that relies on "market forces triggered by technological convergence, business model innovation, product innovation and exponential improvements in both costs and capabilities" (p. 3). This particular report deals with "The second domestication of plants and animals, the disruption of the cow and the collapse of industrial livestock farming."

products outside of their bodies, is that industrial logic is captured within the production of living tissues, eliminating the need for animals and the environments they normally live in. When life *becomes* technology, new forms of technologized vitality, biomaterialized fragments of muscle, and novel categories of living edible tissues proliferate in laboratories and bioreactors, one day ending up on our plates and in our bodies.

Cellular Agriculture: A New Agricultural Narrative

The real or imagined possibilities offered by laboratory-grown meat have ushered in the new food production paradigm known as cellular agriculture (CA), which, over the course of the last decade, has evolved as a response to resource-intensive conventional agriculture, inhumane factory farming, and the environmental impacts of global climate change. The idea is to revolutionize agriculture and to disrupt the food system by creating an animal replacement or post-animal bioeconomy by the middle of this century.

Laboratory-grown meat benefits from a unique historical moment of fast-paced technological convergence, combining scientific advances from a series of disciplines within the life sciences as well as bioprinting, artificial intelligence, and other digital technologies. At the time of writing, the science of producing slaughter-free meat in the laboratory remains in large part experimental; however, the material conditions for its establishment have been carefully put in place, over the last decade, to secure capital flows, regulatory support, and social acceptability. The industry is, in some ways, both in its infancy and maturation phases and the promoters of the technology, assembled under the umbrella of cellular agriculture (CA), promise that this product will be commercialized in the near future, in spite of the fact that major technical and social hurdles remain as discussed in Chap. 5. As cell meat producers announce plans to scale-up production, the industry is also facing some of the same issues and backlash related to the commercialization of Genetically Modified

Organisms (GMOs), a debate that subsists two decades on after their commercialization.

Promoters of cellular agriculture invoke themes of technological determinism, raising questions about whether technological advancements dictate societal change and prompting ethical considerations surrounding the ontological status of life. This revolutionary framing challenges traditional notions of agriculture, proposing a post-anthropocentric view where human intervention can potentially mitigate ethical and environmental concerns associated with conventional meat production. Moreover, it prompts a reevaluation of our moral responsibilities and the potential for biotechnology to redefine what it means to nurture and sustain life within the Anthropocene epoch.

CA can be understood as the "controlled and sustainable manufacture of agricultural products, with cells and tissues without plant or animal involvement" (Coller, 2023). According to the Cellular Agriculture Society (CAS), now rebranded *From Fauna*, an international non-profit organization that supports the development and commercialization of cell-based products, it can be understood as "the process of farming animal products from cells instead of animals" (https://fromfauna.org). This means that most of the production process typically involves engineering and biotechnology techniques using methods such as tissue culture, scaffolding, and bioreactors in order to produce large quantities of edible tissues and proteins from cells which include cultured meat, milk, egg products, as well as other plant and animal materials such as leather and fibers (Barzee et al., 2022; Nyyssölä et al., 2022). This is described in Chap. 3. This signals new levels of complexity and tension when conceptualizing meat as more than flesh from dead animals, interrogating issues surrounding its socio-materiality and symbolic meaning. Food consistently blurs the boundaries of nature and culture; hence, the socio-materiality of synthetic meat, a product of technological convergence, is a subject worthy of attention by social scientists as it attempts to reconcile industrial animal farming with the ethics of producing edible living animal tissue and with the moral imperative of living sustainably while paying attention to human/non-human relations. Indeed, synthetic meat comes with a powerful moral charge and its claims are substantial.

The term "cellular agriculture" encapsulates images and connotations, reflective of its intricate sociotechnical landscape. Semantically, it merges the perceived precision and control of laboratory environments with the age-old practices of agriculture, eliciting a vision of bioreactors and petri dishes juxtaposed against the backdrop of verdant fields and pastoral scenes. In some ways, the traditional view of agriculture ignores the ways in which modern agriculture has been deeply shaped by technological forces. The term came into being in 2015 through a Facebook group called the "New Harvest Community Discussion." Then executive director Isha Datar argued for "cellular agriculture" to describe the industry and the new discipline. The conversation did not, however, specifically steer toward a vision of traditional agriculture. For her, agriculture signified "field growing" or "field cultivation" and, as the conversation thread continues, Datar also suggested that "cell means small room in Latin" which means you could cultivate indoors and in small spaces. As one reads the exchanges, it becomes clear that the term had to include not just in vitro meat but also plant-based and other fermentation-based processes. So, the word play with rural imagery was not lost on the "New Harvest" community, as the name of the non-profit suggests.

In order to illustrate the use of farming metaphors in CA, as a way of navigating between traditional and new agricultural paradigms, the process describing CM chicken production from the company SuperMeat (a CM company based in Tel-Aviv) is very telling:

> How our chicken is made:
>
> 0. It starts with a seed: Cells are taken from a chicken once to form what we call the seed—this contains cells that mature into muscle, fat, and other meat comprising tissues.
>
> 1. Providing nurture: We plant the seed in a meat fermenter, where it is provided warmth, oxygen, and feed. The cells grow and mature into meat tissues just as they would in the animal's body.
>
> 2. The urban harvest: When the meat is mature, we harvest it from the fermenter by removing the remaining liquid feed.
>
> 3. To the kitchen! Once our chicken has been harvested, it is ready to be prepped, cooked and turned into your favourite dish—it can even be processed into any meat application. (SuperMeat, 2024)

The duality between cellular and conventional agriculture, whether intentional or not, underscores the inherent tension between the artificial and the natural, as cellular agriculture seeks to produce products that are ostensibly 'natural'—such as meat and dairy—through inherently artificial means involving extensive biotechnological manipulation and engineering. However, one should be reminded that industrial chicken and cell-based chicken are in perfect continuity (Porcher, 2016) and that the notions of naturalness and realness warrant further critical assessment. The issues surrounding real versus fake meat will be discussed at the end of the book. While proponents champion CA as a sustainable alternative to traditional agriculture, with the potential to significantly reduce land use, greenhouse gas emissions, and animal suffering, critics argue that it represents a further industrialization of the food system, which is often associated with environmental degradation and social inequities.

Furthermore, the term cellular agriculture evokes a profound conflict between tradition and innovation. It challenges the long-standing agricultural practices that are deeply embedded in cultural heritage, offering instead a vision of food production that is technologically advanced and highly controlled. This raises critical questions about the values underpinning our food systems—whether authenticity and the natural variability of terroir are being sacrificed at the altar of efficiency and predictability. Ethically, cellular agriculture foregrounds significant considerations regarding animal welfare, environmental sustainability, and global food security, while simultaneously prompting a discourse on the moral implications of such profound manipulation of life processes. Thus, "cellular agriculture" is a term fraught with conflicting meanings, encapsulating both the promise of a revolutionary approach to food production and the contentious debates surrounding its broader implications for society and the environment. In addition, the term cellular agriculture evokes stories within larger stories and so, narratives of traditional versus cellular farming contain specific forms of understanding about how food production may be radically transformed.

Other terms, like alternative protein (Alt-protein), discussed in detail in the next section, also tell stories. Guthman questions how material promises are discursively and materially enacted:

> scholars writing specifically on new alternative proteins have shown, attempting to enact their specific material promises involves a great deal of discursive, regulatory, and ontological work. Alternative proteins must be positioned as substantially similar to meat, milk, seafood, and eggs, while at the same time shown to be different enough in their production practices to make good on the array of promises they make related to human health, environmental sustainability, climate change, and animal rights and welfare that are their raison d'être. (Guthman, 2022, p. 74)

The promise of the same or better meat competes with new and different meat. Claims that cellular meat is 'real meat' also maintain this tension.

Although the plant-based meat (PBM) analogue industry was already established and expanding, with companies such as Beyond Meat (founded in 2009) and Impossible Foods (founded in 2011) were set to commercialize their meat-free products, the convergence of the factors mentioned above meant that the demand for meat substitutes got an unexpected impetus. The combination of market scarcity, animal diseases, and high prices for conventional meat created demand for 'meat-like' products and as a result, investors, food industry giants, and public interest for these products saw opportunities for these companies to expand beyond vegetarian and vegan markets (Meyer & Terazono, 2020). Today, Beyond Meat products, such as their plant-based chicken, steak, sausages, and burgers, are available in major supermarkets and restaurants globally, as they have partnered with fast-food chains. Impossible Foods also offers plant-based pork, sausage, and chicken nuggets but its most famous product is the Impossible Burger, a burger designed to mimic the taste, texture, and nutritional profile of an animal burger using a key ingredient, a soy leghemoglobin, a product derived from genetically engineered yeast that carries heme, a molecule rich in iron.

The term alt-protein came to be associated with this industry, as it was developing new meat substitutes using a variety of plant-based and other ingredients as well as industrial processes, in order to mimic animal products. These are distinct from cell-based animal products, although they are part of the food tech and alt-protein space, and their development has undoubtedly influenced how cell-based products might be designed and rendered. Their increased presence on supermarket shelves and in fast-food restaurants has certainly been conducive to producing communities of eaters and tasters interested in consuming 'animal as well as non-animal meat products.'

Plant-Based Meats and Meat-Based Plants: The World of Alt-Protein

The food anthropologist Benjamin Aldes Wurgraft (2019) writes, in the introductory chapter of his book entitled *Meat Planet: Artificial Flesh and the Future of Food*, that "protein is protean" (p. 21). What this conveys is that protein, just like meat, is diversified, multifaceted, and ever-changing, ranging from plant-based protein, to cell-based, to insect and other novel meat substitutes and new, often unexpected, sources of protein. Guthman and Biltekoff write:

> Inasmuch as proteins are in fact both ubiquitous in environments and materially protean also presents challenges in making many of these products appear and act as food. (Guthman & Biltekoff, 2023, p. 60)

In fact, the alternative protein paradigm describes the new forms and sources that encompass what is called alt-protein. Protein sources are no longer limited to being plant-based and cell-based and today represent the "leading edge of high-tech food innovation" originating from companies located in Silicon Valley, as their epicenter (Guthman & Biltekoff, 2023).

Meat analogues describe the array of imitation meats or meat substitutes that possess the appearance and characteristics of different types of meat. While animal protein from slaughtered meat provides all the essential amino acids, so-called imitation meats are meats made from non-meat ingredients, usually using soy protein, tempeh, nuts, or wheat gluten, and are made to replace animal protein in the diet. Products like tofu and seitan have been around for a long time, but newer meat analogues tend to use soy, wheat, pea, and fava bean protein isolates and textured vegetable proteins (TVP) (Joshi & Kumar, 2015; Lamas, 2021).

They also undergo a different type of processing in order to mimic the appearance, texture, smell, and flavor of meat, known as high-moisture extrusion processing, in order to alter the protein structure of plants, an important step in achieving the right moisture content, physicochemical properties of the final product (T. McHugh, 2019).

The rise in popularity and market success of plant-based meat, starting and coinciding with the pandemic, when animal meat was in short supply, it was expensive and considered by some to be potentially risky as a vector for diseases. This gave the cellular agriculture and cell meat industry a boost, with investors and large animal corporations investing in start-ups and new processes for laboratory-grown animal products. Plagued by technical and other difficulties, the cultivated meat industry has had to revise its predictions for commercial releases, pushing back release dates for its products. Industry insiders have also pared down their ambition about the potential of cellular agriculture products to revolutionize animal agriculture and food systems. Scientists and entrepreneurs in the industry are constantly perfecting their products, improving systems and finding new ways of making their products affordable and sustainable, all the while focusing on making them totally animal-free.

The rise in meat analogues has played a major role in redefining meat for consumers, be they vegan, vegetarian, or simply meat reducers. These new types of plant-based meat analogues are part of a worldwide trend, partly justified by health, animal ethics, and climate concerns, but also their rise coincides with the COVID-19 pandemic which posed a worldwide challenge to the availability of meat and, as mentioned above, linked meat consumption, with potential infection of zoonotic diseases to humans and a generalized rejection of animal-based foods (Zioga et al., 2022). The new plant-based meat analogues are also called "meat analogues 2.0" or "next generation meat" (Ibid). They also differ from older plant-based analogues in that they are designed to appeal to meat-eaters unlike previous foods like tofu or seitan, also known as vegetarian meats. Because they wish to attract meat-eaters to include these products as part of their diet, they must resemble meat as closely as possible. This is no small feat, as it took six years for Beyond Meat to develop its first Beyond Burger, having gone through several reformulations during that process (Lamas, 2021). Since then, there have been a wide range of plant-based alternatives to animal meat, including fish, bacon, steak, chicken nuggets, meatballs, sausage, etc. The new plant-based meat analogues are part of the alternative proteins space, which includes meat produced from animal cells as a separate set of production processes.

The Good Food Institute (GFI) provides a broad definition of alternative proteins, according to them:

> Alternative proteins are proteins produced from plants or animal cells, or by way of fermentation. These innovative foods are designed to taste the same as or better than conventional animal products while costing the same or less. Compared to conventionally produced proteins, alternative proteins require fewer inputs, such as land and water, and generate far fewer negative externalities, such as greenhouse gas emissions and pollution. Some of these products are available to consumers today, including numerous plant-based and fermentation-derived options. Others, such as cultivated meats, remain primarily in development. (GFI, 2022)

There are three main types and methods for producing alt-protein: plant-based meat, cultivated meat, and fermentation. An important distinction needs to be made at this point because the products obtained from cellular agriculture differ from protein analogues derived from plant, fungal, yeast, insect, micellar, or algal materials in order to mimic meat, milk, or eggs. These products occupy what has been called the "meatless space" or the meat replacement market and have traditionally been aimed at vegetarians or vegans, although they also target consumers known as meat reducetarians or flexitarians. This book focuses on cultivated meat and *animal replacement products* understood as using animal components, cells in the case of CM or proteins and enzymes cloned into yeast and fermented in the production of cellular milk and dairy products. Nevertheless, the development of plant-based alt-protein is important to contextualize because it marks a turning point in people's understanding of meat. Further, because the development of plant-based and cell-based meats is intricately linked, notably as 'blended meat,' it is important to situate plant-based meat within the alt-protein landscape.

According to analysts, the alt-protein market segment is now a mainstream phenomenon; it is no longer a niche market. In 2019 the

U.S. plant-based meat industry was worth $801 million and was the third fastest-growing category of plant-based foods, behind milk and other dairy substitutes with predictions of a net worth of $27.9 billion by 2025, according to researchers at MarketsandMarkets, while the financial firm UBS predicts the industry will grow to $85 billion by 2030 (Siegner, 2019). The GFI plant-based market report shows that in 2023, the U.S. retail plant-based foods' industry value was $8.1 billion, plant-based meats and seafood industries were valued at $1.2 billion, while plant-based milk was worth $2.9 billion. However, they also note a decline of -12% in the case of plant-based meat from the previous year (GFI, 2021). The GFI keeps a database of all the companies involved in plant-based meat, covering 583 companies worldwide.

Companies such as Impossible Foods and Beyond Meat have successfully teamed up with fast-food restaurants such as Burger King and Pizza Hut to popularize their pea protein-based Impossible Burger and the Beyond Burger. Consumers looking for meat alternatives have been quick to embrace these new offerings. In fact, major meat corporations such as Cargill, Tyson Foods, and Smithfields Inc. have entered the alternative protein market, focusing on developing their own lines of plant-based meat analogue products on the strength of their supply chain. Cargill, with more than 8 million cattle harvested and over 8 billion pounds of boxed beef and by-products in North America sold annually, is still betting on animal products. Although it has increasingly invested in diverse protein forms (plant, animal, fungi, and microbe), Food Navigator reports that Cargill has been investing in cultivated meat start-ups like Memphis Meats, Aleph Farms, Wildtype, and Upside Foods (Southey, 2024). At its core, Cargill provides the raw materials for plant-based protein that go into analogue meat and dairy alternatives because it is a conventional meat production company, and they are the world's largest grain trader, with $165 billion in revenue in 2023 (Watts, 2023).

Smithfields Inc. and others at their core see cellular meat as an opportunity to maintain their core business, while also testing the waters in the animal- and plant-based alt-meat industry. Depending on market forces, the distribution networks and infrastructure that these industries control will be a major force toward moving consumers away from conventional animal products. The fact that meat industry giants such as Tyson Foods and Cargill are pursuing opportunities and growing the market for alt-proteins is not surprising. They are venturing into plant-based protein in particular, and it is believed that these new ventures may send mixed signals to vegetarian and vegan consumers. As a way to prevent this, investments in plant-based meat innovations and animal-based laboratory meat have focused on the flexitarian market. Flexitarians or semi-vegetarians occasionally consume meat or fish, however, their main diet consists of vegetarian food. Hence, the demand or the push for cellular meats is not a direct response to veganism, who stay away from animal products. Rather, in the case of cellular meat, this product is clearly aimed at "meat reducers" as they lessen their dependence on traditional animal products, seeking instead novel protein sources or alt-proteins. Flexitarian diets also include fish and seafood products which are also targeted by the alt-protein and cell meat producers.

Laboratory-Grown Meat Today and Food Tech

Cellular agriculture, although difficult to define, can be divided into two categories that Stephens et al. (2018) describe as tissue engineering-based and fermentation-based depending on the production method used. New Harvest, one of the first leading organizations involved in advancing cellular agriculture, uses a nominally different way of describing the two main approaches used to produce animal replacement products in the laboratory as cellular versus acellular agricultural products corresponding to tissue-engineering-based versus fermentation-based, respectively (New Harvest, 2024).

The California-based Good Food Institute (GFI), the alt-protein think tank, categorizes plant-based meat (PBM) products in the following way: (1) *Products similar to meat* which include companies that redefine meat by attempting to emulate the key characteristics of animal-based products from the ground up and by focusing on the "molecular structure and composition rather than the animal origin" of the product (Lagally et al., 2017, p. 4). They include replicas of ground beef, deli meats, sausages and hot dogs, beef burgers, etc. New meats in this category include the Impossible Burger and the Beyond Burger. (2) *Products not similar to meat* which directly replace meat, such as veggie-burgers, plant-based sausages, tofu, seitan, or tempeh, but do not necessarily replicate animal-based products because they target non-meat-eating consumers. (3) *Functional meat products* are plant-based proteins or binding agents used as ingredients of prepared foods or 'heat and serve' products. They do not imitate the taste of meat, but because they "retain their shape, texture and flavor through freezing, thawing, and cooking in a full range of prepared products" (Lagally et al., 2017, p. 4) and provide cheap alternatives to meat. Functional meats include extruded proteins, textured vegetable proteins (TVP), processed and unprocessed plant protein, used in products such as plant-based chicken nuggets served in schools or other institutional settings. (4) *Nature's mimic of meats* describes a class of proteins derived from naturally fibrous wholefoods such as pulses, mushrooms, pressed vegetables, and jackfruit to mimic pulled pork, for instance (Lagally et al., 2017, p. 5). These tend to replicate the taste and mouthfeel of meat but unlike previous meat substitutes, they are not highly processed. Plant-based meats marketed by companies, such as Beyond Meat, Impossible Foods, and Ocean Hugger Foods, represent what the Guardian has called "the rise of ultra-processed fake meat" (Blythman, 2018). These foods are considered to be ultra-processed because they are "manufactured by opaque, high-tech methods using industrial additives and food derived component parts" (Boseley, 2018).

Journalist A.C. Shilton reminds us that "Humans have been manipulating plants into meat-like approximations for centuries—think tofu and tempeh—and today you can get hemp milk foam for your latté, order a barbecue sandwich made of jackfruit, or scoop a ladleful of protein-packed cricket powder into a bulk bag at your organic grocery store" (Shilton, 2019). What these protein trends reveal is a global interest in the alternative protein market, which includes but is not restricted to CM. My focus throughout the book is mostly on in vitro meat with references to the larger alt-protein context as cultural, regulatory, and socioeconomic issues adapt to evolving patterns of food production and consumption and corporate interests invest and realign along "protein platforms." I will include a brief discussion of lab-produced human breast milk as another example of cellular engineering that aims to substitute a food that has resisted scalability (Pfotenhauer et al., 2022; Tsing, 2017) and, to a certain extent, the industrialization process. Human cellular breast milk will be discussed in the following section of this chapter in the context of the Anthropocene as another illustration of how bodies get displaced as attempts are made to replace or substitute naturally produced substances.

In February 2024, a scientific paper published by a team of South Korean researchers in the journal *Matter* described a strategy for growing "a nutrient-rich hybrid food using grains functionalized with nanocoating and bovine cells for a sustainable food system" (Park et al., 2024). In what has been called "beef-rice" or meat-based plant food, the team used rice grains as a scaffold, coating them with fish gelatin and food enzymes, in order to grow organized bovine cells (Ibid). So, this hybrid food is part rice and part beef muscle and fat stem cells, the media photo of the released product depicts a white bowl full of bright pink rice, "which tends to happen when flesh enters the picture" as one journalist commented (Bonk, 2024). The meat cells were grown for 9 to 11 days in a petri dish on the surface of the rice grain as well as inside it, which served as a three-dimensional structure, mimicking a cellular environment (Benke, 2024). One of the researchers is quoted as saying:

> Imagine obtaining all the nutrients we need from cell-cultured protein rice, [...] Rice already has a high nutrient level, but adding cells from livestock can further boost it. (Benke, 2024)

The hybrid beef-rice is said to be relatively easy to produce and researchers see a potential for perfecting their system and adding more nutritious elements to their product. It is said to resemble microbeef sushi (Yeung, 2024). In the future, these scientists aim to produce an immortalized cell line from cows, eliminating the need to harvest cells from live animals. They also hope that this technology will help create a more sustainable food system and prevent food insecurity. For them, this means decreasing our reliance on livestock from meat production. They view their rice hybrid as potentially alleviating problems linked to "health concerns, infectious disease risks, climate change, and resources scarcity" (Ibid). One of the researchers added:

> I didn't expect the cells to grow so well in the rice, […] "Now I see a world of possibilities for this grain-based hybrid food. It could one day serve as food relief for famine, military ration, or even space food. (Yeung, 2024)

So, while cell meat seems to undergo many transformations and as the title of this Atlantic Magazine suggests "Everything Can be Meat" (Tayag, 2024), the future of cellular agriculture lies with the development of hybrid foods, that is, blended products that contain plant- and animal-based ingredients like the addition of lab-grown pork fat cells to fake, plant-based bacon. Mission Barns, the company that produces lab-grown animal fat cells, believes that the addition of fat cells from animals to plant-based products is the key to improving their taste, as consumers often complain about the blandness of plant-based meats. Not only does it add flavor, it also removes the need to add coconut oil, as well as other highly saturated fats, to their final products to enable cooking, thus potentially making these PBMs 'healthier' (Poinski, 2024).

Other Cellular Agriculture Products

Besides meat, other animal replacement products such as milk and eggs are currently being produced and marketed using cellular agriculture. These are different from the non-dairy, plant-based beverages known as 'milks' (soya milk or almond milk, for example) or vegan egg-replacement products. The San Francisco-based companies Perfect Day Foods and the EVERY Company (formerly Clara Foods) use specialized yeast engineer ing to produce milk and egg whites, respectively. Clara Foods aims to produce animal protein without animals, by "creating the protein of tomorrow, today." Perfect Day Foods wants to produce animal-free dairy products. Both companies are using advanced fermentation techniques and yeast engineering techniques. Perfect Day Foods produces "real dairy protein," namely, casein and whey, which are genetically engineered in yeast. In addition to making eggs and egg whites without chickens, the EVERY Company produces protein for several sectors of the food industry (baking, ingredients, nutritional supplements) by "decoupling proteins from the animals that make them, using fermentation." JUST Foods also produces 'eggs,' although they use plant materials to achieve this. Cellular seafood and cellular aquaculture products are also an important part of the industry. Just like meat without animals, their products aim to produce "seafood without the sea." These companies play a significant role in the post-animal bioeconomy, each utilizing distinct cellular production processes and visions. Consequently, I do not examine other replacement products and processes, such as dairy and eggs, nor do I specifically analyze cellular aquaculture, despite its sharing many of the challenges faced by cellular meat.

Another lucrative market for cellular meat is the pet food industry, with companies such as BioCraft Pet Nutrition (formerly called Because Animals), Wild Earth, Meatly (U.K.), and Bond Pet Foods. It is estimated that the pet food industry earns approximately $100 billion U.S. a year and that dogs and cats in the U.S. consume 30% of all meat derived from intensively farmed animals (Oven et al., 2022). The rise in pet ownership since the pandemic is opening up new market demands for human-grade meat or sustainably sourced protein, as opposed to the waste products traditionally used to make pet food (Ibid). In addition, plant-based dog and cat foods are of interest to vegetarians and vegan pet owners. Early in 2024, Meatly announced the production of canned cat food containing cultivated chicken, although not yet approved for sale in the U.K. (Wall, 2024). BioCraft announced, in May 2024, that it achieved price parity with premium pet food for its cell-cultured mouse meat dog and cat food and hopes to market it in early 2026 (Cleaver, 2024). Large pet food companies like Hill's Pet Nutrition are also poised to include cell-cultured animal proteins in their formulations, having just received two metric tons of product from Bond Pet Foods (Ibid). Cellular pet food meat represents a potentially huge market, less technologically challenging than the meat produced for human consumption, since it doesn't have to mimic conventional meat and is less costly as a result (Oven et al., 2022). Moreover, cellular-based pet food does not have to possess the same organoleptic properties, in terms of taste, texture, and smell as meat designed for human palates. It is believed that cellular pet meat could attract pet owners to purchase human-grade cell-based products for themselves, serving as a gateway to the emerging industry. A survey asked consumers (729 respondents) about their willingness to feed their pets with cultured meat, including their own desire to consume this type of product, found that the market for human consumption and for pet food did not necessarily correlate. Some people might be willing to feed it to their pets but did not express interest in eating it themselves. The authors of the study conclude that cell-based pet food may actually have a negative impact on the appeal of human cell-based meat and that the relationship between the two seems to be complex and understudied (Oven et al., 2022). Citing health and safety concerns about pet- and human-grade cell meat, the study also notes that price was an important factor in the survey.

Finally, another relatively new segment of human-animal alt-protein is the production of cultured human breast milk. This product, rather than being revolutionary, perpetuates efforts to replace maternal milk with substitutes like infant formula, in an attempt to closely mimic its natural counterpart. In this case, cell-based breast milk might be considered an example of substitutionism where a natural product is replaced by an industrially produced or manufactured product (i.e. like margarine replacing butter according to Goodman et al.). Until recently, efforts to "humanize" cow's milk involved genetically modifying cows to add or remove the genes responsible for producing proteins that cause allergies in infants. Chinese researchers successfully produced modified human milk in cows by engineering the animals to secrete milk containing lysozyme, a bactericidal protein found in human breast milk, responsible for boosting infant immunity, protecting it from microbial infections (Yang et al., 2011). More recently, a company called BIOMILQ[5] claims to produce "bioauthentic human milk ingredients for better human health" by "combining the unmatched capabilities of human mammary cells with a scalable biomanufacturing process to produce human milk ingredients" (BIOMILQ, 2024). In 2021, the company declared to have "produc[ed] the world's first cell-cultured human milk outside the breast" (Cohen &

[5] BIOMILQ is based in North Carolina and received seed funding from Bill Gate's Breakthrough Energy Ventures in 2020. The company boasts that it is "women-owned and science-led." The idea came about when the CEO had trouble breastfeeding her infant son; at the time she was finishing her post-doc in human physiology using cell culture. Other companies involved in cell-based human milk are Wilk, TurtleTree, and Helaina. These companies are focusing on producing the ingredients for infant formula, but their products such as human Lactoferrin can also be expanded to other food products in order to boost nutrition.

Cassidy, 2022). BIOMILQ has developed a breast milk bank from donations by more than 300 women, which it uses to study the complex science of human lactation and the "intricate composition of human milk" (BIOMILQ, 2024). The startup Wilk uses cells collected from breast reduction surgeries, making a link with the valorization of what would be considered biomedical "waste products" and the development of cell lines and resulting patentable products. Human cell-based breast milk product, like CM, crosses the line between food and regenerative medicine, but like definitions of meat which tend to be complicated by CA, human milk "produced from a human body is notoriously hard to define legally, as it can be thought of alternatively as a food, a tissue, a drug, a relationship, or all of the above" (Cohen & Cassidy, 2022, p. 5). In a piece entitled *Eating the Mother*, philosopher and sociologist Irina Aristarkhova questions the centrality of mothers as food, meat even. She writes:

> But is the mother also a kind of meat? One of archaic definitions of meat is that it is "food of any kind." If the mother is food, therefore, she is meat: it is a simple logical outcome. As solid food (tissue, placenta), she is even "more" meat. However, breast milk, as a liquid type of food, is then a kind of meat (compared to blood, which needs to be cooked into a sausage—for example, a blood sausage, as quoted above—to become "meat"; otherwise it is a "drink of blood"). Here the method of ingestion (and not only the quality of the thing itself) is important: how it feels inside the mouth, whether teeth are engaged or not, and if it is to be drunk, swallowed with or without chewing. Infants, after all, do not have the option of chewing. Their meat needs to be liquid. The mother is also meat because she is a mammal (as opposed to fish or fowl); she is a "red meat" animal, giving her breast milk to the offspring like a cow. (AristarKhova, 2021, p. 46)

This points out to the fact that like the animal's body, the maternal body viewed as meat may also be involved in some aspects of meat politics even though humans are not normally considered 'food.' However, the aim of cell-based maternal milk is not to completely replace human lactation but to augment it, unlike CM. I had originally thought of including lab-produced maternal milk in this book because it opens a discussion about how the degradation of the Earth's ecosystems affects human health and reproduction. But I decided instead to focus my analysis on edible cell meat tissues because lab-produced human milk is not related to agriculture per se, even though there are clear connections. However, cell-based human breast milk is linked to the Anthropocene as both a response to the environmental health challenges of this era and as a symbol of the broader shifts in how humans interact with and rely on nature in an age of unprecedented ecological change. What's more, BIOMILQ breast milk is a product while the act of breastfeeding is a process, involving intimate skin-to-skin communication between mother and infant (Kent, 2020).

As is the case with CM, this is rooted in the ways that technological interventions are developed in response to the environmental crisis. Historically, human breast milk has been considered the epitome of natural nourishment—an essential and uncontaminated life-sustaining substance directly linked to maternal care and the continuity of life. However, as environmental pollutants such as persistent organic pollutants (POPs), heavy metals, and endocrine-disrupting chemicals (EDCs) infiltrate ecosystems, they accumulate in the human body and are passed into breast milk, transforming it from a pure and natural substance into a vehicle for toxins (Boswell-Penc, 2006; Serreau et al., 2024). This contamination problematizes the notion of breast milk as natural, suggesting that in the Anthropocene, even the most intimate aspects of human biology are tainted by industrial and agricultural practices. In this context, cell-based milk emerges as an "unnatural" solution to the unnatural problem of polluted breast milk. The contamination of breast milk with environmental pollutants serves as a stark indicator of broader ecological crises, much like the canary in the mine, as Rachel Carson and others have demonstrated. Feminist scholars highlight the fact that:

> Scientists often use samples of breastmilk to determine the condition of a local ecosystem because breastmilk contains pollutants that are not present in blood and urine samples. Toxic breastmilk thus becomes, not just a tragedy in itself, but an indication of much more wide-ranging devastation…Women will no longer have the choice to nurse their babies, if they want to protect them from the toxins that will have accumulated in the fatty tissues of their bodies. (Boswell-Penc, 2006, pp. 138–139)

This contamination underscores the violation of women's reproductive autonomy, as the ability to safely nurse their children is compromised by the accumulation of toxins in their bodies. The intergenerational transmission of these contaminants, via breast milk, reveals the intimate, embodied connection between human reproduction and environmental degradation. Women's bodies are not only subject to environmental harm but are also rendered complicit in the transmission of these harms to the next generation. In response to these challenges, cell-based breast milk emerges as a move away from embodied, relational forms of care toward a more industrialized, commodified approach to nourishment. In that regard, it is very similar to CM and the commodification of meat. Both represent forms of disembodiment and alienation from "natural" feeding practices. CM and cell-based breast milk, each in their own way, conjure up ideas about the medicalization of eating.

Getting back to alt-proteins, it is important to note that their consumption does not necessarily equate with lesser environmental impacts or production costs. As demand for novel proteins increases, these will require new value chains, which will require that production systems and scalability issues fall in line with sustainability goals. At the moment, large agribusiness companies that control value chains are at an advantage, and they may not only invest in start-ups to keep up with business opportunities but as suppliers of raw materials and other commodities needed in the production of alt-proteins. Some ingredients like soy are known to be resource intensive and tend to be genetically modified. Generally, studies show that plant-based proteins perform better in terms of environmental performance than animal-based food products; however, there is a certain lack of coherence and a high degree of variance among the results obtained, depending on indicators assessed, the functional units selected, and the nature of the food product. In addition, it is my impression that calculation of meat's impact on the environment tends to be based on an aggregate global meat industry, leaving out the farms and ranching operations that raise animals on grass and practice regenerative land management (Niman, 2021). Granted, the footprint or hoofprint of the "Global Meat Complex" (GMC) is considerable. The Institute for Agricultural Trade and Policy (IATP) defines the GMC as a concentrated, vertically and horizontally integrated "web of transnational corporations that controls the inputs, production and processing of mass quantities of food animals" (Sharma, 2018). However, despite the enormous power of the largest meat-producing companies like JBS, Tyson Foods, Cargill, and Smithfield, in 2018, only a small portion of the meat produced in the world was actually traded, since much of it was grown by small producers and tended to supply regional markets (Ibid). Besides, these same companies are also investing in the alternative protein business, including in vitro meat.

One of the major drivers of CM is its potential to have a lesser impact on the climate. It's not clear, however, if it is better for the environment as there is no data available regarding greenhouse gas (GHG) emissions on a commercial scale. Estimates about the climate impact of cellular meat tend to vary considerably since no large-scale operations are currently fully operational. There does not seem to be a consensus on the best way of obtaining an accurate reading of how beneficial CM production models might be, in comparison to the ecological impact of conventional meat. The lack of commercial scale facilities precluding an accurate comparison has meant that most calculations about climate impacts have been modeled differently, often leading to conflicting conclusions or wide arrays of results. Early Life Cycle Analysis (LCA) studies indicate the potential for CM to have "lower carbon footprint, land use and eutrophication effects than most conventional meats" (Sinke et al., 2023). More recent studies show a potential for lower carbon footprints than global averages for all animal meats except for chicken, which will have a higher carbon footprint (Tuomisto, 2022). A study published in 2019, comparing GHG emissions between CM and conventional beef, showed that while CM might have a lower carbon footprint than conventional beef, the long-term effect of CO_2 accumulation of CM on the climate may be higher than for cattle, because the methane produced by cows eventually breaks down into CO_2 (Lynch & Pierrehumbert, 2019). Yet another study published in 2023 found that in a scenario where processes and materials used in the biopharmaceutical industry are used, that near-term CM is likely to have a CO_2 impact "orders of magnitude" higher than conventional beef because of the energy-intensive purification steps to filter out contaminants from the culture medium. The same study, which also compared CM production using non-pharmaceutical grade ingredients or food-grade ingredients, found that CO_2 emissions would be lower than global average emissions from beef cattle (Risner et al., 2023). The food-grade scenario calculated 10–75 kilograms of carbon dioxide emissions per kg of beef while the pharmaceutical-grade study identified levels of emissions between 250 and 1000 kg for every kg of beef. In comparison, the analysis done by Sinke et al. (2023) calculated emissions of between 3 and 14 kg of CO_2 per kg of cultivated meat. However, this study drew criticism due to the fact that it is assumed by GFI scientists that CM will use food-grade components. In any case,

> Cultivated meat will still produce emissions, since energy is required to run the reactors that house the cells as they grow. In the US and most places around the world today, that will likely involve fossil fuels. Renewables could eventually be available widely and consistently enough to power facilities producing cultivated meat. However, even in this case, the reactors, pipes, and all other necessary equipment for production facilities often have associated emissions that are tough to eliminate entirely. In addition, animal cells need to be fed and cared for, and the supply chain involved in that also comes with emissions attached. (Crownhart, 2023)

Until CM is produced at a commercial scale, it will be impossible to obtain an accurate measurement of the climate impact of the industry.

The year 2024 has seen a drop in investment in plant-based protein companies, including meat, seafood, dairy, and eggs. Data compiled by the Good Food Institute (GFI) shows a 28% decline in investments in these companies, from $1.3bn in 2022 to $907m (Crawford, 2024). This coincides with decreases in the sale of plant-based meats possibly signaling the end of the plant-based food revolution. Industry analysts offer possible explanations for this development, namely, the price of plant-based meat has remained relatively high compared to conventional meats. The company Beyond Meat has been experiencing reduced sales, as much as 30.5% less revenue in 2023 than in 2022 and had to cut 200 jobs from its production facility (Flood, 2023). Many brands of PBMs have been a tough sell despite investment, new product launches, and cutting-edge technology. Planterra, the U.S.-based, plant-based meat subsidiary of meat giant JBS, shut down in September 2022, while Maple Leaf Foods let go a quarter of its staff from its plant-based Greenleaf Foods division in the same year (Poinski, 2022).

Convincing consumers to opt for plant-based meats may not be easy. This might explain why companies like Beyond Meat have expanded their product offering to include Beyond Chicken, Sausage, Jerky, and Steak, along with its original Beyond Burger. It appears as though during periods of inflation, consumers are more likely to select more affordable options. A recent industry study showed that repeat plant-based meat buyers are not very high in numbers and these same consumers continue to purchase real or conventional meat. So-called flexitarians will tend to see these products as an option to real meat, not as a replacement and they tend not to be loyal customers (Poinski, 2022).

Yet another factor associated with this loss in enthusiasm may be the taste and texture associated with PBM. Reports of deteriorating perceptions among those who buy it have surfaced in the media. It's also been shown that consumers of plant-based meat may not enjoy the taste and texture of analogues, as ingredient formulations and taste compositions vary. Furthermore, there have been implications surrounding the ultra-processed food status of PBMs as well as the environmental sustainability status of these products. Foods are considered ultra-processed when they go through a series of industrial processes that radically transform their original ingredients. Meat analogues fall into that category, whether plant-based or cell-based. These foods tend to be associated with negative health outcomes, as they tend to contain high levels of salt, sugar, flavor enhancers, and saturated fats in order to make them palatable. There are additional concerns with plant based meats more generally, as these are increasingly consumed. The safety of ingredients such as soy leghemoglobin that provide meat-like flavor and a blood-like appearance in the Impossible Burger has been put into question. Although the FDA considers it Generally Regarded As Safe (GRAS), this ingredient, produced in genetically engineered yeast, has been the subject of controversy because of its potential as an allergen due to the fact that it has never been consumed by humans. A recent study highlights the lack of research on the nutritional and health aspects of these meat products, as well as the fact that the food safety risk data associated with novel proteins is either commissioned by the companies themselves or the organizations that promote them (Bogueva & McClements, 2023). Their analysis arrives at the following conclusion:

> Currently, there is insufficient information about the potential safety and nutritional health impacts of consuming plant-based meat alternatives with different nutritional profiles to animal-based ones, and further research is clearly needed. In addition, there are new safety concerns associated with incorporating next-generation plant-based foods into the human diet, which arise from the potential presence of allergens, chemical contaminants, antinutritional factors, and pathogenic microorganisms. The long-term health effects of regular PBMA consumption are also not yet comprehensively assessed, and further investigation is clearly required. (Bogueva & McClements, 2023, p. 22)

There are also concerns associated with the health and safety status of lab-grown meat. It is believed that, in addition to nutritional issues, there are potential health and safety risks with the production process for achieving the final product. According to the FAO and WHO report on the food safety of cell-based meat, hazards can be introduced during four stages of production: cell selection, cultivation and manufacturing, harvesting and processing. The first is during the sourcing of cells and their storage, where either bacterial contamination, viral pathogens, parasites, or zoonotic infectious diseases that could originate from the animals used for biopsies. The production phase, including cell cultivation and manufacturing, can also be a source of microbial contamination from the animal-derived serum and other components used in growth media. There could be yeast, spores, and other fungal contaminants at that point. Another potential risk is:

> At the level of the cell lines used, there is a risk of (epi)genetic drift in cell lines due to constant sub-culturing, where mutations build up over time that may eventually cause changes in phenotypes (Soice & Johnston, 2021). The use of quality-controlled cell banks of cryopreserved cell lines is a way to mitigate the risk of losing cell-line fidelity to genetic drift, as well as protecting against the presence of viruses, bacteria, yeast and Mycoplasma spp. (FAO & WHO, 2023, p. 26)

The next level of risk can occur during harvesting, which removes a large portion of the chemical and biological components and residues from the media used for cell cultivation. Common ingredients found in culture media are mixtures of salts, sugars, vitamins, amino acids, organic acids, growth factors, and hormones. It is believed that the process of harvesting may introduce enzymes and chemicals used to separate the cells from the microcarriers that support the cells during growth. The FAO/WHO notes that "Of special attention is the use of biological components such as growth factors and hormones from animal (serum) or non-animal origin, as these biologically active molecules might interfere with the metabolism or have been associated with the development of certain cancers" (FAO and WHO, 2023, p. 27).

In terms of the processing phase and formulation of the final product that turns it into meat or nuggets,

> To process cultivated cells or tissues into cell-based food products for consumption, they are formulated with other ingredients and additives to improve e.g. the structure, texture, taste, colour or shelf-life of the end product…This can be edible and biocompatible microcarrier or scaffolding material that was already included for cell proliferation and differentiation, or e.g. binders, flavor agents and preservatives that are added after harvesting. These ingredients may exert an allergenic effect, and the allergenicity of the components of the cell-based food products is therefore an important factor for assessment. The cell line used might also have allergenic potential by itself, which in particular is the case for cell lines from fish or shellfish...Moreover, ingredients added to improve product characteristics, such as wheat gluten/hydrolysates, soy protein or milk components might be the cause of allergic reactions. All additives, ingredients, nutrients and all other substances added will need to be approved for application (e.g. considered to be safe and allowed for the specific cell-based food) and all applicable food labelling requirements will apply (including allergen labelling). As in the other stages of cell-based food production, potential microbiological hazards also exist during food processing, which should be minimized using GHPs. (FAO and WHO, 2023, p. 27)

Meat from animals normally contains vitamin B12, also known as cobalamin, which plays an essential role in red blood cell formation, cell metabolism, DNA synthesis, and other vital functions. Vegetarians and vegans normally take vitamin B12 supplements because it is not readily available from plants and the human body does not produce it. From a nutritional standpoint, for cultured meat to be considered as a substitute or equivalent to conventional meat, it must contain vitamin B12. An article published in 2020 entitled *Sensorial and Nutritional Aspects of Cultured Meat in Comparison to Traditional Meat: Much to Be Inferred* discusses the challenges of producing cultured meat that is biologically equivalent to traditional meat (Fraeye et al., 2020). The authors, collaborators from agricultural science, base their study on the available scientific literature about the production process of cultured meat since insufficient quantities of meat are available for a complete assessment. Nevertheless, they raise important questions about the post-mortem biochemical changes that occur when muscle becomes meat, after an animal has been slaughtered and attempt to correlate how these processes might occur in cultured meat. The process of becoming meat involves complex biochemical post-mortem transformations which alter animal flesh, including the production of lactate, which influences the properties of meat. In addition, animal meat undergoes a period of aging for tenderization which produces additional flavors. They question the "intracellular conditions" of cellular meat which may involve different steps for tenderization and for flavor development. Generally, the article argues that:

> Based on the available state of the art regarding production processes, it can be inferred that cultured meat currently differs significantly from traditional meat in its technological, sensorial and nutritional properties. Revealing the extent to which postmortem processes occur in cultured meat is crucial to understand its impact on sensorial and technological properties. Production of cultured meat resembling fresh, unprocessed meat entails the biggest challenges with respect to texture, color, flavor as well as nutritional composition ...With regard to nutritional value, we illustrated the long trajectory of additional research that is needed before the composition of cultured meat could resemble traditional meat, as well as the complexity of the medium composition needed to achieve this. This will not only add to the cost of the medium, but also increase the environmental footprint of the entire process. (Fraeye et al., 2020, p. 5)

Besides questioning the texture, color, and flavor of cell meat, the authors also discuss its nutritional composition. Without going into specific details, at issue are: the exact protein content of cell meat; the nutritional status of the scaffolding material used to support the tissues during growth; the level of fat; the lack of mineral content such as zinc, iron, and selenium and vitamin B12 which will have to be supplemented post-culture; other bioactive ingredients like taurine, essential for metabolic processes in humans but also for dogs and cats, as well as creatine, which they believe should be added to cell culture media. In terms of the levels of adverse products found in traditional meat, such as carcinogenic proteins and hemes, it remains unclear if they might be "lowered in cultured meat without compromising sensorial and nutritional aspects" (p. 5).

As a result, the challenge to manufacture meat products that are nutritionally equivalent to or superior to conventional meats is unlikely to be met, not only because of the findings of Fraeye et al.'s study but also because some of the same issues associated with consuming red meat will persist with cellular meat. Secondly, the cheapest kind of cellular meat currently available, the animal- and plant-based hybrids, like GOOD Meat's chicken nuggets, probably contains higher levels of additives, artificial colors, artificial flavors, and chicken fat than conventional meat.

Finally, in terms of safety risks, the FAO/WHO report suggests that cultured meat will have to meet the limits of harmful bacteria, allergens, residues like antibiotics, growth hormones, and other chemicals, just like any other standards established for other foods (Jones, 2023). Until cell meat products reach the market, it is impossible to know how safe and nutritious they actually are. It is also difficult to ascertain how comparable they are to conventional meat. Only time will tell.

Bibliography

AristarKhova, I. (2021). Eating the Mother. In Meat! A Transnational Analysis. Chatterjee and Subramaniam Eds. (pp. 39–60). Duke University Press.

Attard, G. (2022). Robots in Livestock Management. In Q. Zhang (Ed.), *Encyclopedia of Smart Agriculture Technologies*. Springer International Publishing.

Barzee, T. J., El Mashad, H. M., Cao, L., Chio, A., Pan, Z., & Zhang, R. (2022). Cell-Cultivated Food Production and Processing: A Review. *Food Bioengineering, 1*(1), 4–25. https://doi.org/10.1002/fbe2.12009

Benke, K. (2024, February 14). *By Growing Animal Cells in Rice Grains, Scientists Dish Up Hybrid Food.* EurekAlert! Retrieved from https://www.eurekalert.org/news-releases/1033922

Bennett, C. E., Thomas, R., Williams, M., Zalasiewicz, J., Edgeworth, M., Miller, H., Coles, B., Foster, A., Burton, E. J., & Marume, U. (2018). The Broiler Chicken as a Signal of a Human Reconfigured Biosphere. *Royal Society Open Science, 5*(12), 180325. https://doi.org/10.1098/rsos.180325

BIOMILQ. (2024). *Home | BIOMILQ | Human Milk Ingredients | Research Triangle Park, NC, USA.* BIOMILQ. Retrieved from https://www.biomilq.com

Blaustein-Rejto, D. (2022). *Livestock Don't Contribute 14.5% of Global Greenhouse Gas Emissions.* The Breakthrough Institute. Retrieved from https://thebreakthrough.org/issues/food-agriculture-environment/livestock-dont-contribute-14-5-of-global-greenhouse-gas-emissions

Blythman, J. (2018, February 12). The Quorn Revolution: The Rise of Ultra-processed Fake Meat. *The Guardian.* Retrieved from https://www.theguardian.com/lifeandstyle/2018/feb/12/quorn-revolution-rise-ultra-processed-fake-meat

Bogueva, D., & McClements, D. J. (2023). Safety and Nutritional Risks Associated with Plant-Based Meat Alternatives. *Sustainability, 15*(19), 14336. https://doi.org/10.3390/su151914336

Bonk, L. (2024, February 15). *You've Tried Plant-Based Meat, but here Come Meat-Based Plants.* Engadget. Retrieved from https://www.engadget.com/youve-tried-plant-based-meat-but-here-come-meat-based-plants-163654564.html

Boseley, S. (2018, February 14). Ultra-processed Foods may be Linked to Cancer, Says Study. *The Guardian.* Retrieved from https://www.theguardian.com/science/2018/feb/14/ultra-processed-foods-may-be-linked-to-cancer-says-study

Boswell-Penc, M. (2006). *Tainted Milk: Breastmilk, Feminisms, and the Politics of Environmental Degradation.* State University of New York Press.

Cleaver, L. (2024, May 30). *BioCraft's Cultivated Mouse Meat Reaches Price Parity with Premium Pet Food.* PetfoodIndustry. Retrieved from https://www.petfoodindustry.com/ingredient-issues/proteins/news/15676448/biocrafts-cultivated-mouse-meat-reaches-price-parity-with-premium-pet-food

Cohen, M., & Cassidy, T. (2022). Milk from Mars. The Challenges of Regulating Lab-Produced (Human) Milk. *Food and Drug Law Journal*, 77.

Coller, J. (2023). *Council Post: How Business Innovation Can Turn The Food Crisis Into A New Green Revolution*. Forbes. Retrieved from https://www.forbes.com/sites/forbesbusinesscouncil/2023/07/10/how-business-innovation-can-turn-the-food-crisis-into-a-new-green-revolution/

Crawford, E. (2024, April 29). Plant-based 3.0: Industry insiders share strategies for pushing alternative proteins forward. *Foodnavigator.com*. https://www.foodnavigator.com/Article/2024/04/29/Plant-based-3.0-Industry-insiders-share-strategies-for-pushing-alternative-proteins-forward

Crownhart, C. (2023). Here's What We Know About Lab-grown Meat and Climate Change. *MIT Technology Review*. Retrieved from https://www.technologyreview.com/2023/07/03/1075809/lab-grown-meat-climate-change/

Devyver, A. (2022). Freedom and Dialectics: On the Critical Theory of Moishe Postone and Theodor Adorno. *Philosophy & Social Criticism*, 01914537221137842. https://doi.org/10.1177/01914537221137842

FAO, & WHO. (2023). *Food Safety Aspects of Cell-Based Food*. FAO; WHO. Retrieved from https://openknowledge.fao.org/handle/20.500.14283/cc4855en

Fish, K. (2013). *Living Factories: Biotechnology and the Unique Nature of Capitalism*. McGill-Queen's University Press.

Flood, E. (2023, July 31). *From Prediction to Production: A Timeline of Cultivated Meat*.

Fraeye, I., Kratka, M., Vandenburgh, H., & Thorrez, L. (2020). Sensorial and Nutritional Aspects of Cultured Meat in Comparison to Traditional Meat: Much to Be Inferred. *Frontiers in Nutrition, 7*, 35. https://doi.org/10.3389/fnut.2020.00035

Galusky, W. (2022). *Protein Machines, Technology, and the Nature of the Future*. Springer International Publishing. https://doi.org/10.1007/978-3-031-08717-2

GFI. (2021, January 19). *Plant-Based Retail Market Overview | GFI*. Retrieved from https://gfi.org/marketresearch/

GFI, G. F. I. (2022). *2022 Cultivated Meat State of the Industry Report*. Retrieved from https://gfi.org/wp-content/uploads/2023/01/2022-Cultivated-Meat-State-of-the-Industry-Report-2-1.pdf

Gillespie, K. (2018). *The Cow with Ear Tag #1389*. The University of Chicago Press.

Goodman, D., Wilkinson, J., & Sorj, B. (1987). *From Farming to Biotechnology: A Theory of Agro-industrial Development*. Basil Blackwell. Retrieved from https://trove.nla.gov.au/version/45809245
Guthman, J. (2022). The CAFO in the Bioreactor. *Environmental Humanities, 14*(1), 71–88. https://doi.org/10.1215/22011919-9481440
Guthman, J., & Biltekoff, C. (2023). Agri-food Tech's Building Block: Narrating Protein, Agnostic of Source, in the Face of Crisis. *BioSocieties, 18*(3), 656–678. https://doi.org/10.1057/s41292-022-00287-3
Hannan, J. (2020). *Meatsplaining: The Animal Agriculture Industry and the Rhetoric of Denial*. Sydney University Press.
Haraway, D. (2015). Anthropocene, Capitalocene, Plantationocene, Chthulucene: Making Kin. *Environmental Humanities, 6*(1), 159–165. https://doi.org/10.1215/22011919-3615934
Haug, T. (2018). The Capitalist Metabolism: An Unachieved Subsumption of Life under the Value-form. *Journal for Cultural Research, 22*(2), 191–203. https://doi.org/10.1080/14797585.2018.1461352
Hayek, M. N., & Miller, S. M. (2021). Underestimates of Methane from Intensively Raised Animals Could Undermine Goals of Sustainable Development. *Environmental Research Letters, 16*(6), 063006. https://doi.org/10.1088/1748-9326/ac02ef
Hribal, J. (2003). "Animals are Part of the Working Class": A Challenge to Labor History. *Labor History, 44*(4), 435–453.
Jones, N. (2023). Will Cultured Meat Replace The Real Thing? *Nature, 619*, 22–24.
Joshi, V., & Kumar, S. (2015). Meat Analogues: Plant based Alternatives to Meat Products—A Review. *International Journal of Food and Fermentation Technology, 5*(2), 107. https://doi.org/10.5958/2277-9396.2016.00001.5
Kent, G. (2020). Why make Fake Breastmilk? *World Nutrition, 11*(3), 2–11. https://doi.org/10.26596/wn.20201132-11
Knorr-Cetina, K. (1999). *Epistemic Cultures: How the Sciences make Knowledge*. Harvard University Press.
Lagally, C., Clayton, E. R., & Specht, L. (2017). Plant-based meat mind maps. *The Good Food Institute*. https://gfi.org/wp-content/uploads/2021/02/PBMap.pdf
Lamas, M. (2021). *How Scientists make Plant-Based Foods Taste and Look More Like Meat*. The Conversation. Retrieved from http://theconversation.com/how-scientists-make-plant-based-foods-taste-and-look-more-like-meat-156839

Landecker, H. (2007). *Culturing Life: How Cells Became Technologies* (First paperback ed). Harvard Univ. Press.
Lewis, S. L., & Maslin, M. A. (2015). Defining the Anthropocene. *Nature, 519*(7542), 171–180. https://doi.org/10.1038/nature14258
Lynch, J., & Pierrehumbert, R. (2019). Climate Impacts of Cultured Meat and Beef Cattle. *Frontiers in Sustainable Food Systems, 3*. https://doi.org/10.3389/fsufs.2019.00005
Malm, A. (2015). The Anthropocene Myth: Blaming All of Humanity for Climate Change lets Capitalism off the Hook. *Jacobin Magazine*. Retrieved from https://jacobin.com/2015/03/anthropocene-capitalism-climate-change/
Malm, A., & Hornborg, A. (2014). The Geology of Mankind? A Critique of the Anthropocene Narrative. *The Anthropocene Review, 1*(1), 62–69. https://doi.org/10.1177/2053019613516291
McHugh, T. (2019, October 1). How Plant-Based Meat and Seafood Are Processed. *Food Technology Magazine, 73*(10). Retrieved from https://www.ift.org/news-and-publications/food-technology-magazine/issues/2019/october/columns/processing-how-plant-based-meat-and-seafood-are-processed
McMichael, P. (2013). *Food Regimes and Agrarian Questions*. Fernwood Publishing.
Meyer, G., & Terazono, E. (2020, April 26). *Pandemic Accelerates Shift to Meat Substitutes*. Retrieved from https://www.ft.com/content/0127984a-6def-4040-9bca-002b6ffd4e0a
Moore, A. (2016). Anthropocene Anthropology: Reconceptualizing Contemporary Global Change. *Journal of the Royal Anthropological Institute, 22*(1), 1. https://doi.org/10.1111/1467-9655.12332
New Harvest. (2024). *What is Cellular Agriculture*.
Niman, N. H. (2021). *Defending Beef: The Case for Sustainable Meat Production*. Chelsea Green Publishing.
Nyyssölä, A., Suhonen, A., Ritala, A., & Oksman-Caldentey, K.-M. (2022). The Role of Single Cell Protein in Cellular Agriculture. *Current Opinion in Biotechnology, 75*, 102686. https://doi.org/10.1016/j.copbio.2022.102686
Oven, A., Yoxon, B., & Milburn, J. (2022). Investigating the Market for Cultivated Meat as Pet Food: A Survey Analysis. *PLoS One, 17*(12), e0275009. https://doi.org/10.1371/journal.pone.0275009
Park, S., Lee, M., Jung, S., Lee, H., Choi, B., Choi, M., Lee, J. M., Yoo, K. H., Han, D., Lee, S. T., Koh, W.-G., Bang, G., Hwang, H., Lee, S., & Hong, J. (2024). Rice Grains Integrated with Animal Cells: A Shortcut to a

Sustainable Food System. *Matter, 7*(4), 1677. https://doi.org/10.1016/j.matt.2024.03.007
Pfotenhauer, S., Laurent, B., Papgeorgiou, K., & Stilgoe, J. (2022). The Politics of Scaling. *Social Studies of Science, 52*(1), 3–34. https://doi.org/10.1177/03063127211048945
Poinski, M. (2022). How can Plant-Based Meat Get Back to Growth? *Food Dive*. Retrieved from https://www.fooddive.com/news/plant-based-meat-back-to-growth/636834/
Poinski, M. (2024, June 8). *Why Mission Barns Believes Cultivated Fat is the Key to Better Plant-Based Meat*. Food Dive. Retrieved from https://www.fooddive.com/news/mission-barns-cultivated-fat-plant-based-meat/624960/
Polanyi, K. (1944). *The Great Transformation: The Political and Economic Origins of our Time*. Beacon Press.
Porcher, J. (2016). Demain, une agriculture sans élevage? *Pour, 231*(3), 255. https://doi.org/10.3917/pour.231.0255
Postone, M. (2020). Capital and Temporality. In *The Marx Revival: Key Concepts and New Critical Interpretations* (p. 157). Cambridge University Press.
Reisman, E., & Fairbairn, M. (2021). Agri-Food Systems and the Anthropocene. *Annals of the American Association of Geographers, 111*(3), 687–697. https://doi.org/10.1080/24694452.2020.1828025
Risner, D., Kim, Y., Nguyen, C., Siegel, J. B., & Spang, E. S. (2023). *Environmental Impacts of Cultured Meat: A Cradle-to-gate Life Cycle Assessment* [Preprint]. Ecology. https://doi.org/10.1101/2023.04.21.537778
Serreau, R., Terbeche, Y., & Rigourd, V. (2024). Pollutants in Breast Milk: A Scoping Review of the Most Recent Data in 2024. *Healthcare, 12*(6), 6. https://doi.org/10.3390/healthcare12060680
Sharma, S. (2018). *Mighty Giants: Leaders of the Global Meat Complex*. Retrieved from https://www.iatp.org/blog/leaders-global-meat-complex
Shilton, A. C. (2019, March 18). *Protein of the Future*. The Hub. Retrieved from https://hub.jhu.edu/magazine/2019/spring/lab-grown-alternative-meat-protein-ethics/
Shiva, V. (2019, January 20). *A New Report Sustains Unsustainable Food Systems*. Seed Freedom. Retrieved from https://seedfreedom.info/poison-cartel-toxic-food-eat-report/
Siegner, C. (2019). *Plant-Based Meat Market Forecast to Reach $85B by 2030, Report Says*. Food Dive. Retrieved from https://www.fooddive.com/news/plant-based-meat-market-forecast-to-reach-85b-by-2030-report-says/559170/

Sinke, P., Swartz, E., Sanctorum, H., van der Giesen, C., & Odegard, I. (2023). Ex-ante Life Cycle Assessment of Commercial-Scale Cultivated Meat Production in 2030. *The International Journal of Life Cycle Assessment, 28*(3), 234–254. https://doi.org/10.1007/s11367-022-02128-8

Soice, E., & Johnston, J. (2021). Immortalizing Cells for Human Consumption. *International Journal of Molecular Sciences, 22*(21), 11660. https://doi.org/10.3390/ijms222111660

Southey, F. (2024, March 6). *From Meat to 3D Printed Plants: Why Cargill is Diversifying its Protein Investments*. Foodnavigator.Com. Retrieved from https://www.foodnavigator.com/Article/2024/03/06/Why-Cargill-is-diversifying-its-protein-investments

Stafford, B. M. (2007). From Genetic Perspective to Biohistory: The Ambiguities of Looking Down, Across, and Beyond. In E. Kac (Ed.), *Signs of Life: Bio Art and Beyond* (pp. 373–386). The MIT Press.

Steffen, W., Broadgate, W., Deutsch, L., Gaffney, O., & Ludwig, C. (2015). The Trajectory of the Anthropocene: The Great Acceleration. *The Anthropocene Review, 2*(1), 81–98. https://doi.org/10.1177/2053019614564785

Steinfeld, H. (2006). *Livestock's Long Shadow: Environmental Issues and Options*. Food and Agriculture Organization of the United Nations.

Stephens, N., King, E., & Lyall, C. (2018). Blood, Meat, and Upscaling Tissue Engineering: Promises, Anticipated Markets, and Performativity in the Biomedical and Agri-food Sectors. *BioSocieties, 13*(2), 368–388. https://doi.org/10.1057/s41292-017-0072-1

SuperMeat. (2024). *Manufacturing—SuperMeat*. Retrieved from https://supermeat.com/manufacturing/

Tayag, Y. (2024, March 8). Everything Can Be Meat. *The Atlantic*. Retrieved from https://www.theatlantic.com/health/archive/2024/03/lab-grown-meat-beef-rice/677689/

Tsing, A. L. (2017). *The Mushroom at the End of the World: On the Possibility of Life in Capitalist Ruins* (First paperback printing). Princeton University Press.

Tubb, C., & Seba, T. (2020). *RethinkX: Disruption, Implications and Choices. Rethinking Food and Agriculture 2020–2050*. RethinkX.

Tuomisto, H. L. (2022). Challenges of Assessing the Environmental Sustainability of Cellular Agriculture. *Nature Food, 3*(10), 10. https://doi.org/10.1038/s43016-022-00616-6

Wall, T. (2024, March 19). *Cell-cultured Meat for Pet Food Nearing Market Appearance*. PetfoodIndustry. Retrieved from https://www.petfoodindustry.

com/blogs-columns/news-hound/article/15666679/cellcultured-meat-for-pet-food-nearing-market-appearance

Watts, J. (2023, May 4). Grain Trader Cargill Faces Legal Challenge in US over Brazilian Soya Supply Chain. *The Guardian*. Retrieved from https://www.theguardian.com/environment/2023/may/04/cargill-faces-legal-challenge-us-over-brazilian-soya-supply-chain-deforestation

Willett, W., Rockström, J., Loken, B., Springmann, M., Lang, T., Vermeulen, S., Garnett, T., Tilman, D., DeClerck, F., Wood, A., Jonell, M., Clark, M., Gordon, L. J., Fanzo, J., Hawkes, C., Zurayk, R., Rivera, J. A., De Vries, W., Majele Sibanda, L., et al. (2019). Food in the Anthropocene: The EAT–Lancet Commission on Healthy Diets from Sustainable Food Systems. *The Lancet, 393*(10170), 447–492. https://doi.org/10.1016/S0140-6736(18)31788-4

World Economic Forum. (2018). *Meat: The Future. Time for a Protein Portfolio to Meet Tomorrow's Demand* [White Paper]. Retrieved from https://www3.weforum.org/docs/White_Paper_Meat_the_Future_Time_Protein_Portfolio_Meet_Tomorrow_Demand_report_2018.pdf

Yang, B., Wang, J., Tang, B., Liu, Y., Guo, C., Yang, P., Yu, T., Li, R., Zhao, J., Zhang, L., Dai, Y., & Li, N. (2011). Characterization of Bioactive Recombinant Human Lysozyme Expressed in Milk of Cloned Transgenic Cattle. *PLoS One, 6*(3), e17593. https://doi.org/10.1371/journal.pone.0017593

Yeung, J. (2024, February 15). *Move over, Plant-Based Beef: Hello, Beef-infused Rice*. CNN. Retrieved from https://www.cnn.com/2024/02/15/asia/south-korea-beef-rice-study-intl-hnk-scn/index.html

Zioga, E., Tøstesen, M., Kjærulf Madsen, S., Shetty, R., & Bang-Berthelsen, C. H. (2022). Bringing Plant-Based Cli-meat Closer to Original Meat Experience: Insights in Flavor. *Future Foods, 5*, 100138. https://doi.org/10.1016/j.fufo.2022.100138

3

Cultivating Meat from Cells

Rather than the end of industrial animal agriculture or its ultimate form, cellular meat represents what French sociologist Jocelyne Porcher calls the anthropological rupture of the relation between humans and animals in farming (Chapelle, 2014). In Marxist terms, this means that rather than proposing an alternative to animal farming, laboratory-grown meat represents the ultimate stages of industrial agriculture, not its antithesis (Porcher, 2010). It is its continuation of the industrialization trends observed since the nineteenth century and it constitutes animal exploitation and factory farming by other means.

The production of cultivated meat relies on complex technical processes that transcend conventional methods of animal farming and slaughter. This section aims to explore these processes, highlighting the key steps involved in creating muscle tissues from stem cells in order to understand the arguments developed throughout the book. Advances in biotechnology and tissue engineering allow for the controlled growth of animal cells in a laboratory environment, resulting in a substance that shares certain characteristics with traditional meat. Lab-grown meat thus exemplifies how technology can be mobilized to recompose living organisms, transforming conventional approaches to food production and redefining our material relationship with biological resources.

E. Abergel, *Dead Meat*, https://doi.org/10.1007/978-981-97-9049-4_3

The successful transfer of cellular reproduction outside organic bodies represents, according to Hannah Landecker, one of the most significant advances in biotechnology (2007). According to her, cell culture necessitates the creation of a new type of "body" in which cells can be maintained and manipulated in a laboratory in an artificial growth medium. Additionally, cryopreservation allows cells to exist "out of biological time." These advances make biological materials more plastic, enabling greater control and flexibility in terms of the human manipulation of cells. According to Landecker, liberating cells from the spatial and temporal constraints of the body constitutes a new modality of exploiting living organisms, guided by human intent (Landecker, 2007). These advances in cellular technologies and the new attitudes they generate toward biological matter are now applied to the production of synthetic meat (Galusky, 2010).

Manufacturing Cultivated Meat: Biofabrication and Bioassemblage

The production of in vitro meat begins with the collection of cells capable of forming muscle, which are then expanded in a bioreactor (Post, 2012). Myosatellite cells, known as primary cells, are adult muscle stem cells taken from a living or recently slaughtered animal. It is also possible to extract them from a fertilized egg (Mulvaney, 2023). These cells can develop and survive for several days in the laboratory. If provided with appropriate environmental conditions and "fed" with a nutrient-rich growth medium, these primary cells will proliferate as they do in an animal. Primary cells grow in a simulated in vivo growth environment. However, even under optimal conditions, most cells will undergo a limited number of divisions before experiencing senescence, aging, and death. The use of myosatellites in cultured meat production requires successful induction of their differentiation from an undifferentiated state into myoblasts, which will subsequently fuse to form muscle fibers (Bach et al., 2003; Datar & Betti, 2010). Some cells can divide between 30 and 50 times before a new biopsy is necessary.

Myogenic satellite cells have been successfully isolated from the skeletal muscle tissue of several species used for meat production, including cattle (Dodson et al., 1987), chickens (Yablonka-Reuveni et al., 1987), fish (Powell et al., 1989), pigs (Blanton Jr et al., 1999; Wilschut et al., 2008), and turkeys (McFarland et al., 1988). However, Post (2012) notes that isolating, culturing, and maintaining a dynamic proliferative state of myosatellite cells is possible but remains a challenge.

An important consideration for manufacturing is having access to an infinite cell line by achieving cell immortality. What this means is creating a cell population "of non-senescent cells that escape cell cycle restriction and proliferate indefinitely *in vitro*" (Guo et al., 2022). The immortalization of primary cells offers the possibility of obtaining cell lines through genetic manipulations capable of proliferating indefinitely, without the need to renew biopsies, thus indicating the potential to eliminate dependence on living animals and achieve an infinite supply of meat (Jönsson, 2016). Cell lines become immortal when they undergo alterations in their cell cycle regulatory pathways, bypassing the natural process of senescence where cell division would normally cease. This ability to avoid senescence is often associated with specific genetic mutations, such as alterations in tumor suppressor genes or oncogenes, which disrupt the delicate balance between cell division and programmed cell death regulation. The processes of cell line immortalization involve using genome modification techniques, such as viral genes, mutations, or tools like CRISPR (Clustered Regularly Interspaced Short Palindromic Repeats). However, cell lines intended for food production present particular technical challenges and remain very difficult to immortalize (Soice & Johnston, 2021). Knowledge about stem cells for certain animal species such as marine animals, invertebrates, and mollusks is lacking, and few cell lines currently exist.

The production of cultured meat generates an extraordinary demand for cells that are both highly pure and capable of rapid proliferation. Most companies surveyed by the GFI in 2023 had a strong interest in obtaining cell lines from both terrestrial and aquatic species. Despite the well-established production and quality procedures in the cell and gene therapy (CGT) industry, the substantially higher cellular requirements for food production necessitate the development of large-scale culture methods and advanced cryopreservation solutions, even at the earliest stages of production. Different cell types are used in the production of cultured meat, including adult stem cells (ASC), embryonic stem cells (ESCs), induced pluripotent stem cells (iPSCs), and immortalized cell lines. Furthermore, the high costs associated with current cell banking practices pose significant barriers to their direct application in the food industry. Consequently, issues of scale and cost present substantial challenges for cell banking, within the context of cultured meat production (Martins et al., 2024). The industrial scale of cell cryopreservation to meet the needs for cell banking for cultured meat is unprecedented, especially if the goal is to produce cell meat commercially, replace conventional meat, or contribute significantly to the meat economy.

The choice of cell type used to produce meat depends on the bioprocessing design used by the company. Embryonic stem cells, for use in cell meat, come from embryonic stem cell lines, derived from an early embryo formed a few days after fertilization. Currently, half of the companies making cell meat depend on slaughtered animals, as their source of cells, but express the desire to use live animals for more cell lines in the future (Ravikumar & Powell, 2023). This seems rather ironic in light of the ethical standards the industry proclaims, but heavy reliance on slaughtered animals persists due to problems accessing live materials and stabilized cell lines.

Ultimately, the cell lines used in cultivated meat production are crucial for determining many factors later in the process. Cells that can self-renew and differentiate into various meat tissue cells, such as myofibers, adipocytes, fibroblasts, chondrocytes, and endothelial cells, are particularly desirable. This necessitates starting with stem cells. Each type of stem cell possesses varying abilities to differentiate into multiple cell types. For instance, embryonic stem cells can become any cell type from the three developmental layers (ectoderm, mesoderm, and endoderm), while adult stem cells are generally more specialized and limited to producing cells within the same layer or organ type (GFI, 2024). The use of induced pluripotent stem cells or iPSCs enables the properties of embryonic stem cells, without having to source them from an embryo. This means that using cellular reprogramming can be performed on adult somatic cells like skin fibroblasts or white blood cells (Ibid). According to the GFI, iPSCs have been produced from most agriculturally important species except for marine species which remain elusive. For instance, the company, Future Meat, uses fibroblasts as their starting cell line. However, the use of adult stem cells remains the most common for culturing the various components of meat, fat, and muscle. One category of ASC is myosatellite or satellite cells obtained from a small muscle biopsy, under local anesthesia or from animals recently slaughtered, and purified in the laboratory. Satellite cells lie alongside myofibers under the basal lamina of the muscle tissue, where they remain quiescent until activated upon injury or stress. These cells are the most direct way to obtain skeletal muscle tissue in vitro but are not considered an ideal starting cell type for the creation of other cellular components of meat. Immortalized cell lines for meat production would present obvious advantages, but this goal remains elusive. Generally, it is believed that development of reliable cell lines is critical for the industry. According to Specht et al.:

> While cultivation of primary cells is possible for small-scale clean meat applications, ultimately the process is best suited for robust cell lines that exhibit consistent performance over many production cycles, rather than primary cell harvests, which introduce variability and increased risk of contamination from the isolation protocol. Long-term cultivation is critical for reducing costs for clean meat because it allows for continuous cultivation of the seed train with only the final scaling and maturation stages of the production process requiring batch processes. It is likely that clean meat cell lines will be utilized in a similar fashion to fermentation strains in brewing: cultures can be used continuously for some number of generations, but periodically they are re-started from frozen stocks to avoid genetic drift. (Specht et al., 2018, p. 164)

The establishment of cell lines for cultured meat will necessitate the development of cell banks in order to “enhance their stability, reproducibility and long-term maintenance” (Ibid). There are several projects underway to develop cell banking of farm animals and for obtaining agriculturally important animal cell lines. Cell banking would provide the industry with consistent and reproducible cell lines for bioprocessing. As mentioned before, it is an important tool of tissue economies because it creates an institutional infrastructure and a site for storing, organizing, identifying, collecting, circulating, and distributing biological materials. Animal cell lines for food purposes present enormous costs. There is currently a limited availability of publicly accessible cell lines, from agriculturally significant species, suitable for cell-based meat production and privately held cell lines for specific types of animals are currently the norm. Nevertheless, there is an urgent need to develop new cell lines that can be preserved and made available to researchers in both academic and industrial settings. These biorepositories could be established in a manner akin to those used for preserving tissues and cell lines for endangered species conservation, like the *Frozen Ark* in the U.K., or large-scale distribution networks, such as the American Type Culture Collection (ATCC) that could supply ex situ cryopreservation of animal farm tissue and DNA. The availability of diverse cell lines from various animal species will be critical for advancing the research necessary for the long-term success of the cultivated meat industry. To address this need, the Good Food Institute has initiated the *Frozen Farmyard* project, in a dedicated biorepository for cultivated meat research (GFI, 2024). The aim of the Frozen Farmyard project is to create a centralized biobank of farm animal cells. The research group collects samples from farms across Europe, which will be used to create stem cell lines to grow muscle and fat. It is said that the samples can be obtained from ear tagging so as to avoid harming the animals but, as in the case of Waldby and Webster's study of tissue economies, the materials collected will rely on donations by farmers which is reminiscent of the gift economy, that is, human tissues and organ donation. Collected animal biopsies will eventually be used to make pluripotent stem cells (iPSCs), which can be used indefinitely. In contrast, the majority of cell-based meat companies today are working with stem cells from adult animals, which requires new animal biopsies for continuous production. The high cost of growing iPSCs remains another barrier to the industrialization of cellular agriculture. One of the limiting factors adding to the cost of producing iPSCs, according to researchers, is the high price of growth factors and other molecules used to grow stem cells into differentiated adult animal cells. The same goes for cell feed or the materials used to grow the cells for meat and allow them to differentiate which means that many companies are still relying on fetal bovine serum as a major source of growth factors. Hence, finding affordable and non-animal alternatives for optimizing cell growth remains an important priority to make the entire process of producing edible tissues cost-effective. A recent Nature article described the situation as of July 2023:

> GFI has compiled a catalogue of known useful cell lines (52 so far, from animals from cow to quail), and has partnered with the reagent company Kerafast in Boston, Massachusetts, to store and sell frozen samples of four cell lines so far. The catalogue contains several fish species, such as Atlantic mackerel (*Scomber scombrus*). (Jones, 2023a, p. 24)

A study published by GFI scientists reveals that companies are striving to obtain cell lines for the most commercially desirable terrestrial and aquatic animals. However, the majority of cell meat producers still rely on myoblasts, fibroblasts, and mesenchymal stem cells (MSCs) because of difficulties associated with accessing potentially superior cell types.

Challenges and Technical Hurdles

One of the major concerns surrounding cultivated meat involves the health risks associated with consuming the various cell lines used. Currently, consumers are not accustomed to eating cultivated animal cell lines, and these may potentially express oncogenes, necessitating confirmation that food products made from these cells do not pose a risk of tumor development (Soice & Johnston, 2021). Although there is limited evidence suggesting that genetically modified plant cell DNA can integrate or be transferred into human somatic cells or intestinal microflora, an in-depth study of genetically modified animal cells remains a scarcely discussed aspect by in vitro meat proponents (Ibid). Generally, it is important to understand that in vitro meat represents a technological convergence of cellular manipulations that remain controversial in the public sphere, such as the creation of GMOs or the use of gene editing tools such as CRISPR-Cas9. However, a comprehensive government regulatory framework has yet to be developed for overseeing the safety of cell lines used in the food industry.

As was just mentioned, the majority of animal replacement products consists of laboratory-grown tissues originating from biopsies used to isolate the stem cells of live animals such as chickens, cows, pigs, turkeys, ducks, fish, and seafood. Two companies (JUST and Memphis Meats) start from cells obtained from the feathers of a chicken. In order to comply with Kosher and Halal religious laws, the stem cells must be harvested from recently slaughtered animals, provided the tissue is viable (Balasubramanian et al., 2021). Others, like Aleph Farms, insist on providing details about the provenance of their cells by identifying the animals its steaks are derived from, as a way to make the process of cell isolation more transparent and possibly as a way of building consumer trust by providing the product's origins. Moreover, meat traceability is often associated with food safety.

Once isolated, cells could undergo a process of immortalization in order to produce a starter cell line to be used in perpetuity. This would enable the elimination of the donor animal as a source of biological materials for meat production, as starter cells could be sourced from cell banks.

The main objective of this type of tissue engineering is to "mimic regeneration of muscle after trauma and/or embryonic muscle myogenesis" or the process of muscle formation (Stephens et al., 2018). Muscle as well as fat cells are cultured in the laboratory using a mix of nutrients, growth factors, and hormones to allow division and maturation into muscle tissue. Mature muscle cells (myoblasts) can differentiate into muscle fibers or myocytes which can then adhere to a 3D structure, mimicking its "natural" corporeal environment. Scaffolding is required to not only allow proper adhesion but also for the proliferation of the cells for tissue development. Scaffolding replicates the cellular environment, known as the extracellular matrix, which is an interconnected network of fibrous proteins, such as collagen and other essential substances necessary for cell function. Materials for constructing scaffolding for 3D skeletal muscle formation are mainly sourced from animals due to their compatibility with the in vivo environment. The proteins in the extracellular matrix provide signals to cells, prompting them to either proliferate, differentiate into specialized cells like muscle or fat, or develop a specific structure. Some laboratories are investigating the use of food-safe plant-based, algal, pectin, modified cellulose (Specht et al., 2018) and combinations of biomaterials for cell support. The choice of scaffolding type and material can vary depending on the final product: microcarrier beads are used for producing ground meat; decellularized plants and fungal materials can be used to grow thicker tissues; finally, bioprinting can create custom shapes and complex structures by layering cells using a computer-aided design (CAD) process. One important consideration for the industry is whether the scaffold should be edible or whether the cells should be removed from the scaffold, either as it biodegrades or in order to reuse it. There are other issues concerning the extracellular matrix called hydrogel, which is used to support and mimic the natural tissue. Some of the things that may limit the success of the tissue production include the use of non-native constituents for the composition of the hydrogels, which may prevent the anchoring of the cells and affect replication and scalability; the contraction or shrinking of the hydrogel causing cells to grow on the edges of tension lines rather than at the center where they should be forming fibers; inconsistencies in fiber alignment and production; tissue thickness and scaffold degradation rates, etc. (ibid: 159–160). The point is that cell growth-enabling tissue formation must be optimized and scaled, and research to understand the processes of cell proliferation and differentiation is still ongoing. To date, successful muscle constructs are a few microns in thickness and multiple sheets of tissues have to be assembled to form large enough layers for meat fabrication.

When transferred to a bioreactor, cells multiply rapidly, and starter cell cultures are attached to a scaffold structure to enable maturation into muscle, fat, and connective tissue. The bioreactor, a large tank, contains nutrient-rich medium that recreates the environment inside the animal's body. The culture medium is referred to as feed for cells, and it contains all the components found in animal blood that sustain cell life, proliferation, and differentiation into the various cell types that constitute meat.

One of the limiting factors in the production of an edible and marketable in vitro meat product is the use of animal-derived products used in the cell growth media and for the scaffold material. Cells grow better in media that contain collagen, cellular factors, and blood serum all derived from animal byproducts in slaughterhouses. Although fetal bovine serum, harvested from pregnant cows during slaughtering, makes for the best culture medium, research is underway to find plant-based or non-animal substitutes. A few companies claim to have produced a serum-free medium. Growth media components can be derived from plant and fungus extracts or synthesized through fermentation, similar to the production methods for many vitamins and medicines. As they grow and mature, the cells need to attach to a scaffold to give them some structure, usually made of gelatin and/or collagen materials, sourced either from animals or from genetically engineered microbes and yeast. In order to make large quantities of clean meat, scientists need to develop efficient ways of feeding and growing cells as the culture medium represents 80% of the cost of producing the end product. The culture medium will likely differ for fish or cow cells as these have specific nutritional as well as biological needs. Scalability remains a huge challenge to the clean meat industry as cells need to attach to a structure which may limit the use of current bioreactors which tend to be around five liters. The goal is to use bioreactors such as the 2000-liter ones used in the biomedical field or larger ones for industrial purposes. As a result, new meat bioreactors will have to be designed specifically to accommodate how cells grow (using 3D scaffolds) or the process may be designed using several steps. This will also depend on the final product, whether the company is producing hamburger meat, nuggets, or steak (as Aleph Farms intends to do).

In fact, FBS has been largely eliminated from large-scale biopharmaceutical manufacturing precisely because of its potential for virus and bacterial contamination. Replacing this product or other animal sera for the production of cultured meat might not be such an easy proposition, even though companies claim to have developed animal-free growth media or are in the process of developing it (Humbird, 2021). In a scale-up scenario, involving large bioreactors, one can contrast the relatively slow growth of animal cell growth with the rapid growth of bacteria. According to David Humbird[1] (quoted in Fassler, 2021):

> bacteria grow every 20 minutes, and the animal cells are stuck in 24 hours. You're going to crush the culture in hours with a contamination event. (Fassler, 2021, p. 17)

[1] David Humbird is a chemical engineer commissioned to study the potential of cell meat for Open Philanthropy in 2018. He published an extensive techno-economic analysis (TEA) of the potential costs of industry scale-up.

Humbird contends that this type of event will produce a microbial culture instead of an animal cell culture in a very short time, leading to economic losses but also potential safety problems. The medium needs to include high-quality food-grade components, and this represents a major hurdle that needs to be overcome. Non-animal-based growth factors cost more than using fetal bovine serum. Recently, it was announced that scientists have engineered bovine muscle stem cells to produce their own growth factors (Fibroblast Growth Factor 2 or FGF2) (Stout et al., 2024). The experiment was described as proof of concept. This finding opens the way for potentially reducing the cost of growth cell media components but also signals another way to increase the biovalue of cell lines.

Once the production time has been completed, after five to seven weeks depending on the species, tissues can be harvested. Production timelines for cellular meat start from the time cells are placed in a bioreactor until harvest. Larger systems will probably require more time for cells to grow. The post-harvest phase is food processing into meatballs, nuggets, patties, or filets in the case of fish. The final product undergoes post-processing steps to make the product look, taste, and feel like real meat.

Yet another challenge faced by the industry is to control the costs involved in transforming cultured meat into a commodity, considering that the first in vitro burger, produced by Mark Post, costs around $325,000 U.S. In 2018, Aleph Farms, an Israeli meat start-up, claimed to be able to produce a steak for $50 U.S.; in 2021, the same company presented the first lab-cultivated rib-eye cut, "complete with all the flavor and texture of regular meat" using 3D bioprinting (Winer, 2021). The cost of growing cell meat using a large-scale facility

> has been estimated at $600M U.S. and the depreciation costs of these facilities will be a major component of the final cost of goods. While interesting, other technology that is being developed for 'CBM,' like edible cell-scaffolds and 3D printing with multiple cell types, just add to the challenge of scaling and cost of goods. The development of blended products with plant-based material will help to reduce costs and stabilize formulation. (Wood et al., 2023, p. 70)

The limitations of manufacturing cellular meat will be discussed further when we examine the results of a techno-economic model study. Suffice it to say that making CM affordable on a scale large enough to compete with conventional meat products remains highly uncertain, despite the enthusiastic predictions of the industry. As the above quotation suggests, and industry insiders believe:

> In the meantime, companies can justify large venture capital investments by pursuing cheaper products that combine cultivated and plant-based components. (Yaman, 2023)

Bioprinting Cellular Meat

The project of 3D bioprinting meat represents a profound intersection of biotechnology, material science, and food technology, promising to reshape food production and consumption. Because of the mechanical nature of the tissue assembly, 3D bioprinting is being explored to perform "scalable extrusion of cultured cells for use in forming three-dimensional tissue structures" (Modern Meadows quoted in Stephens et al., 2018). Harvard researchers claim to have sustained a printed cell culture for weeks (Ibid); however, many hurdles remain before it can be used in a manufacturing setting.

Building on the principles of 3D printing and tissue engineering, this technology involves the layer-by-layer deposition of bioinks, composed of living cells, growth factors, and biomaterials to construct tissue structures that resemble traditional meat. This process utilizes advanced 3D printing techniques adapted for biological materials, allowing for the precise control of cellular placement and tissue architecture. Bioprinters are robotic devices that receive information from computer software. The technology operates under the principles of additive manufacturing, wherein each layer of bioink is meticulously deposited to build complex, three-dimensional tissue constructs (Guo et al., 2024). Guo et al. describe in detail the three main approaches involving 3D bioprinting which include extrusion-based, inkjet, and laser-assisted jetting and light-assisted printing (Handral et al., 2022). The perceived advantage of bioprinted cultured meat is to automate the manufacturing process of biofabrication, a term that describes the automated creation of biologically active products with organized structures such as tissues, using living cells, bioactive molecules, biomaterials, or hybrid cell-material constructs which includes tissue maturation processes. Biofabrication is often used to describe the process of bioassembly. This automation would be beneficial for large-scale manufacturing of cultivated meat but also would facilitate the customization of meat in the future, allowing the programming of specific structural designs of the final product. Customization might also involve improving the nutritional profile of cultured meat, including adjusting the fat to protein ratio for instance (Cadloff, 2022).

The founders of the first cell meat company, Modern Meadow, Gabor, and Andras Forgacs, worked on 3D bioprinting of meat, from the expertise they developed in their other venture, a 3D bioprinting biomedical company called Organovo. Organovo develops bioprinted structures for medical purposes, having successfully 3D bioprinted functional blood vessels in 2010 (Moskovitch, 2013). Today, Modern Meadow is no longer involved in cellular meat, but is focused on the production of biomaterials such as lab-grown leather (Anusci, 2015; Stephens et al., 2019). In the lab, Modern Meadow scientists were able to produce 20,000 filaments grown from bovine stem cells which were inserted into a biocartridge and were combined by successive layering of the various strands in a desired shape in order to form living tissue (Anusci, 2015). According to Dr. Gabor Forgacs, bioprinting meat presents several challenges: "We are printing live material – the cells are alive when we are printing them" (Moskovitch, 2013). He describes the difference between bioprinting organs or living structures for regenerative medicine compared to producing meat. He claims that printing meat is much simpler because unlike the complexities of organs or tissues, hamburgers consist in a basic 2D shape. Printing organs and tissues for biomedical use involves the creation of biomimetic structures that are used for drug testing, tissue repair, and, in the future, it is believed, organ transplantation. In this case, the bioprinted tissue needs to remain active and functional during printing and cultivation in order to be clinically functional. For cellular meat on the other hand, "cells only need to keep active during product manufacturing to enable proliferation and differentiation" (Guo et al., 2024). Once created into the final product, it no longer remains a living structure. Cell-based meat is, as Fogacs called it, a "post-mortem tissue" because the tissue construct is eventually "killed" after it reaches the final stages of maturation into the final structure—nuggets, hamburger, etc. Printing cultured meat creates a unique challenge, compared to gastronomic printing of meat in the food industry, which is aimed at producing and designing meats and meat combinations with specific shapes, textures, flavors, etc. from existing materials because it constitutes an extension of livestock production (Guo et al., 2024).

The 3D bioprinting of cell meat also differs from the process of producing cell meat in the traditional way, with no printing. Cell meat produced without printing entails a different series of manual manipulations and a process which consists of attaching proliferated muscle cells to a scaffold or a carrier such as microbeads and transferring it to a bioreactor containing growth media. It seems that there are two approaches to bioprinting cellular meat: one which uses bioinks to print out desired shapes that are essentially living structures and ultimately become cultured meat; the second approach is scaffold-based and prints out "biocompatible scaffolds" onto which animal cells are seeded in order to expand into specific edible tissue (Guo et al., 2024). In the 3D printing of cell meat, muscle cells, fat cells, and even extracellular matrix (ECM) within the scaffold can all be printed in order to support cell growth and proliferation (Handral et al., 2022). The bioink used in the extrusion cartridges contain both animal cells and biomaterials that enable the scaffold structures to be fabricated. As for non-printed meat, these materials must be edible, food-grade level and maintain their shape and size after frying. Scaffolds must also be edible and absorbed by muscle cells, unlike the case of biomedical research where scaffolds are used as support for cell clusters (Su et al., 2023).

The future of 3D bioprinted cell meat will depend on the ability to produce multicomponent cultured meat that is made up of muscle cells, fat (marbling), vascular structure (tendons), and possibly bone structure, in a way that makes it comparable to conventional meat. This includes designing bioinks and scaffolds that meet scalability, cost-effectiveness, shelf-life, and printability requirements, while also replicating the complexities in architecture, safety, taste, and flavor of conventional meat. According to experts, meat bioprinting is still in infancy and is limited to laboratory scale (Guo et al., 2024; Handral et al., 2022).

In 2019, Aleph Farms, the Israeli start-up, announced its experiment in the International Space Station to grow the first-ever cultured meat in space, using Russian 3D bioprinting technology in zero gravity. In 2021, the company produced the first rib-eye steak using 3D bioprinting. In 2022, Aleph Farms conducted further experiments in space, as part of the Rakia Mission, on the International Space Station, to explore "the effects of reduced gravity on the growth and maturation of cow cells" (Aleph Farms, 2024). They claim that their true goal is to be able to grow meat anywhere whether on Earth or in space. Another Israeli company, Steakholder Foods, produced a 104-gram cultivated beef steak in 2021 and a cultivated "fish filet" in 2022 using bioprinting. It plans to continue producing complex cultured meat structures, including hybrid meats, using its bioprinting technology. The company has developed a fish bioprinter as well as a meat bioprinter to create its products. Interestingly, it seems the company offers the possibility to bioprint tissues for interested customers, at a cost. It also markets its proprietary bioprinters which begs the question about how these companies position themselves in the market, as cell meat companies—each with their own brand of bioinks—or as bioprinting companies offering services to other industry players. A study exploring the social, legal, and geographical dimensions of 3D bioprinting in regenerative medicine analyses the patentable products and technologies within the field, as well as the formation of a legal discursive network, through legal tools and the patent documents filed by actors (Bicudo et al., 2021). Not only are bioprinters patentable, so are software packages, bioinks, and final products. One of the findings of the study is that a specialized bioprinting market might focus attention on key areas of research and that collaborations with academic researchers might lead to more concrete applications of bioprinting. It is thought that the experimental phase might lead to more translational research. In the bioprinting meat space, companies like Steakholder Meats or Aleph Meats might license their bioprinting technologies to other players in the field, selecting key players and products. Their use of the technology and overcoming the obstacles of bioprinting meat on a large scale might make bioprinting part of the landscape of expectations and imaginaries surrounding cellular meat. The fact that bioprinting involves automation and standardization, in the production of cultured meat, is seen as a way of accelerating its path to market (Heine et al., 2024). It is also viewed as a solution for producing food 'anywhere' even when resources might be lacking as Aleph Farms claims. In describing their Space Program, the start-up discusses its mission to make cellular meat:

> Prolonged exploration in space, such as establishing human bases on the Moon and Mars, is limited by the ability to provide quality nutrition to astronauts. We are taking our food security vision to the next stage by developing a versatile technological platform for the local production of Aleph Cuts in space. This self-sustaining food system would provide astronauts continuous access to fresh food, which is essential for their physical and mental wellbeing.
>
> The constraints of space—such as fast-changing temperatures, limited access to natural resources, and zero gravity conditions—challenge us to create new frontiers in our processes. The knowledge gained from our exploration and space-based technologies, where we work under these extremely limiting conditions could help us build resilient food systems on Earth:
>
> - Adapting our food systems to cultivate quality, delicious products anywhere, from remote rural areas to dense city centers
> - Building resilient food systems that are less vulnerable climate change, to food supply disruptions from crises like war, disease, and natural disasters
> - Producing quality, delicious products more efficiently to meet the protein demands of a rapidly growing global population
> - Producing quality, delicious products with less land, water, and emissions in order to preserve critical natural resources for future generations. (Aleph Farms, 2024)

Here, the people behind Aleph Farms are making a reference to NASA's early involvement in the creation of cellular meat, with Benjaminson's experiments with goldfish tissues. They are also playing on the notion of multiplanetary imaginaries of humans no longer being tied to Earth but becoming multiplanetary (Tutton, 2018), presenting space colonization as a fact, an inevitable fate, in light of terrestrial environmental and resource depletion. Richard Tutton qualifies these imaginaries as utopian, constructing specific visions of the future through the lens of technoscientific innovation and its projected possibilities. The question of global food security is another important part of their messaging.

Clearly, space colonization constitutes an important recurring element in CM discourse and, in particular, as it relates to the ability to print meat. Bioprinting soft tissues in space constitutes an image of a possible future, presented as providing in situ practical solutions for resource-limited, human-sustaining systems, both for biomedical and food purposes (Kang et al., 2021; Rezapour Sarabi et al., 2023). The idea of working toward self-sustaining systems is reminiscent of the ambitions of Biosphere II and also links with Anthropocene narratives that view space colonization as a long-term option for humanity (Ćirković, 2019).

Key to bioprinting is the relationship between the biological and the digital and how these are conceptualized to work within the parameters of bioprinting. Bioprinting also creates expectations about the way cells and other biological materials should behave in order to conform to the parameters of the machine. This highlights conflicting paradigms between engineers and biologists, as my co-authors and I have shown. In our study, involving the bioprinting of skin for therapeutic and cosmetic purposes, the biologists were constantly having to adapt their protocols and practices in order for the cells to "work" in the printer's bioink cartridges (Lafontaine et al., 2021). Hence, a hierarchy of knowledge was established between what is perceived as the accuracy and precision of bioprinting engineering and the messiness and unpredictability of biological processes and products. Our conclusion was that the bioprinting of tissues and organs represents the near realization of the machine-based ideal of life, achieved through the technologizing of biological systems. Bioprinting embodies this ideal by transforming the inherently organic process of tissue growth into a purportedly controlled, reproducible, and scalable procedure driven by the mechanistic paradigm of predictability, automation, and computer programming. The development of bioinks, which are made up of biological parts (cells, fat, scaffolds, biomaterials, etc.), as components of cell meat that can be modularized in order to then produce live flesh is reminiscent of machine parts. This approach underscores the convergence of engineering and biological sciences but also exemplifies the use of additive manufacturing as a way to construct complex tissues and organs in a way that makes these constructible in a context of organ shortages or food insecurity, under conditions of climate change. The implications of this are that engineering solutions can address these social problems. It is important to question the ethical implications of merging living materials with technologies, such as bioprinting, because bioprinted human and animal tissues might be considered mere mechanical constructs, further removed from their biological origins, raising the question of their social and legal status within tissue economies.

Vital matter is produced using the techniques of tissue engineering and in vitro cell and tissue culture, to reconstruct three-dimensional tissues, usually skeletal muscle, from the cellular level up, to then produce edible tissues. As mentioned, this involves recreating the conditions of dematerialized bodies. According to Melinda Cooper:

> The field of tissue engineering, as its name suggests, has fostered all kinds of cross-disciplinary alliances between biologists, materials scientists, and engineers (chemical, mechanical, and even electrical), but the specific problems it encounters have also gone hand in hand with a rethinking of traditional mechanistic models of engineering. (Cooper, 2008, p. 105)

Cooper studies organogenesis in biomedicine but similar principles apply in the case of cellular meat. She proposes an understanding of tissue engineering through the lens of morphogenesis of form as process, particularly when conceptualized through the lens of mathematics of topological space. Hence, the technical possibility of reproducing the morphogenesis of the body outside of "metric and genealogical" time, in other words suspended in time, means that the field of tissue engineering must contend with recreating bodily space, as well as its architectural form. For Cooper this implies that, in tissue engineering, morphogenesis is now considered a process that is biologically modulable using various techniques that use bioreactors and computer-assisted modeling, as in the case of bioprinting live tissues. The processes of cell differentiation in the bioreactor involve subjecting tissue constructs to various stimuli that mimic function (contraction for skeletal muscle tissues). This means creating a virtual space in which the "entire field of bodily transformations can be materialized *in vitro*" and where tissues can be "modulated, deformed, continuously remoulded" (Ibid). In terms of molding tissues and creating complex structures that replicate meat from bioprinting, it is clear that the extreme "mutability, unexpected recalcitrance and peculiar generativity of the living biological" may be at odds with the "stability, reproducibility, and standardized form" demanded by a mechanistic view of the body.

Bibliography

Aleph Farms. (2024). *Space Meat—The Future of Astronaut Food*. Aleph Farms. Retrieved from https://aleph-farms.com/aleph-zero/

Anusci, V. (2015, February 17). *I'm the Future of 3D Printing, Nice to "Meat" You*. All3DP. Retrieved from https://all3dp.com/meat-3d-printing/

Bach, A. D., Stern-Straeter, J., Beier, J. P., Bannasch, H., & Stark, G. B. (2003). Engineering of Muscle Tissue. *Clinics in Plastic Surgery, 30*(4), 589–599. https://doi.org/10.1016/S0094-1298(03)00077-4

Balasubramanian, B., Liu, W., Pushparaj, K., & Park, S. (2021). The Epic of In Vitro Meat Production—A Fiction into Reality. *Foods, 10*(6), 1395. https://doi.org/10.3390/foods10061395

Bicudo, E., Faulkner, A., & Li, P. (2021). Patents and the Experimental Space: Social, Legal and Geographical Dimensions of 3D Bioprinting. *International Review of Law, Computers & Technology, 35*(1), 2–23. https://doi.org/10.1080/13600869.2020.1785066

Blanton, J. R., Jr., Grant, A. L., McFarland, D. C., Robinson, J. P., & Bidwell, C. A. (1999). Isolation of Two Populations of Myoblasts from Porcine Skeletal Muscle. *Muscle & Nerve, 22*(1), 43–50. https://doi.org/10.1002/(SICI)1097-4598(199901)22:1<43::AID-MUS8>3.0.CO;2-O

Cadloff, E. B. (2022, February 28). Steak's Next Frontier: Customized and Lab-Grown to Order. *Modern Farmer*. Retrieved from https://modernfarmer.com/2022/02/custom-cell-grown-meat/

Chapelle. (2014, November 18). *Et si le but ultime de l'industrie agroalimentaire était de se débarrasser des animaux d'élevage?* Basta! Retrieved from https://

basta.media/jocelyne-porcher-industrie-agroalimentaire-disparition-animaux-elevage-viande-de-synthese
Ćirković, M. M. (2019). Space Colonization Remains the Only Long-term Option for Humanity: A Reply to Torres. *Futures, 105*, 166–173. https://doi.org/10.1016/j.futures.2018.09.006
Cooper, M. (2008). *Life as Surplus: Biotechnology and Capitalism in the Neoliberal Era*. University of Washington Press.
Datar, I., & Betti, M. (2010). Possibilities for an In Vitro Meat Production System. *Innovative Food Science & Emerging Technologies, 11*(1), 13–22. https://doi.org/10.1016/j.ifset.2009.10.007
Dodson, M. V., Martin, E. L., Brannon, M. A., Mathison, B. A., & McFarland, D. C. (1987). Optimization of Bovine Satellite Cell-derived Myotube Formation *In Vitro*. *Tissue and Cell, 19*(2), 159–166. https://doi.org/10.1016/0040-8166(87)90001-2
Fassler, J. (2021, September 22). *Lab-grown Meat is Supposed to be Inevitable. The Science Tells a Different Story.* The Counter. Retrieved from https://thecounter.org/lab-grown-cultivated-meat-cost-at-scale/
Galusky, W. (2010). Playing Chicken: Technologies of Domestication, Food, and Self. *Science as Culture, 19*(1), 15–35. https://doi.org/10.1080/09505430903557874
Good Food Institute. (2024). *Alternative Protein Company Database* [Dataset]. Retrieved from https://gfi.org/resource/alternative-protein-company-database/
Guo, D., Zhang, L., Wang, X., Zheng, J., & Lin, S. (2022). Establishment Methods and Research Progress of Livestock and Poultry Immortalized Cell Lines: A Review. *Frontiers in Veterinary Science, 9*, 956357. https://doi.org/10.3389/fvets.2022.956357
Guo, X., Wang, D., He, B., Hu, L., & Jiang, G. (2024). 3D Bioprinting of Cultured Meat: A Promising Avenue of Meat Production. *Food and Bioprocess Technology, 17*(7), 1659–1680. https://doi.org/10.1007/s11947-023-03195-x
Handral, K. H., Hua Tay, S., Wan Chan, W., & Choudhury, D. (2022). 3D Printing of Cultured Meat Products. *Critical Reviews in Food Science and Nutrition, 62*(1), 272–281. https://doi.org/10.1080/10408398.2020.1815172
Heine, S., Ahlfeld, T., Albrecht, F. B., Gelinsky, M., & Kluger, P. J. (2024). How Biofabrication can Accelerate Cultured Meat's Path to Market. *Nature Reviews Materials, 9*(2), 83–85. https://doi.org/10.1038/s41578-024-00650-9
Humbird, D. (2021). Scale-up Economics for Cultured Meat. *Biotechnology and Bioengineering, 118*(8), 3239–3250. https://doi.org/10.1002/bit.27848

Jones, N. (2023a). Lab-grown Meat: The Science of Turning Cells into Steaks and Nuggets. *Nature, 619*(7968), 22–24. https://doi.org/10.1038/d41586-023-02095-6

Jönsson, E. (2016). Benevolent Technotopias and Hitherto Unimaginable Meats: Tracing the Promises of In Vitro Meat. *Social Studies of Science, 46*(5), 725–748. https://doi.org/10.1177/0306312716658561

Kang, D.-H., Louis, F., Liu, H., Shimoda, H., Nishiyama, Y., Nozawa, H., Kakitani, M., Takagi, D., Kasa, D., Nagamori, E., Irie, S., Kitano, S., & Matsusaki, M. (2021). Engineered Whole Cut Meat-like Tissue by the Assembly of Cell Fibers using Tendon-Gel Integrated Bioprinting. *Nature Communications, 12*(1), 5059. https://doi.org/10.1038/s41467-021-25236-9

Lafontaine, C., Wolfe, M., Gagné, J., & Abergel, E. (2021). Bioprinting as a Sociotechnical Project: Imaginaries, Promises and Futures. *Science as Culture, 30*(4), 556–580. https://doi.org/10.1080/09505431.2021.1977264

Landecker, H. (2007). *Culturing Life: How Cells Became Technologies* (First paperback ed). Harvard Univ. Press.

Martins, B., Bister, A., Dohmen, R. G. J., Gouveia, M. A., Hueber, R., Melzener, L., Messmer, T., Papadopoulos, J., Pimenta, J., Raina, D., Schaeken, L., Shirley, S., Bouchet, B. P., & Flack, J. E. (2024). Advances and Challenges in Cell Biology for Cultured Meat. *Annual Review of Animal Biosciences, 12*(1), annurev-animal-021022-055132. https://doi.org/10.1146/annurev-animal-021022-055132

McFarland, D. C., Doumit, M. E., & Minshall, R. D. (1988). The Turkey Myogenic Satellite Cell: Optimization of *In Vitro* Proliferation and Differentiation. *Tissue and Cell, 20*(6), 899–908. https://doi.org/10.1016/0040-8166(88)90031-6

Moskovitch, K. (2013, January 10). Modern Meadow Aims to Print Raw Meat using Bioprinter. *BBC News*. Retrieved from https://www.bbc.com/news/technology-20972018

Mulvaney, K. (2023, July 19). De la viande cultivée en laboratoire commence à être servie aux États-Unis. *National Geographic*. Retrieved from https://www.nationalgeographic.fr/sciences/alimentation-revolution-ecosysteme-de-la-viande-cultivee-en-laboratoire-commence-a-etre-servie-aux-etats-unis

Porcher, J. (2010). La production de viandes in vitro, stade ultime? *Revue Politique et Parlementaire, 1057*, 97–104.

Post, M. J. (2012). Cultured Meat from Stem Cells: Challenges and Prospects. *Meat Science, 92*(3), 297–301.

Powell, R. L., Dodson, M. V., & Cloud, J. G. (1989). Cultivation and Differentiation of Satellite Cells from Skeletal Muscle of the Rainbow Trout Salmo gairdneri. *Journal of Experimental Zoology, 250*(3), 333–338. https://doi.org/10.1002/jez.1402500314

Ravikumar, M., & Powell, D. (2023). *Cell Line Development and Utilisation Trends in the Cultivated Meat Industry.*

Rezapour Sarabi, M., Yetisen, A. K., & Tasoglu, S. (2023). Bioprinting in Microgravity. *ACS Biomaterials Science & Engineering, 9*(6), 3074–3083. https://doi.org/10.1021/acsbiomaterials.3c00195

Soice, E., & Johnston, J. (2021). Immortalizing Cells for Human Consumption. *International Journal of Molecular Sciences, 22*(21), 11660. https://doi.org/10.3390/ijms222111660

Specht, E. A., Welch, D. R., Rees Clayton, E. M., & Lagally, C. D. (2018). Opportunities for Applying Biomedical Production and Manufacturing Methods to the Development of the Clean Meat Industry. *Biochemical Engineering Journal, 132,* 161–168. https://doi.org/10.1016/j.bej.2018.01.015

Stephens, N., King, E., & Lyall, C. (2018). Blood, Meat, and Upscaling Tissue Engineering: Promises, Anticipated Markets, and Performativity in the Biomedical and Agri-food Sectors. *BioSocieties, 13*(2), 368–388. https://doi.org/10.1057/s41292-017-0072-1

Stephens, N., Sexton, A. E., & Driessen, C. (2019). Making Sense of Making Meat: Key Moments in the First 20 Years of Tissue Engineering Muscle to Make Food. *Frontiers in Sustainable Food Systems, 3.* https://doi.org/10.3389/fsufs.2019.00045

Stout, A. J., Zhang, X., Letcher, S. M., Rittenberg, M. L., Shub, M., Chai, K. M., Kaul, M., & Kaplan, D. L. (2024). Engineered Autocrine Signaling Eliminates Muscle Cell FGF2 Requirements for Cultured Meat Production. *Cell Reports Sustainability, 1*(1), 100009. https://doi.org/10.1016/j.crsus.2023.100009

Su, L., Jing, L., Zeng, X., Chen, T., Liu, H., Kong, Y., Wang, X., Yang, X., Fu, C., Sun, J., & Huang, D. (2023). 3D-Printed Prolamin Scaffolds for Cell-Based Meat Culture. *Advanced Materials, 35*(2), 2207397. https://doi.org/10.1002/adma.202207397

Tutton, R. (2018). Multiplanetary Imaginaries and Utopia: The Case of Mars One. *Science, Technology, & Human Values, 43*(3), 518–539. https://doi.org/10.1177/0162243917737366

Wilschut, K. J., Jaksani, S., Van Den Dolder, J., Haagsman, H. P., & Roelen, B. A. J. (2008). Isolation and Characterization of Porcine Adult Muscle-Derived Progenitor Cells. *Journal of Cellular Biochemistry, 105*(5), 1228–1239. https://doi.org/10.1002/jcb.21921

Winer, S. (2021, February 10). Higher Steaks as Israeli Company Makes First Lab-Cultivated Rib-Eye Cut. *The Times of Israel.* Retrieved from https://www.timesofisrael.com/higher-steaks-as-israeli-company-makes-first-lab-cultivated-rib-eye-cut/

Wood, P., Thorrez, L., Hocquette, J.-F., Troy, D., & Gagaoua, M. (2023). "Cellular Agriculture": Current Gaps Between Facts and Claims Regarding "Cell-Based Meat". *Animal Frontiers, 13*(2), 68–74. https://doi.org/10.1093/af/vfac092

Yablonka-Reuveni, Z., Quinn, L. S., & Nameroff, M. (1987). Isolation and Clonal Analysis of Satellite Cells from Chicken Pectoralis Muscle. *Developmental Biology, 119*(1), 252–259. https://doi.org/10.1016/0012-1606(87)90226-0

Yaman, R. (2023). *Is Cultivated Meat For Real?—Asterisk.* Asterisk Mag. Retrieved from https://asteriskmag.com/issues/02/is-cultivated-meat-for-real

4

Meat by Design: Cellular Economies and the Engineering of Meat

Post-Animal Bioeconomies

New Harvest, one of the leading organizations involved in the research and promotion of cell-based meat, described its mission as "Our vision is a strong foundation of accessible, public, fundamental cellular agriculture research, upon which we can build a *post-animal bioeconomy*, where we harvest animal products from cell cultures, not animals, to feed a growing global population sustainably and affordably" (Datar et al., 2016).

The bioeconomy concept is a neologism, originally introduced in the 1970s by the Romanian mathematician and economist Nicholas Georgescu-Roegen. His concept also known as bioeconomics was rooted in an ecological perspective on economics that considers the biological origin of economic processes, arguing that economic activities should be understood through their impact on the physical environment and resource availability. In addition, Georgescu-Roegen emphasized the entropic nature of all economic activities leading to environmental degradation and resource scarcity (Vogelpohl & Töller, 2021). As a result, he was critical of traditional economic growth models, advocating for economic systems that respect ecological limits. He is believed to be one of the original thinkers of the degrowth movement. Among the leading

E. Abergel, *Dead Meat*, https://doi.org/10.1007/978-981-97-9049-4_4

institutions using the concept of bioeconomy are the Organization for Economic Cooperation Development (OECD) and the German Bioeconomy Council (BGC). They outlined their bioeconomy framework over a decade ago,[1] consisting of a vast policy-driven framework aimed at enhancing economic growth—which some might say contradicts Georgescu-Roegen's vision—in line with sustainability and job creation. Georgescu-Roegen's bioeconomy is broad and philosophical, addressing fundamental economic and ecological relationships. The OECD's approach is more practical and sector-specific, involving agriculture, biotechnology, and policy frameworks to foster economic development and stimulate industrial growth.

The OECD's definition of the bioeconomy is focused on the integration of biological resources and biotechnology into the economy, using novel forms of capitalism based on a biotechnological innovation and a biomass economy. The biotechnological economy is associated with the industrial exploitation of biotechnological knowledge defined as "an understanding of DNA, RNA, proteins and enzymes at the molecular level; of ways to manipulate cells, tissues, organs or whole organisms; and of bioinformatics for analysis of genomes and proteins" (Datar et al., 2016). Strategies surrounding the development of a bioeconomy usually promote biotechnological innovation for economic output and "public benefit" in order to meet global challenges (Pavone & Goven, 2017). The bioeconomy emphasizes the role of biotechnology and other advanced technologies in transforming biological resources into economic outputs, highlighting innovation and commercial viability. According to Pavone and Goven, this vision of the bioeconomy has been criticized for promoting:

> "an overtly neoliberal model of economic development" (Parry 2007), for "treat[ing] the economic aspects of biotechnology as central while tending to treat risks and ethical issues as secondary concerns" (Hilgartner 2007), and for representing complex socio-political problems as technological

[1] In the 1970s, Nicholas Georgescu-Roegen coined the term "bioeconomics" to "make us bear in mind continuously the biological origin of the economic process and thus spotlight the problem of mankind's existence with a limited store of accessible resources, unevenly located and unequally appropriated" (Georgescu-Roegen, 2011:103).

challenges solvable through technological innovation alone (Goven and Pavone 2015). (2017:4)

The interest in the bioeconomy approach is its links to biotechnology and speculative practices and claims. In addition, within bioeconomic discourse, animal science and agriculture define specific ways in which new technologies such as animal biotechnologies relate to ideas of sustainable agriculture and are negotiated within this policy framework (Twine, 2007). Therefore, it is a perfect lens through which to explore CA. The promissory character of life sciences research intersects with speculative finance through the active construction of assets and the securing of venture capital, creating what Rajan calls biocapital (Rajan, 2006), heralding new forms of bioeconomic activities and expectations coming from patients in the case of medical technologies or ethical consumers, in the case of cellular agriculture.

The promissory bioeconomy of clean meat gives rise to the prospects of future foods, projecting the production of ecological, yet highly technologized, animal-friendly meat, thus articulating new forms of capitalism. Stem cell science has responded to health and ecologically conscious consumers by promising safe, sustainable, ethical meat. The speculative claims of potentially achievable futures affect investment decisions, consumer attitudes, and willingness to purchase as well as governmental regulations. All this relies on the promise of generating *economic* value, as the work of Cooper demonstrates (2015). The bioeconomic vision of biotechnology and cell meat contributes to new forms of capital accumulation based on the appropriation of the living attributes of all life forms. Climate change or the Anthropocene define new ways for embedding market logic into ecological relations and creating opportunities for new modes of value-added technological appropriation of living systems for food production. The Anthropocene infuses new economic impetus for restructuring capital relations within the food system (Abergel, 2011). As such, the deployment of tech foods and the emphasis on alt-protein is seen in economic and social terms as a way to demonstrate how disruptive technologies and products like cell meat can benefit the environment while also enabling the expansion and power of the biotechnology industry.

Cellular meat, global climate change, and Anthropocene diets intersect to create a series of imaginaries that become enmeshed in the cultured meat narrative and its promissory economy. Sociotechnical imaginaries and other categories of imaginaries expressed via environmental narratives about the Anthropocene and food futures provide a multi-layered understanding of the ways in which the anthropogenic destruction of the Earth's living systems ushers in the production of new edible life forms in the laboratory, conceived as part of a sustainable future. How will we eat at the end of Nature? Continuity and rupture and extinction and regeneration are played out in a perpetual cycle of utopian and dystopian visions, seen through the lens of end-of-nature narratives. This type of analysis informs how cell-based technologies are used and developed for different ends, while interrogating "how altering biology changes what it is to be biological" on the one hand, while problematizing on the other, how laboratory practices that engage with the potentialities and limits of biological matter "characterize contemporary attitudes to the living" (Landecker, 2004), all the while signaling a potential way out of the current ecological crisis.

Tissue Economies, Biocapital, and the Creation of Value

Because it lays the groundwork for CM, it is important to first look back at the tissue economy and its connection to the bioeconomy. The bioeconomy provides the context for the insertion and commodification of biological materials and processes into "product-making" and "profit-seeking" networks and policy infrastructures. At its base, the bioeconomy creates different socioeconomic systems of production, circulation, and exchange, with different societal implications, including distinct institutional arrangements. This section deals with the idea of tissue economies as first described in the book *Tissue Economies: Blood, Organs, and Cell Lines in Late Capitalism,* by Catherine Waldby and Robert Mitchell (2006). Their research describes the ways the accelerated circulation of human tissues developed via bioscientific research and medical technologies, and which include skin, bones, embryos, stem cell lines, and other substances, become the subject of ethical and societal concerns concerning the distinction between the gift economy and commercial transactions which structure a tissue market for profit. The gift economy refers to donating or receiving organs, blood, breast milk, ovules, sperm, tissues, and other biological materials; it is associated with the notion of "corporeal generosity" which frames donation around intercorporeal and interpersonal exchange (Shaw, 2008). In most cases, these donations occur within a non-commercialized context and are altruistic, although the U.S. allows gamete donations to be commercialized. The work of Waldby and Mitchell highlights the mechanisms underpinning various types of tissue exchanges, from the inception of cell banks for embryo donations, to the creation of economic value from medical 'waste products' such as umbilical cord blood. They analyze the production of biovalue, as new types of biocommodification are created in the circulation of tissues inside and outside of bodies and laboratories. A definition of biovalue is that:

> Biovalue refers to the yield of vitality produced by the biotechnical reformulation of living processes. Biotechnology tries to gain traction in living processes, to induce them to increase or change their productivity along specified lines, intensify their self-reproducing and self-maintaining capacities. This intensification or leveraging of living process typically takes place not at the level of the body as a macro-anatomical system but at the level of the cellular or molecular fragment, the mRNA, the bacterium, the oöcyte, the stem cell. (Waldby, 2002b, p. 310)

The authors refer to the circulation of tissues and cells, including stem cells, transferred from one body to the next as well as from body to laboratory as a key feature of the logic of tissue economies. Waldby also states that one of the key features of biotechnology in this process is to find "insertion points between living and non-living systems...where new and contingent forms of vitality can be created, capitalizing life" (Waldby, 2002a, 2002b). Before going further into the implications of tissue economies, it is important to connect what is going on in agricultural biotechnologies and the biomedical sciences. Authors like sociologist Richard Twine, writing about animal biotechnologies, argue that:

> if one was to base an analysis solely on either agriculture or biomedicine without any reference to their interconnection, one might well come up with a particularly partial representation of the biotechnological shaping of human/animal relations. (Twine, 2007, p. 100)

The implications of this are that there are clear ontological connections between biomedicine and animal biotechnologies and because separating these two domains finds its origins in the human/non-human dualism (Ung-Lanki, 2014).

Following up on this work, Melinda Cooper and Catherine Waldby wrote *Clinical Labor: Tissue Donors and Research Subjects in the Global Bioeconomy* (2014) where the labor question in biomedical production is examined. The analysis focuses mostly on human cells and tissues and the material conditions that create markets, consisting of "extensive technical systems for donation, circulation, analysis, and transplantation of human tissues" (Waldby & Cooper, 2014, p. 6). The concerns, raised by the authors, focused on the production of value within what they call new forms of embodied labor, which include "surrogacy, the provision and sale of body tissues, participation in clinical trials," and constitute important aspects of the growing bioeconomy but which tend to be less visible when thinking about the creation of biovalue. Their analysis considers how novel forms of living matter become the material base for biocommerce via the extension of labor. According to Waldby and Mitchell, the patenting of multicellular entities, cell lines, embryonic stem cell lines, and genetic sequences have transformed biomedical research into a lucrative area of investment for the increasingly mobile forms of finance and venture capital that have dominated the global economy since the seventies. Novel forms of living matter thus become the material base for "highly intensive kinds of biocommerce, and companies" (Waldby & Mitchell, 2006, p. 23). In the same way, the body, whether human and animal, is now an open source of free biological material for commercial use (p. 24).

These ideas are taken up in Kaushik Sunder Rajan's work (2006) regarding the various strategies of capital accumulation. He describes this process as "the overdetermination of the life sciences by the capitalist political economic structures within which they emerge" (Rajan, 2006,

p. 5). Rajan's analysis builds on studies which conceptualize tissue economies within the processes of capitalist circulation and see biocapital as a system of exchange. Building on these, he writes, "Biocapital is a study of the systems of exchange and circulation involved in the contemporary workings of the life sciences, but is also a study of those life sciences as they become increasingly foundational epistemologies for our time" (Rajan, 2006, p. 12).

Hence, the possibility offered by cellular meat and its creation comes as a result of successive and complex mediations between scientists, entrepreneurs, policy experts, big agricultural and food industries, cultural influencers like chefs and people concerned about climate change, animal welfare, and agricultural sustainability. The notion of overdetermination, first described in Freud's work and later taken up by Althusser and then by Rajan, as quoted above, is very important because it describes how epistemic developments in the biosciences are not the direct cause of specific political economic formations but rather that these emerge as part of a total productive process where, at any point in time, every element within that process is determined by all the others, a kind of cumulative internalization which enables the possibilities of scientific research to emerge as potentially realizable within a bioeconomic framework. Viewing the rise of the bioeconomy and tissue economies solely as the outcome of capitalist relations results in an overly simplistic cause-and-effect analysis, neglecting the complex interplay of multiple underlying determining factors. Such a deterministic view would hinder our understanding of how the emergence of cellular meat results from the complex interplay of interconnected and opposing forces. These forces include a diverse array of political, economic, social, and cultural actors and themes. By focusing only on one aspect, we overlook the nuanced and often contradictory influences that drive this innovation, including regulatory policies, market dynamics, technological advancements, ethical debates, and shifting consumer preferences. This includes considering modernity's philosophical themes such as progress, rationality, and the relationship between humans and nature.

Recognizing this complexity is essential for a comprehensive analysis of the development and impact of cellular meat. This enables an analysis that combines the material aspects as well as the immaterial and "imaginary" dimensions of scientific work. The sociology of expectations first identified how scientific and technological changes are shaped by social and other types of expectations. As science and technology scholars have shown, technoscientific research is by definition future-oriented and its potential comes about in the form of visions, imaginings, and expectations (Borup et al., 2006). It follows that these are "generative and guide activities, provide structure and legitimation, attract interest and foster investment" (Ibid, p. 286). In essence, they are the engine of innovation, but, moreover, they tend to manifest within a particular context that Borup et al. call "future-oriented understandings about the future." The concrete aspects of scientific and technical change can be thought of as a dynamic process of expectations that mobilizes forces and resources, at the macro (national policy, regulation, funding), meso (innovation networks and sectors), and micro (research groups, individual laboratories, and scientists) levels and articulates these different levels vertically and horizontally. The innovation process can also be fear-inducing as it signals social anxieties brought about by technological change. These are well documented and historically specific, but the use of genetically modified crops in agriculture or the establishment of 5G cell phone networks come to mind. Besides the fact that expectations are constitutive, in the sense that they structure and are structured by material scientific and technological activity, they are also performative, as they tend to perform possible futures by tracing specific paths toward their realization. This explains why the immaterial aspects of expectations have been of great interest to social scientists. But as the sociology of science and technology suggests, the actors involved in technological change and their values are literally and materially scripted into technologies and sociotechnical systems. This highlights the inseparable nature of science and values.

Rather than viewing biocapital within a static framework of capitalism, Rajan seeks to analyze developments in the biosciences in the context of capitalism's evolving nature, highlighting both the complementarities and tensions among its various forms. This provides a complex understanding of how speculative forms and commodity productive forms of capitalism coexist within a biotechnology firm, to create value. Hence, the process of valuation in the biotech industry relies on speculative capital, even when the system of commodity production is not yet in place, expressing its imaginary existence. In this way, the possibilities of cell-based meat result from lucrative synergies between tech investors, scientists, and others using Silicon Valley as a model for the cross-fertilization of ideas, technological convergence, and potentialities for enterprises. Even as cell-based meat has not yet reached, and may not reach, the commodity production and manufacturing stage, its valuation rests not only on a tangible material product but also on less tangible forms based on a capitalization of the future. These forms of capitalization often take the form of patents on biological processes or products which may be seen as forms of appropriation of the properties of living systems, in order to ensure the generation of capital even before innovations become operational (Abergel, 2011).

Rajan's book extends work regarding the emergence of tissue economies and their relation to recent developments in the biosciences associated with new biological possibilities in stem cell research and synthetic biology, which rely on the articulation and integration of different sciences and technologies, such as engineering, physics, chemistry, etc. (Rajan, 2006, p. 5). According to Webster, the changing boundaries of the biosciences in terms of disciplinarity and biological possibilities also extend to our social understanding of the boundaries between nature, the body, and identity, including the creation of novel forms of hybrids in the laboratory as "bio-objects" which require new types of state and society regulation and governance (p. 2). Stemming from regenerative medicine, initial controversies focused on the creation of human-animal hybrids but later extended to thinking about the impact of tissue economies, in relation to "the longstanding history of managed reproduction and the manipulation of life itself" (p. 2).

The term tissue here is used in "a generic sense to include blood, organs, and any other kind of living matter taken from the body" (p. 6). Similarly in the case of cell-based meat, tissues are composed of a complex array of biological and physical (sometimes inert) materials, which produce living matter used for food. Lab-grown edible animal tissues correspondingly refer to the final product at the end of a long process, which goes from the harvesting via donation or biopsy of somatic stem cells from an animal to a series of in vitro transformations. This is discussed in the next chapter, but the important point is that CM is a "tissue."

According to Waldby and Cooper, the circulation of animal tissues carries affective significance, the same way human tissues might generate complex ideas and sentiments about human identity and community (2014, p. 6). Of course, in the case of human tissue, the authors are referring to humans donating tissue or whole organs to others. In the case of animal tissues, the situation is different, as the circulation and proliferation of cells and tissues are not for biomedical purposes but for food production. However, the ways in which animal and human tissue economies intersect are that the same technologies are used in biomedicine for developing transplantable tissues and organs. The same life processes are at work, the creation of biovalue from cells to other components of living matter apply.

From Tissue Economies to Cellular Meat

Clearly, the logic of human tissue economies used for reproductive and therapeutic purposes is very similar to cellular meat, as far as the globalization of cellular meat research and the commodification of cells and tissues. They both operate within the "technical framework of biomedicine and the commodity frameworks of biotechnology capital" (Waldby, 2002a). There are, however, some marked differences. While the biological production of tissues might be similar in terms of the scientific processes used to grow them in vitro, cellular meat operates within the context of the food economy. The science of tissue culture and its historical and material developments for human medical science are in fact what inspired the possibilities of cellular meat. Several CEOs of alt-meat startups are medically trained and gained their expertise in biomedical health research. Clearly, the sourcing, circulation, and productive capacity of cells and living tissues follow similar lines of inquiry as Waldby and Webster, in terms of the global mobility of animal tissues, the development of immortalized cell lines that can be patented, the commodification of these tissues within food systems (as opposed to health systems), intellectual property rights of the final products (meats) and of proprietary biological processes involved in production and biomanufacturing, and regulatory issues from one country to another to name a few. One of the main insights from their work relates to a description of the processes leading to the production of patented cell lines from stem cells. They describe how rendering cell lines patentable works within tissue economies, further disentangling stem cells from their bodies of origin or corporeal entities. This commodification of cells, tissues, and organs develops through commodified circuits of value creation and expands as venture capital investments in biotechnology increase. Biobanks signal the institutionalization of biological storage systems but, more importantly, they act as repositories for donor materials for scientific research and development purposes, as well as significant infrastructures for the commodification and circulation of human tissues.

Nonetheless, cellular meat derived from cultured animal cells does not pose the same type of ethical questions raised in *Tissue Economies* because, unlike human tissues used for therapeutic or reproductive purposes in a biomedical context, lab-grown edible tissues are meant to be ingested and circulated through bodies as foods. Furthermore, the tissues undergo a series of processing and cooking steps before being consumed. As such, they are maintained functionally alive until cooked, at which point they become edible. As comestible tissues, they become incorporated into the body but the level and type of incorporation contrast from biomedical processes that involve the transfer of functional biological materials like sperm, ova, blood, tissues, etc. Cellular meats, on the other hand, are digested through the gut and define a different type of tissue transfer. In that sense, the degree of transcorporeality involved in eating might be considered vastly different from the experiences of tissue transfer between donors and recipients in terms of "embodied identity" and the social and relational aspects of what constitutes personhood (Waldby, 2002a). Beyond the technical aspects of tissue transfer, including ways in which the body of recipients is altered to accept 'foreign' tissue, Waldby argues that:

> contemporary developments in biomedicine and biotechnology provoke a reconsideration of the significance of introceptivity and visceral materiality as social and relational sites. (Waldby, 2002a, p. 241)

It is my contention that despite the substantial differences between the experiences of donors and recipients in human tissue transfers, Waldby's exploration of the social and relational aspects of transcorporeality is pertinent to the context of consuming cellular meat. It has to do with the boundaries of and between bodies and the social negotiations that ingesting novel foods such as cellular meat might pose for eaters. This will be discussed in the chapter covering eating bodies and biosocialities. I further contend that the experience of eating cellular meat might reflect the idea of tissue transfer.[2] This is due to the incorporation of animal-derived, biologically synthesized matter into the body as food, while simultaneously highlighting the lack of donor identity. This supposes an indirect form of corporeal exchange between humans and the donor cells or the immortalized cell lines derived from the bodies of animals. What this entails is a further distancing of humans from their own animality, as the tissue fragments they are consuming do not emanate from an integral living subject but from cells that must acquire animal identity as 'beef,' for instance, and that must be constructed into becoming and behaving as 'muscle' during the tissue growing process. However disembodied and impersonal industrial meat may be, the absent referent as it were, comes from animals that were once alive. If cellular meat becomes a commercial reality, eaters will face new types of material and relational exchanges when deciding which meat to consume. They may need to negotiate the tension between the exploitative practices of intensive animal agriculture and the biomanufactured, disembodied nature of lab-grown meat, balancing ethical considerations with personal preferences and societal impacts.

Tissue economies intersect with food and health economies, they also differ in fundamental ways: as the cell meat project advances, the transfer of tissues and cell lines will follow particular trajectories of power and wealth; the global mobility of lab-grown edible animal tissues will necessitate regulatory oversight from one country to another and introduce new market complexities for enabling the sale of animal tissues; the biotechnology and agribusiness will play a role as international brokers of animal tissues—having established the systems and infrastructures for agricultural value chains; what will be their role in shaping food policy?

[2] Cultured meat is considered as edible tissues. A distinction is made with meat obtained from a once living animal and it is assumed that consuming conventional meat does not constitute a form of tissue exchange because it is not produced as the product of extreme biological fragmentation or tissue biofabrication in vitro generated within the context of biotechnological capital.

There is a dichotomy between the gift model for obtaining human tissues and their subsequent commodification and between their donation and their sale across jurisdictions. In the case of cell meat, animals are property, so these questions don't get asked about the status of animal bodies, they are already commodities. However, the gift economy comes into play, when farmers participate in donating animal cells from their herds to assist in the banking of animal cell lines and contribute to building the cellular economy.

In her important work, Hannah Landecker suggests that as we proceed in transforming cells into technologies, we're not just "mining" living organisms for value through these experimental interventions and the creation of living objects in the laboratory. She claims that these practices establish much more, as they serve to shape how humans relate to the living world.

Vital Economies: The Intersection of Lively Capital and Cellular Agriculture

While the question of defining life and its complexities is not the subject of this book, it suffices to argue that cell-based meat is a technoscientific development that necessitates critical thinking about life and vitality. Additionally, it provides a platform to explore the ramifications of the social project that is inherent to the various narratives used to assert its quasi-inevitability. To a certain extent, cell-based meat embodies the promise of creating new forms of life and sustenance, as we navigate an increasingly uncertain future. Nevertheless, if we transition to cell-based meat production and cease to rely on farm animals, the ontological question of being and beings will remain unresolved, including the social transformations that such a transition portends.

However, the concept of cell-based meat goes beyond its potential to address the food and environmental crisis. It also embodies both a material and symbolic representation of life and vitality, challenging conventional understandings of the natural world (Thacker, Eugene, 2010). The philosophical implications of this are significant and aspects of the question have been explored in academic literature. The ethical implications of cell-based meat are that it represents a shift in our understanding of what it means to be alive, as well as a rethinking about our relationship with animals. Cell-based meat signifies a significant transition toward an increasingly technologically mediated comprehension of life. It represents the ability to engineer and manipulate life within a laboratory setting, challenging the boundaries of our scientific prowess to shape it according to our desires. This carries significant implications for our understanding of what it means to be alive, as well as the ethics and governance of emerging biotechnologies. Nevertheless, irrespective of its revolutionary narratives, the types of vitality and life inherent in cell-based meat emanate from a specific vision of the future, where the conventional understanding of life as exclusively natural and biological is contested.

As a result, cellular meat is not simply about the "future of meat" or even planetary survival, but its significance rests on a multitude of interacting temporalities, including biological, economic, and geological timescales, which are discussed in later chapters. What lab-grown meat also interrogates is the unsettled relationship between "the temporality of biological life [and living matter] and the temporality of value" as the forces of neoliberalism and the life sciences join (Cooper, 2015). The commodification process of these includes adding biovalue, where the value of living organisms is determined not just by their natural existence but also by their potential for economic profit and manipulation, namely, through the applications of biotechnologies. This integration of life sciences with market forces further complicates ethical considerations and challenges traditional notions of value and of life's intrinsic worth. Amid an intensifying climate crisis, the engineering of cells into meat proceeds within a largely depoliticized discourse that fails to examine how this technology intersects with existing economic structures, such as the corporate control of food production and farm labour, the urban-rural divide, and broader society dynamics, including issues of inequality and environmental justice. The narrow framing of "climate-friendly meat" as a sociotechnical project deflects attention from the urgent need for systemic transformation of the global food system, transformations that are achievable in the short term.

The entanglement of the life sciences and market economies means that biological value is bound to the technological possibilities associated with cellular processes, revealing opportunities for industrial-scale advancements in the fields of biomedicine, materials science, and agriculture. As the entrepreneurial turn and corporatization of the life sciences as Sunder Rajan writes became increasingly naturalized over the past four decades, the domains of economics/capitalism and the life sciences/biotechnology evolved common goals dealing with "similar types of problem-spaces" (Sunder Rajan, 2012). The concept of lively capital explores the ways in which the practices and knowledges of the life sciences are increasingly tied to markets, building on the work of Melinda Cooper, who demonstrates the overlap between biotechnology and capitalism and analyzes how security discourse is framed around the biological and life processes in the neoliberal era.

In her work, Cooper focuses on the U.S. government's dual strategy of financialization and debt creation as means of reviving capitalism. The speculative and promissory natures of the life sciences provide the essential tools for guaranteeing value creation, returns on investments and profits, at a moment when economic growth is no longer achievable in traditional industrial sectors. By focusing on emerging threats—biological and ecological, local and global, human-made or natural, etc.—as disruptive social and biological forces which must be tamed, the life sciences offer a means of producing surplus value, where patents on life processes "generate and capture production itself, in all its emergent possibilities" and bioproduction signifies "the accumulation of biological futures rather than the extraction of non-renewable resources and the mass production of tangible commodities" (Cooper, 2008, p. 24). Referring to the concept of lively capital, Rajan et al. revisit value creation and systems of valuation in the life sciences, in terms of the articulation between political economy and ethics not as distinct domains but that considers regimes of ethics, implying that "ethics contains its own political economies and capacities for being institutionalized and made corporate" (Sunder Rajan, 2012, p. 19). This signifies that ethics—altruism, tissue and cell donation, the gift economy, etc.—exist in parallel with regimes of technoscience and capital. Both Cooper and Rajan provide insights into the entangled temporalities of life and capital, the speculative moment and promissory futures involved in the technological re-creation and re-design of life as a function of damaged, ruined, and fast depleting presents. As a result, Cooper suggests that "the delirium of contemporary capitalism … is intimately and essentially concerned with the limits of life on Earth and the regeneration of living futures—beyond the limits" (p. 20).

These ideas are encapsulated in the Organization for Economic Cooperation and Development's (OECD) 2021 report on science and technology, entitled "Times of crisis and opportunity." The report acknowledges the role of the COVID-19 pandemic as having played a central role in stimulating biomedical research and innovation systems, thus reifying a belief in science and scientific research to mitigate existing and emerging threats. The report states: "The pandemic has underscored more than in other recent crises the importance of science and innovation to both preparing for and reacting to upcoming crises" (OECD, 2021). In particular, the development of vaccines within the span of several months has been a key achievement for the life sciences as has been the use of rapid technology-driven collaborations and convergence, including the acceleration of public response and regulatory approvals in biomedical research. COVID-19 has been a catalyst for new technologies and value creation. This has also facilitated new technology adoption; the coronavirus pandemic created an emergency situation which was essential in promoting technological innovations in cellular agriculture and the alt-protein industries. It also gave rise to new research frameworks for converging technologies (the life sciences' use of big data analytics, artificial intelligence, machine learning, Internet of Things (IoT), bioprinting, smart robots, etc.), as well as novel models for private-public partnerships and funding.

Cellular agriculture positions itself at this point in time as a perfect illustration of these processes. CA's emphasis on securing the future on Earth and beyond, as it creates opportunities for space exploration and colonization, rests on renewed investments in technoscientific solutions and infrastructure. Regardless of our perspectives on the Anthropocene, this concept radically changes our view of nature and the relationship we as humans entertain with other animals and living organisms (Tønnesen et al., 2016). Ideas relating to the Anthropocene and food narratives help us to understand the social dynamics that drive social change and produce knowledge about edibility. More importantly, I contend that agricultural innovation and technoscientific framings of the ecological crisis are not simply about the creation of novel meats or new foods for humans to consume, they can also be read as foreshadowing extreme forms of manipulation and alteration of cellular matter (Landecker, 2002), cast as our food future. This is indicative of a particular understanding of human evolution whereby changes to the biological become socially, scientifically, ethically, economically, and culturally negotiated as lively capital. The industrial scale of cells as biologicals means that they are manipulated and exploited "both as a source of materials or as a potential goal for materials being developed" (Landecker, 2002). Hence, technology's ability to produce flesh in the laboratory hinges on harnessing the plasticity and temporality of biological life as it exists in vivo but managing it in vitro, producing new kinds of vitality via biotechnological means.

Bibliography

Abergel, E. (2011). Climate-ready Crops and Bio-capitalism: Towards a New Food Regime? *The International Journal of Sociology of Agriculture and Food, 18*(3), 260–274. https://doi.org/10.48416/IJSAF.V18I3.248

Borup, M., Brown, N., Konrad, K., & Lente, H. V. (2006). The Sociology of Expectations in Science and Technology. *Technology Analysis & Strategic Management, 18*(3–4), 285–298. https://doi.org/10.1080/09537320600777002

Cooper, M. (2008). *Life as Surplus: Biotechnology and Capitalism in the Neoliberal Era*. University of Washington Press.

Cooper, M. (2015). In Vivo Economies: Temporalities of Life and Value. *Journal of Cultural Economy, 8*(3), 257–259. https://doi.org/10.1080/17530350.2015.1039461

Datar, I., Kim, E., & d'Origny, G. (2016). New Harvest: Building the Cellular Agriculture Economy. In B. Donaldson & C. Carter (Eds.), *The Future of Meat without Animals* (pp. 121–131). Roman and Littlefield International.

Georgescu-Roegen, N. (2011). *From bioeconomics to degrowth: Georgescu-Roegen's 'new economics' in eight essays*. Routledge.

Landecker, H. (2002). New Times for Biology: Nerve Cultures and the Advent of Cellular Life In Vitro. *Studies in History and Philosophy of Science Part C:*

Studies in History and Philosophy of Biological and Biomedical Sciences, 33(4), 667–694. https://doi.org/10.1016/S1369-8486(02)00026-2
Landecker, H. (2004). Living Differently in Time: Plasticity, Temporality and Cellular Biotechnologies. *Culture Machine, 7*(0). Retrieved from http://svr91.edns1.com/~culturem/index.php/cm/article/view/26
OECD. (2021). *OECD Science, Technology and Innovation Outlook 2020: Science and Innovation in Times of Crisis*. OECD. https://doi.org/10.1787/75f79015-en
Pavone, V., & Goven, J. (2017). *Bioeconomies—Life, Technology, and Capital in the 21st Century*. Palgrave Macmillan. Retrieved from https://www.palgrave.com/gp/book/9783319556505
Rajan, K. S. (2006). *Biocapital: The Constitution of Postgenomic Life*. Duke University Press.
Shaw, R. (2008). The Notion of the Gift in the Donation of Body Tissues. *Sociological Research Online, 13*(6), 41–50. https://doi.org/10.5153/sro.1832
Sunder Rajan, K. (2012). *Lively Capital: Biotechnologies, Ethics, and Governance in Global Markets*. Duke University Press.
Thacker, Eugene. (2010). *After Life*. The University of Chicago.
Tønnesen, M., Armstong Oma, K., & Rattasepp, S. (Eds.). (2016). *Thinking about Animals in the Age of the Anthropocene*. Lexington Books.
Twine, R. (2007). Animal Genomics and Ambivalence: A Sociology of Animal Bodies in Agricultural Biotechnology. *Genomics, Society and Policy, 3*(2), 99. https://doi.org/10.1186/1746-5354-3-2-99
Ung-Lanki, S. (2014). Constructing the Biotech Nonhuman Animal: Instrumentalism and Ambivalence. *Society & Animals, 22*(5), 439–458. https://doi.org/10.1163/15685306-12341349
Vogelpohl, T., & Töller, A. E. (2021). Perspectives on the Bioeconomy as an Emerging Policy Field. *Journal of Environmental Policy & Planning, 23*(2), 143–151. https://doi.org/10.1080/1523908X.2021.1901394
Waldby, C. (2002a). Biomedicine, Tissue Transfer and Intercorporeality. *Feminist Theory, 3*(3), 239–254. https://doi.org/10.1177/146470002762491980
Waldby, C. (2002b). Stem Cells, Tissue Cultures and the Production of Biovalue. *Health: An Interdisciplinary Journal for the Social Study of Health, Illness and Medicine, 6*(3), 305–323. https://doi.org/10.1177/136345930200600304
Waldby, C., & Cooper, M. (2014). *Clinical Labor: Tissue Donors and Research Subjects in the Global Bioeconomy*. Duke University Press.
Waldby, C., & Mitchell, R. (2006). *Tissue Economies: Blood, Organs, and Cell Lines in Late Capitalism*. Duke University Press.

5

Making and Eating Meat in/for/Against the Anthropocene

A Brief History of In Vitro Meat and Synthetic Flesh Imaginaries

In vitro meat thus resonates with a century of science fiction scenarios, political visions, and scientific hopes about a world where meat is produced independently of animal bodies. Today, many researchers place significant anticipation in this industry, even seeing in vitro meat as the inescapable future of humanity (Bhat et al., 2015), or the logical conclusion of tissue culture research (Jönsson, 2016). Although it represents a radically new way of thinking about and producing meat (Verbeke et al., 2015), some see it as the logical evolution of the meat industry (Hocquette et al., 2015), given that the production capacity of conventional livestock farming has practically reached its limits (FAO, 2006, 2009, 2011). This evolution has been described, following hunting and livestock farming, as a new agricultural revolution, used by humans in meat production.

The idea of cultivating meat outside of the animal body is not new. In the case of in vitro meat, indeed, the literature identifies a number of individuals, as well as cultural creations, as foundational to the sociotechnical imaginaries working prior to and/or in parallel with its development. The following section provides key moments in the history of

E. Abergel, *Dead Meat*, https://doi.org/10.1007/978-981-97-9049-4_5

cultured meat, together with fictional accounts of synthetic flesh in science fiction in order to show that there is a consistent interplay between art, science fiction, and science. The significance of this interplay is its power to shape collective imaginaries and the ways in which each of these envisions the future. In addition, these dynamic interactions actively participate in shaping public perceptions and ethical considerations about the future of cellular meat. Each of these categories contextualizes synthetically produced flesh within broader speculative narratives. This section does not provide a comprehensive history of in vitro meat, nor does it describe a timeline of events, rather it focuses on particular moments in the development of this nascent science and industry. Neither does it present a complete inventory of all the science fiction and visual arts productions that feature cell meat. My short exploration into these themes aims to juxtapose artistic and fictional explorations of 'meat' in light of scientific advancement in the field. Imaginaries, like utopias, are not simply passive reflections but the product of active and rational critiques of our times and they often challenge us to envision what the future might hold. It is also a way of telling new stories about existing stories, by mixing the factual and the fictional (Tsing, 2017), by using a number of methods for thinking through things, as Donna Haraway states:

> It matters what matters we use to think other matters with; it matters what stories we tell to tell other stories with; it matters what knots knot knots, what thoughts think thoughts, what descriptions describe descriptions, what ties tie ties. It matters what stories make worlds, what worlds make stories. (Haraway, 2016, p. 12)

These intersecting narratives contrast the critical and dystopian perspectives found in cultural productions against the utopian aspirations of scientific research toward industrialized, slaughter-free, and environmentally sustainable meat production. They can be important for understanding the complex relationship between science and society. Louis Lemkow studies how science fiction depicts cell meat. He writes:

> In vitro meat (IVM) is as much part of the science fiction universe as spaceflight, extraterrestrial life, robots and artificial intelligence. Science fiction

is replete with synthetic meat. In scenarios of the far future, IVM is grown in the laboratory-factories of mammoth sized starships, providing protein for human colonists that traverse light years to conquer new worlds. In the not so distant future, outposts on the Moon or Mars sustain their inhabitants through hydroponic farms and lab meat (as IVM is sometimes referred to in science fiction). Closer to home, in the dystopian imaginary of Earth's dismal future, we find foul tasting and poorly textured IVM providing a last-ditch solution to gross overpopulation, natural resource depletion and animal protein shortages, temporarily saving us from world hunger and malnutrition.

More optimistic narratives project IVM as a cheap solution to human protein requirements in a crowded world, where fortunately, technological fixes arise for every societal challenge. Today, hype often surrounds IVM; supporters of this technology insist that it is a key to our future wellbeing, and media outlets like the Guardian newspaper report that it is no longer a futuristic fantasy: "Is science fiction on the verge of becoming fact?…Two scientists on opposite sides of the world both claim to be on the verge of serving up the first lab-grown hamburger and saving the planet in the process. The new reality is so close, you can almost taste it". (Lemkow, 2015, p. 45)

Sergey Brin, who funded the first "test-tube burger," shared his thoughts in a video released shortly after the product's public unveiling. The video was posted on CulturedBeef.net, a website created by University of Maastricht biologist Mark Post and his team for the lab-grown burger project. He said:

> Some people think this is science fiction—it's not real. It's somewhere out there…I actually think that's a good thing. If what we are doing is not seen by some people as science fiction, it's probably not transformative enough. (Boyle, 2013)

This intersection of cellular meat and science fiction, where once-imagined futures begin to take shape, finds its roots in the early scientific breakthroughs, such as the discovery of cell culture media, that laid the foundation for the first experiments in lab-grown meat. As early as 1882, Sydney Ringer developed the first saline solution that closely mimicked the composition of bodily fluids. This groundbreaking solution enabled Ringer to sustain the beating of frog hearts, even after they had been excised from the animals. By maintaining the proper ionic balance, Ringer's solution was able to preserve the physiological function of the heart tissue outside the living organism. This work was pivotal in the field of experimental physiology, demonstrating that organs and tissues could be kept viable ex vivo under the right conditions. The development of Ringer's solution laid the foundation for future advancements in organ preservation and tissue culture, influencing a wide array of biomedical research and applications (Yao & Asayama, 2017). Ringer's innovative approach involved using a balanced mixture of sodium chloride, potassium chloride, calcium chloride, and other electrolytes, which approximated the natural environment of the body's extracellular fluid. This precise formulation was crucial in maintaining the contractility of the heart muscle cells, allowing for sustained physiological activity in a controlled laboratory setting. His work provided critical insights into the essential conditions required for the survival of isolated tissues, significantly advancing the understanding of cellular physiology and paving the way for later developments in tissue engineering and regenerative medicine.

Alexis Carrel, Nobel Prize laureate in Physiology or Medicine in 1912, is renowned for his experiments on the preservation of blood vessels and living organs in vitro for potential transplantation purposes. Carrel maintained a "living" chicken heart for an estimated duration of between 28 and 37 years, significantly exceeding the typical lifespan of a chicken, which is about 5 years. His work was pioneering because it raised questions about the lifespan of living organisms and, in so doing, interrogated the process of cell mortality as an internal process. Carrel's experiment began in 1912, when he successfully cultivated a fragment of embryonic chicken heart in vitro. His method involved keeping the tissue alive by regularly providing it with a fresh nutrient solution. This experiment was intended to explore the potential for long-term preservation and even transplantation of tissues, an idea that was revolutionary at the time. The longevity of the chicken heart tissue raised fundamental questions about the nature of cellular aging and the intrinsic factors that govern cell death (Landecker, 2007). In addition, he pioneered the techniques of cell culture by devising a glass flask still used today (hence the term in vitro, literally, in glass to describe the glass apparatus used to perform these experiments).

Carrel's work, however, has not been without controversy. In part because generations of researchers found his results challenging to replicate, but mostly because it was revealed in the 1960s by Leornard Hayflick[1] that his immortalized chicken heart culture consisted of normal cells with a set number of replications, approximately 50 doublings or 35, according to Hayflick (Rasko & Power, 2015). Hayflick concluded that when Carrel's assistant refreshed the nutrient medium, new viable cells were probably added, maintaining the illusion that the original cells were still alive. Nonetheless, his contributions laid the groundwork for the field of tissue engineering and had a profound impact on subsequent biomedical research. Carrel's experiment demonstrated the potential for in vitro cultivation of tissues, paving the way for later developments in regenerative medicine and synthetic biology. It is interesting to note how:

> Carrel's work now becomes seen as a precursor to current food production experiments, and what was essentially a failed experiment (chicken tissue cannot be held alive for nearly as long as Carrel suggested) is rendered inspirational. (Jönsson, 2016, p. 11)

A few years later, in the 1930s, the lawyer and conservative politician Frederick Edwin Smith (1872–1930) wrote that it would soon no longer be necessary to raise cattle to eat steak; according to him, it would be possible to cultivate it from a single steak carefully selected for its qualities. One of Smith's closest friends, Winston Churchill (1874–1965), published an essay in 1931, *Fifty Years Hence*, in the work *Thoughts and Adventures*, where the statesman described a world where farmlands are transformed into parks and gardens, as it is no longer necessary to raise chickens to produce wings and thighs. His statement and the quotation have been a fixture in the historical accounts of cell culture meat. This quotation forms an integral part of cell meat narratives and imaginaries:

[1] Hayflick is known today because he introduced the notion of the Hayflick limit, that is, the number of times a cell population can divide until it reaches senescence (Shay & Wright, 2000).

> We shall escape the absurdity of growing a whole chicken in order to eat the breast or wing, by growing these parts separately under a suitable medium.—Fifty Years Hence, Winston Churchill (1931)

Later, in 1943, the French writer and journalist, René Barjavel (1911–1985), published the science fiction novel *Ravage*, in which he depicted a post-apocalyptic world where a large Parisian house produces in vitro steak in its basement, explicitly following the work of Alexis Carrel. In 1927, J.B.S. Haldane and Carl A. Price wrote an essay entitled *The Future of Biology* contained in his works *Possible Worlds and Other Essays* where they imagine ways of keeping a killed animal's heart and liver alive for a day or more and making beefsteak, grown out of tissue culture of muscle with no nervous system and hormones, to keep it growing (Haldane & Price, 2001). Much more recently, van der Weele and Tramper (2014) imagined villages where pigs are raised as pets in people's backyards while their cells proliferate in local factories.

The idea of using tissue culture to produce "meat" belongs to Willem van Eelen, a Dutch researcher trained in medicine and an entrepreneur who is dubbed as "the godfather of *in vitro* meat" (van Mensvoort, 2015). On the surface, it appears to be a simple proposition, one that van Eelen came up with while he was a prisoner of war, in Japanese camps during World War II (WWII) in Indonesia where he starved for five years (Hickman, 2009). According to media accounts, due to van Eelen's experience of extreme hunger and his observations of the harsh treatment inflicted on animals by his captors (Specter, 2011; Wurgaft, 2019), he became deeply concerned about animal cruelty and vowed to work against unnecessary animal treatment, although he apparently never became a vegetarian.

Decades later, the science of tissue culture caught up to van Eelen's imagination with the discovery of mouse stem cells in the early 1980s, which could differentiate into different types of tissues. Recognizing the potential of this discovery, he encouraged the Dutch government to invest in research about cell cultured meat which led to Mark Post's first lab-grown burger in 2013 (Wurgaft, 2019). van Eelen's training as a medical doctor inspired him to work on his ideas of producing in vitro meat throughout his career by the end of which he had filed a series of patents in the U.S. and in the Netherlands. He has held a U.S. patent and an international patent for *the industrial production of meat using cell culture methods* since 1999. As a medical student in 1948, he witnessed researchers experimenting with stem cells to cultivate skin for burn victims. This sparked an idea in him. As he later told his daughter, "The only thing I saw was a piece of meat." This moment ignited a lifelong mission: filing patents, making unannounced visits to potential investors to secure funding, and collaborating with others who also held patents related to lab-grown meat (Seamons, 2017). Together with John Vein, they launched a dedicated effort to secure financing and attract more researchers. However, van Eelen passed away in 2015 without seeing his vision come to fruition. After his death, at the age of 91, Josh Tetrick, the CEO and co-founder of Eat Just—a San Francisco-based cellular meat company originally called Beyond Eggs and Hampton Creek, founded in 2011—purchased all of van Eelen's patents. Tetrick contacted his daughter, Ira van Eelen, expressing interest in her father's patents. He acquired the patents in 2017, around the time van Eelen's daughter joined the company's board, it was described as a "done deal" (Seamons, 2017). van Eelen's patents, starting with the first filed in 1999, consisted of:

> extracting cells from the embryonic muscle or somatic cells through a biopsy. It also dealt with constructing a thin scaffold that would enable cell growth, which the researchers claimed would result in an edible product. In 2006 another invention by Van Eelen concerned a meat product from in vitro cultivation of non-human cells in a culture medium free of dangerous substances, capable of producing a three dimensional animal muscle tissue. (Fernandes et al., 2022, p. 3)

Prior to this, several works of fiction involving synthetic meat had been imagined which reinforced the idea that cultured flesh has been a long-standing motif in science fiction and in popular imagination. In 1897, the German writer and scientist Kurd Lasswitz (1848–1910) published the science fiction novel *Two Planets*, in which Martians who had established a base at the North Pole introduced "synthetic meat" to Earth. Later, in 1939, George Orwell's novel *Coming Up for Air* featured a modern, disenchanted narrator disgusted by a fake sausage, made from a rubber casing filled with fish. In 1954, Frederick Pohl and Cyril Kornbluth co-authored a novel titled *The Space Merchants* where they describe an overpopulated planet governed by advertisers, where an artificial meat called "Chicken Little" is manufactured. In 1973, the dystopian film *Soylent Green*, based on Harry Harrison's novel, published a few years earlier, depicts New York City in a post-apocalyptic world where, due to the near-total absence of arable land and ocean life, humans are forced to consume a synthetic food produced by the multinational Soylent. It is revealed at the end of the film that Soylent Green is made from human corpses, turning humans into cannibals. More recently, in 2003, the Canadian novelist Margaret Atwood wrote an equally dark speculative fiction, *Oryx and Crake*, in which she describes a post-apocalyptic world where transgenic animals on the loose threaten Snowman, the last human. In her novel, Atwood describes the Pigoon Project:

> The goal of the pigoon project was to grow an assortment of foolproof human-tissue organs in a transgenic knockout pig host – organs that would transplant smoothly and avoid rejection, but would also be able to fend off attacks by opportunistic microbes and viruses, of which there were more strains every year. A rapid-maturity gene was spliced in so the pigoon kidneys and livers and hearts would be ready sooner, and now they were perfecting a pigoon that could grow five or six kidneys at a time. Such a host animal could be reaped of its extra kidneys; then, rather than being destroyed, it could keep on living and grow more organs, much as a lobster could grow another claw to replace a missing one. That would be less wasteful, as it took a lot of food and care to grow a pigoon. A great deal of investment money had gone into OrganInc Farms. (Atwood, 2013, pp. 22–23)

These various fictional accounts provide a dystopian discourse in parallel to science and its imagined potentialities. They are a provocation and, as such, incite larger ethical inquiries about bioengineering animals for food as a technoscientific project (Castle, 2022). The Pigoons are reminiscent of genetically engineered pigs, produced as sources of organs for human transplants, through the process of xenotransplantation. In 2022, the *New England Journal of Medicine* published a paper describing an experiment whereby pig kidneys, genetically edited to reduce the chances of organ rejection and of infection, were transplanted into brain-dead human recipients as a way to test their function (Montgomery et al., 2022). In 2014, the company Organovo printed liver tissue for the first time in their newly developed Bioprinter. The tissue functioned as a real liver for 42 days and researchers later attempted to bioprint fully functional human kidney tissues (Listek, 2019). The same technology is applied today to produce organoids in the biomedical field. These consist of cell cluster structures that are formed by cultivating stem cells in a 3D environment in vitro in order to simulate the living environment of a cell in vivo (Nie et al., 2021). These models are used to screen drugs for toxicity and exist for different types of human tissues. Imagine these being used for food, like the pigoons which

> quietly get added to the human food chain as both waste products of research and development for the biomedical industry and welcome additions to a depleted meat-production stream. But the cleverness of these pig-shaped genetic hybrids proves another unintended consequence of developing this animal as a biotech fix to human medical problems, only far more deadly. (McHugh, 2010, p. 193)

Works of fiction like Atwood's novel become metaphors for thinking about the "perplexing forms of species and social life already embedded deeply in our own and other tissues" and, as such, are not "so much fantastic extensions as practical examinations of these lived conditions" (Ibid, p. 196).

It would also seem that science fiction is mobilized as a discourse in the technosciences. Nora Castle analyzes CM and science fiction using Atwood's depiction of ChickieNobs, a near depiction of bioengineered chickens, found in *Oryx and Crake*. She views this example as the movement between non-human animals as animals toward non-human animals as biotechnology. She sees fictional accounts of cultured meat as establishing the sociocultural framing for CM in public and corporate discourse. More specifically, she asserts that:

> In fact, IVM and other "New Food" companies themselves mobilize speculative narratives and sf tropes to contextualize their products. Sf acts as the "discursive imaginary" (Hollinger 149) through which these companies construct their products narratologically. In claims that it will "disrupt" (a buzzword in startup culture) existing animal agriculture, IVM is simultaneously an actual technology that can produce "animal" flesh, and an exaggerated novum for the stories which New Food companies want to tell, stories that gain them access to the speculative finance of venture capital, even without a sellable product. (Castle, 2022, p. 157)

The Genesis of Cellular Agriculture: 1998 to 2014

The early 2000s marked the genesis of cellular agriculture, with pioneering research focusing on the basic science of growing animal cells in vitro. Advances in stem cell biology played a crucial role, as scientists explored using stem cells to produce muscle tissue. In their review of the first 20 years of tissue engineering for making cell meat, Stephens et al. describe the development of CM, from idea to possible reality, as follows:

> In general, the first wave CM community was typified by university-based projects (sometimes with industrial contacts) driven by environmental and animal ethics concerns but overwhelmingly hitting a wall when trying to attract funding from the traditional institutions that support university work. They generally saw their work as basic science, focused on building a knowledge base for subsequent innovators to develop products. The majority of the researchers involved came from biomedical backgrounds, often based in biomedical departments, and were fully aware of how some colleagues attributed an oddness to their pursuit of tissue engineered meat. (Stephens et al., 2019, p. 4)

Much of the scientific work done to produce edible animal tissue tended to be as side experiments in biomedical laboratories. Attracting research funding and generating general interest is also a feature of this period, as several funding bodies in different countries rejected scientific proposals (Ibid).

During this period, patents began to be filed for methods and technologies related to lab-grown meat. The Tissue Culture and Art Project, a collective of bioartists, produced animal tissue for the purpose of creating food in 1996. As part of an exhibition called *Disembodied Cuisine*, the artists created frog 'meat' from culturing cells. They served the frog tissues at the end as part of a meal, the final feast, where members of the public were invited as guests. Chapter 7 details this particular event in light of Mark Post's first lab-grown hamburger served in 2013. It should be noted that this event came as a desire to show the public "a proof of concept" of what the industry and the technology were capable of doing, but it was predominantly a response to the Dutch government's failure to provide funding for the national in vitro meat consortium, because of their claim that the public and private sector showed little interest (Jönsson, 2016).

Between 1998 and 2001, NASA showed interest in a project that had struggled to secure financial backing, appointing scientist Morris Benjaminson to lead a research team aimed at producing fresh meat in the laboratory for space exploration. Benjaminson's initial experiments involved extracting pieces of tissue from living goldfish and placing them in a nutrient medium composed of fetal calf serum. After a week, he observed that the fish cells had proliferated, resulting in a 15% increase in the initial volume of the tissue samples. Although Benjaminson's experiment did not produce edible meat and did not progress further, his work demonstrated the feasibility of cultivating living tissue outside of the animal body (Benjaminson et al., 2002). His research was groundbreaking in the field of cellular agriculture and tissue engineering, as it provided proof of concept for the in vitro cultivation of animal tissues. The use of fetal calf serum, which contains essential growth factors and nutrients, was crucial in supporting cell proliferation and tissue growth. This experiment underscored the potential of biotechnological approaches to address the challenges of food production in extreme environments, such as space.

The significance of Benjaminson's work lies in its pioneering demonstration that animal tissues can be maintained and grown outside the living organism. This laid the foundation for subsequent research in the field, inspiring further exploration into alternative methods of meat production that could reduce reliance on traditional livestock farming. The implications of his findings extend beyond space exploration. Upon inspection, the goldfish explants were considered "acceptable as food" (Stephens et al., 2019).

At the time of Benjaminson's goldfish experiments, in the early 2000s, a number of experiments were conducted on in vitro meat:

> with key figures including Vladimir Mironov, then of the Medical University of South Carolina (MUSC), Douglas McFarland of South Dakota State in the US, and Henk Haagsman of Utrecht University in Europe. Mironov and McFarland co-authored the first academic CM review article with Edelman et al. (2005). Edelman had previously written a review article in 2003 during a science journalism course in Wageningen, the Netherlands. Matheny had been inspired by the NASA work and was in the process of establishing a non-profit supporting the sector, which came to be the now well-known New Harvest, which at that time gave 100% of donations to research at universities. (Stephens et al., 2019, p. 3)

In 2005, as a result of van Eelen's advocacy, the Dutch government funded research into IVM. According to Stephens et al. (2019) this was unexpected since the grant was obtained for a program in sustainable food systems which weren't typical funders for this type of research. The project ended in 2009. One of the researchers was Dr. Mark Post, a medical doctor and professor of physiology at Maastricht University, and Peter Verstrate, a food technician. They are known for creating the famous $325,000 lab-grown hamburger unveiled in 2013. Today, Post is the co-founder and Chief Scientific Officer at Mosa Meat, a Maastricht-based food technology company that produces beef from animal cells. He is also the co-founder and Chief Scientific Officer of Qorium, a cell-based leather company. Mosa Meat was incorporated in 2016 and has been developing new techniques for growing muscle as well as having defined an animal-free medium for growing cells. In 2023, the company unveiled its new Center for Advanced Meat Production, Upscaling and Sustainability (CAMPUS). During this event Mark Post was quoted as saying: "When we introduced cultivated meat to the world, we predicted it would take 10 years to create a consumer product. Now, almost exactly 10 years later, we have a consumer product that we can start making in larger quantities and that we can start serving to consumers in Singapore, pending regulatory approval" (Edible Planet Ventures, 2023). In 2020, Singapore granted regulatory approval to one company, Eat Just or GOOD Meat, for their cell-based chicken meat, despite the long list of companies having filed to sell their products there. Since 2015, Maastricht has hosted the annual International Scientific Conference on Cultured Meat (ISCCM) which has been a gathering place for cell meat advocates, scientists, investors, and others. Prior to these meetings, conferences were organized as early as 2008 with the intent to "bring people together and symbolically mark new stages in the field's development" (Stephens et al., 2019).[2] These gatherings and subsequent CM events have been key to the emergence of the industry.

Starting in 2004, a new organization dedicated to promoting the development of IVM started advocating for the industry. New Harvest founder Jason Matheny was an animal rights activist and a graduate student in utilitarian philosophy (McHugh, 2010) when he got inspired by the powers of cell meat to stop animal violence. Not only was he inspired by the NASA experiments conducted by Benjaminson, he also became aware of the bioart collective *Tissue Culture and Art Project*'s work on frog meat, a tissue created while artists were in residence in a laboratory. The frog tissue later became part of an art exhibition which will be discussed in detail in Chap. 7. Information about the history of New Harvest from their website explains his involvement with the Dutch government from 2004 to 2013 advocating on behalf of the emerging industry. The New Harvest website went online in 2004 and that year, the non-profit organization was invited to discuss cultured meat at the PROtein Foods, Environment, Technology and Society (PROFETAS) conference in Wageningen, where Jason Matheny met with the Dutch agriculture Minister to discuss funding for cultured meat research (Datar et al., 2016).

In 2015, Isha Datar, New Harvest's first employee, started a Facebook thread within the New Harvest Community Discussion that gave birth to the term "cellular agriculture." In her Facebook message, Datar asks the "community" to come up with a sexy name in order to "establish a

[2] Stephens et al. (2019) provide detailed accounts of these meetings and the key figures who influenced the field of CM.

new academic discipline and an industry focused on 'animals products without animals' (Zassenhaus, 2021). We will return to the phrase "cellular agriculture" and what it conjures up in the broader context of agriculture and eaters, when we discuss meat imaginaries. Essentially, the promise articulated by New Harvest at the time was that the in vitro production of meat without animals opened the way for advancing meat analogues that would be palatable to meat eaters (McHugh, 2010).

The establishment of organizations like New Harvest, in 2004, played a pivotal role in promoting research and development in cellular agriculture, shifting from public to private funding for research and development. The creation of annual New Harvest conferences, starting in 2016 in San Francisco, served as catalysts to share the latest scientific developments in cellular agriculture, for promoting research into CM and as a way to attract investors, the public, and government officials to believe in the future of the industry. These meetings mark a shift "away from the university-based disciplinary activity of the first wave to the start-up culture," according to Stephens et al. (2019).

These organizations started to create networks of researchers and began advocating for the potential of lab-grown meat. Besides offering research fellowships and grants, the organization linked researchers with non-profits, philanthropists, and governments. This led to the establishment of the National Institute for Cellular Agriculture based at Tufts University, in 2021. The years following the creation of New Harvest in the U.S. and in the wake of the research performed as part of the in vitro Meat Consortium in Europe, PETA (People for the Ethical Treatment of Animals) launched its "*in vitro* chicken contest," offering a $1 million prize to the first laboratory capable of producing commercially viable chicken meat from cells. The contest was announced in 2008 and ran until 2014, as no candidates came forward. The rules stipulated that the contestants had to produce commercially significant quantities of product that could be sold at the comparable market price of boneless chicken breast. While still largely confined to academic and scientific communities, the concept of lab-grown meat started to gain attention. Media coverage was sporadic and often skeptical, but the foundational ideas began to spark interest among forward-thinking investors, philanthropists, and scientists. The PETA contest contributed to spreading the idea of cell culture in line with animal rights. Ironically, at that time, the technology relied heavily on animals and animal products, so it was interesting that PETA defended this approach against animal cruelty. Besides, as McHugh notes in her work, up until then goldfish, chickens, sheep, and toad cells were sacrificed to advance the technology, being kept alive with serum derived from other animals but always with the promise of "meat's transcendence from animal life" in the future (2010). More on this contest and its aftermath is discussed in the chapter on building cr/edibility. This decade laid the foundational scientific groundwork but offered little in terms of commercial progress or public engagement.

The annual Maastricht and New Harvest events spawned a series of other similar events in the U.S. and in other countries. The Good Food Institute launched its own series of conferences in the San Francisco Bay area in 2018. The development of a CM community outside of academia was instrumental for laying the groundwork for making CM a "physical technology" with promise. This enabled funding and interest to come from third-sector investors, as well as fostering an entrepreneurial culture reflected in the emergence of start-ups.

High Hopes and Industrial Ambitions: 2015–2020

Having gone through an identity crisis and found a name, it was time for the industry to start showing concrete results. The early 2010s marked a turning point with the unveiling of the world's first lab-grown burger by Dr. Mark Post, in 2013, as mentioned before. This event demonstrated the practical potential of cultured meat and sparked significant interest in the field but did not receive the expected governmental financial support. During this period, researchers focused on refining cell culture techniques, improving the taste and texture of lab-grown meat, and reducing production costs. Innovations in bioreactor design and growth media optimization were central to these efforts.

A number of cultured meat companies were founded as start-ups. One of these was Perfect Day Foods, co-founded with New Harvest, in 2014.

The company used fermentation to produce dairy proteins without animals. Despite the scientific advancements, the industry still faced significant technical challenges. Start-ups like Memphis Meats (now UPSIDE Foods), Mosa Meat, and JUST (now Eat Just) emerged, attracting attention and funding from venture capitalists, tech moguls, Hollywood stars, and traditional food companies. However, commercial products remained elusive. UPSIDE Foods and Eat Just are headquartered in the San Francisco Bay area known for its strong tech and biotech sectors. In addition, the Silicon Valley region benefits from a robust venture capitalist ecosystem and a culture of innovation. The nascent industry struggled with scaling up production and making the process cost-effective. Regulatory bodies began discussing frameworks for cultured meat, but substantial progress in this area was still required. Public acceptance was growing, but the path to market-ready products was fraught with technical and logistical hurdles.

The peak years for the greatest number of cell meat start-ups emerged from 2015 to 2019. During this period, there was a significant surge in new companies entering the field, driven by increased media attention, successful demonstrations of lab-grown meat products, and a growing interest in sustainable and ethical food production.

The rise of these start-ups was facilitated by tech accelerators such as IndieBio which play a crucial role in providing early-stage start-ups with essential resources, mentorship, and funding. By providing a network of industry experts and successful entrepreneurs, tech accelerators were instrumental for CM developers. Described as exemplifying a second wave in CM research, Stephens et al. (2019) explain how IndieBio's approach worked to fulfill the technological potential of the industry and foster a collaborative space for scientists:

> The teams spend an intensive period of usually 3–4 months in residence at the accelerator, during which time they are given physical space, technical resources, seed funding and networking opportunities in exchange for an equity stake in the company. The programmes usually culminate in a "demo day" where the teams showcase their prototypes and pitch their business potential to an invite only audience of investors, business personnel and media. The model was developed within and for the information

technology industry, but has been extended to other sectors, in the case of CM most notably by IndieBio. (Stephens et al., 2019, p. 6)

The guidance offered also covered business strategy as well as market positioning. Several start-ups in the cell-based meat industry, such as Memphis Meats (now Upside Foods), Mosa Meat, and Aleph Farms, benefited from participation in tech accelerators. Among the distinctive approaches emanating from Silicon Valley Food Tech culture was for start-ups "pitching their ventures [...] to identify the explicit problem(s) that their innovation will purportedly solve" (Ibid). Global climate change was identified as a major issue CM would solve and, in particular, one likely to appeal to mission-driven investors. However, it appears that global food security has been a new focus, reflecting the interests of venture capitalist firms (Ibid).

The establishment of organizations like the Good Food Institute in 2016 further catalyzed the growth of start-ups by providing support and advocacy for the industry. The Cellular Agriculture Society (CAS) was founded in 2017, in the U.S., by Kristopher Gasteratos,[3] as a lobby organization and a non-profit whose mission is to spread awareness about cultivated meat. Recently renamed *From Fauna* in 2024, their purpose is to educate about cultivated meat, but most importantly, they produce projects that "help people imagine what a future with cultivated meat could look and feel like" (From Fauna, 2024). We will discuss two of their future visions for CM in the section on visual narratives.

Many cell meat start-ups were located in the U.S., mostly in the Silicon Valley, but Mosa Meats, in the Netherlands, and Higher Steaks (rebranded as Uncommon in 2023),[4] in the U.K., were early leaders in the European cell agriculture industry. Other countries like Israel, benefiting from strong biotech research sectors, a dynamic start-up culture and a supportive regulatory environment, became an important player in the industry.

---

[3] He coined the term "neomnivore" to describe someone who only consumes animal products from cellular agriculture, although the term never really caught on (Gasteratos, 2017).

[4] The name change came about because the company focuses on pork belly and bacon products and not on steaks. Interestingly, over the years since CA has developed, many companies have changed names as a way to reposition their products in the public mind but also to refocus their "brand strategy."

Companies like Aleph Farms (2017), SuperMeat (2015), and Future Meat Technologies, also known as Believer Meats (2018), are at the forefront of the industry and the country remains a global hub for research and development in the field of cellular agriculture. In 2018, the founder and CEO of Aleph Farms, Didier Toubia, stated that Israel is a start-up nation (Niamh, 2018). The FAO and WHO's study on the safety of cell-based foods reports that Israel's start-ups represented 36% of global investments in cell-food products in 2021 (FAO & WHO, 2023). This company intended to commercially release its cultured beef steaks, Aleph Cuts, by 2022, pending regulatory approvals. The product called "Cultivated Petit Steak" was approved for sale in Israel in January 2024. The prevalence of vegetarianism and veganism in Israel is often cited as a reason for the strong support of the alternative meats industry but its limited farmland might also be a factor.

In Asia, Singapore positioned itself as a global player in food innovation, becoming the first country in the world to approve the sale of lab-grown meat in 2020. Companies like the recently merged Shiok Meats (2018) and UMAMI Bioworks (2020) specialize in cellular aquaculture and seafood.

The period saw several key milestones that reinforced the sense of progress. Start-ups began demonstrating their prototypes and achieving significant technological milestones. For example, Memphis Meats produced the first cultured meatball, in 2016, and companies like Mosa Meat and Aleph Farms showcased their developments in cultured beef and steak, respectively. These milestones created a perception of rapid advancement toward market readiness.

This second phase of CM was characterized by inflated expectations in the cellular agriculture industry driven by a combination of technological optimism, significant investment, growth in the number of start-ups, as well as the beginning of some high-profile media coverage. The rapid pace of scientific progress and the belief that these innovations could soon lead to commercial viability fueled a lot of excitement and very high expectations. Several of the CEOs announced the impending release of their products; however, some technical hurdles (scaling up, optimal media formulation, and cost effectiveness mostly), issues of consumer acceptance, and the lack of a regulatory framework for the human consumption of cell meat remained critical challenges. Media coverage during this period intensified, portrayals of cellular meat tended to be either neutral or positive, a lot of publications relayed information from the industry players. Visual imagery of CM for promotional purposes or media exposure also tended to rely on images of conventional meat rather than what was being produced in the laboratory.

In 2018, the World Economic Forum published a White Paper in its series *Meat: The Future* where they made the following prediction:

> Current industry watchers suggest large technology, IP sharing and regulatory hurdles mean it will take at least five years for cultured meat to arrive consistently in high-end restaurants and perhaps a decade to reach mass-market. It is also not yet clear exactly what scale of environmental benefits a mass-produced cultured meat sector could actually deliver. As with the technologies for renewable energy or electric vehicles, perhaps it might take the equivalent of a major economy such as Germany or China to help pump-prime the development of the global consumer market for new protein portfolios, as they did for solar energy a decade or so ago? (WEF, 2018, p. 11)

The temporal horizon of the commercial release of cell-based meat was a major issue of discussion, with various timelines projected by CM advocates. While some organizations like the WEF predicted 2023 for supplies of the product in high-end restaurants, others predicted at least a decade or two until the large-scale release of cell-based products.

By the end of 2020, there were 40 cultivated meat companies, compared to 5 in 2015, according to the GFI, and investors had spent $350 million into those ventures, double what had been invested into the industry since its beginning. At this point, the investment climate was positive and the industry was filled with enthusiasm and excitement at the prospects of expanding the business and nearing its goal of producing cell meat.

The Long Road Ahead to Scalability 2020–2024

At the beginning of 2021, Forbes reported that over $250 million went into funding cell meat start-ups. Some of the investors included celebrities and Silicon Valley tech innovators like Bill Gates, Richard Branson, Kimball Musk, Sergey Brin, Leonardo de Caprio, and Peter Thiel (Philpott, 2021). In this phase, the industry faced the realities of technical and regulatory challenges, leading to a more grounded approach. Despite setbacks, progress was made in scaling production and securing regulatory approvals, such as Eat Just's cultured chicken in Singapore. The annual cell meat conferences in the U.S. and Maastricht continued to facilitate collaboration and innovation. The merger of Shiok Meats with UMAMI Bioworks, in 2024, highlighted strategic consolidations aimed at overcoming hurdles and achieving commercial viability—Shiok meat faced financial difficulties but prior to 2024 was an important player in the cellular seafood space. The focus shifted toward sustainable growth, refining technologies, and building consumer acceptance, setting the stage for broader market entry in the near future. While important technical milestones were achieved in terms of the science of cell-based meats and regulatory approvals were secured in some markets, the costs and challenges of scaling up meant that the technology's potential and limitations were becoming apparent.

During that period, the idea of describing cellular agriculture as a "moonshot" project was prevalent. In the cell meat conferences I attended, speakers would often refer to this notion, which describes large-scale interventionist government programs believed to "reboot their post-pandemic economies and to attain bold targets such as sharply reducing and eventually eliminating CO_2 emissions" (Henrekson et al., 2024). Originating with John F. Kennedy's idea of putting a man on the moon within a decade, Barack Obama's 2016 Cancer Moonshot initiative, revived in 2022 by Joe Biden, similarly launched a vast program to eliminate cancer. Another moonshot program is also the Green Deal in Europe and the U.S. This idea comes from the work of Mariana Mazzucato who argues that capitalism needs to be restructured around missions developed through private-public partnerships, mostly by expanding the role of governments in the creation and the steering of markets (Henrekson et al., 2024). Drawing inspiration from the Apollo moon landing, Mazzucato advocates for government-led missions that leverage public funds and policy frameworks to stimulate innovation and address issues like climate change, health care, and technological advancements. Moonshots fall under mission-oriented industrial policies, and they originate from

> third-generation innovation policy, which posits that governments should not only provide basic research and contribute to the commercialization of it but also guide innovation efforts in specific directions. According to this approach, it is no longer enough for the government to increase positive knowledge externalities by supporting R&D activities, nor is it enough to provide targeted support or platforms strengthening the links between diverse actors such as universities, start-ups, and incumbent firms. The purposeful direction of these activities and proactive intervention in the marketplace is deemed necessary. A critical element distinguishing the mission-oriented approach is therefore directionality. (Henrekson et al., 2024, p. 9)

Moonshot capitalism relies on active involvement by the state in capital accumulation and innovation. This positions the state as a critical actor in driving economic growth, which is counter to neoliberal ideologies that advocate for minimal state intervention in the market. Without proper mechanisms in place, the benefits of publicly funded innovations such as cellular agriculture could disproportionately be captured by private corporations and interests, reinforcing existing inequalities. In Mazzucato's model, governments are not the sole value creators. She includes altruistically motivated NGOs and non-profit organizations (such as New Harvest or the Good Food Institute as well as venture capitalist firms and third-sector investors in the case of cellular agriculture) which work with governments, often "entangling their work in political mire" (Ibid, p. 192).

In the context of cellular agriculture, government funding and support are essential for advancing research and development (R&D) in lab-grown meat and other biotechnological products. Such investments lay the groundwork for turning scientific discoveries into viable commercial applications. This aligns with the principles of biocapital (Rajan, 2006), where biological research is transformed into economic assets, facilitating the emergence of new markets and industries, based on the commodification of biological entities and processes. Biocapital also refers to the commodification and capitalization of biological materials and processes, wherein life sciences become pivotal to economic growth and technological innovation (Rose, 2007; Waldby & Mitchell, 2006). Innovations in the development of cellular meat represent more than scientific progress and a desire to resolve humanity's pressing needs. They are economic opportunities aimed at the commercialization of life. In this instance, commodification of living processes is a crucial aspect of biocapital accumulation that ensures the monetization of biological innovations and proprietary technologies through patents and intellectual property rights. The fact is that large animal intensive industries like Cargill, Tyson Foods, and others have invested in the technology because they see a potential to continue profiting from industrial agriculture. Some believe that large corporations rebranding themselves as protein companies view lab-grown meat as a bridge between real meat and plant-based products.

In the U.S., the federal government plays a pivotal role in the bioeconomy. In a speech inaugurating the launch of the National Biotechnology and Biomanufacturing Initiative in 2022, President Biden explained:

> Analyses and facts suggest that before the end of the decade, engineering biology holds the potential to be used in manufacturing industries that account for more than one third of global output. That's equivalent to almost $30 trillion in terms of value… The investments, programs, and partnerships in this initiative will advance research and development in engineering biology and biomanufacturing, expand and strengthen domestic biomanufacturing capacity and resilient supply chains. It will help bring bioproducts to market that will help support American farmers and reduce risk through biosecurity innovation, as well as enhance workforce training and education programs for the next generation of biotechnologists… We're also looking to improve food security and drive agricultural innovation, including through new technologies that protect crops from disease,

> enhance seeds and fertilizers and foods made with cultured animal cells. (The White House, 2022)

In the realm of cellular agricultural technology, a moonshot refers to a highly ambitious, exploratory, and innovative project undertaken without a guaranteed short-term return on investment or benefit, often initiated without comprehensive evaluation of potential risks and benefits (Awati et al., 2022). The idea is to solve problems *completely*, as many cell agriculture advocates claim. Bruce Friedrich of the Good Food Institute interviewed by Ezra Klein, in an article entitled *Ending the Age of Animal Cruelty*, illustrates this point:

> It's interesting to see these companies reconstituting themselves as protein companies and excited to move in this direction because it's a more efficient way of producing meat, so it will be more profitable over time, but also excited because human beings want to be part of doing good in the world and this is a way to do even more good in the world. (Klein, 2019)

This wish is echoed by altruistic investors and venture capitalists advocating for the end of animal cruelty. Is a market-based solution such as cellular agriculture going to solve the problems of animal exploitation in our food system?

Moonshot approaches seek to gain government support in terms of providing the conditions for market entry of cellular agriculture products, including providing R&D funding, setting up an infrastructure for teaching and research in the field, developing a favorable regulatory climate for industrialization, adopting safety standards, patents, and intellectual property safeguards, and setting up the conditions in the marketplace for consumer acceptance (whether it is in terms of naming the products or labeling, etc.). So, the moonshot is there to solve a large number of problems that the nascent cell ag community cannot resolve on its own. While the initial phases of product development in cellular agriculture` focused on entrepreneurial start-up culture and the disruptive potential of the field, it is now evident that bringing cell-based meat to market requires creating a new market for unprecedented products. This endeavor necessitates not only convincing the public to embrace cellular agriculture but also persuading bureaucracies, administrative bodies, and evaluation systems to adapt and support this innovation. It also means overcoming technical as well as other barriers before these products become reality. In its State of the Industry 2023 report the GFI writes:

> the year was not without its challenges. Funding constraints, difficulties with scaling, and the spread of misinformation and disinformation all posed obstacles for cultivated meat. Consumer education, understanding, and adoption are still in the early stages. And the work to secure increased government and private investment continues. (Good Food Institute, 2023, p. 2)

The disinformation about cell meat includes efforts to ban cell meat in some countries like Italy and in several states in the U.S. These will be discussed in the conclusion.

By the middle of 2021, it was clear that the promises of cellular meat for dinner were not to be realized in the short or long term. In a rather critical piece published in Mother Jones, journalist Tom Philpot produced an interactive chart where the timelines of the predicted product releases, by each cell meat company, can be visualized. For instance, Mosa Meat predicted it would sell its beef product in 2020. Most of the companies, according to this chart, expected to have released products between 2020 and 2024. With a few exceptions, companies like BluNalu and Finless Foods (cell-based seafood companies) indicated 2029 as the possible date for reaching commercial status (Philpott, 2021). In 2024, the prospects of large-scale cultivated meat remain far off, even with the FDA granting full approval for the sale of cultivated meat in the U.S. (Flood, 2023). The year 2022 is also when consensus formed in the cellular agriculture community around the term "cultivated meat" to describe animal without animal products.

Wurgaft speaks of cultured meat being "bound up with a culture of prediction," he explains that according to Isha Datar, the head of New Harvest at the time:

> the money for vat meat research is principally philanthropic, because the venture capitalists who support companies need to see returns in far shorter increments of time than twenty or ten years. (Wurgaft, 2019, p. 14)

So it seems, as the venture capital investments decrease, the timelines for cell meat extend. Expectations now are that products won't roll out of the meat factories for at least another decade or two, potentially. It is also during this period that nomenclature, what to call CM, becomes important, as the prospects of near to long-term commercialization open up.

The various forums and conferences I attended during this period included a discussion about nomenclature. The name of the product evolved as did thinking about how to position it as meat in consumers' minds. The pros and cons of various formulations such as 'clean meat,' 'cultured' versus 'cultivated,' 'cellular,' etc. were debated among attendees and several studies published focused on the issue of what to call it, mostly from a marketing perspective. A study conducted by the opinion firm Embold Research commissioned by the GFI found that the term "cultivated meat" was more attractive to consumers and was better for differentiating the product with conventional meat. Industry stakeholders agreed to adopt this nomenclature (GFI, 2023).

From in vitro to cultivated, the names attributed to this meat are indicative of a wider set of discursive politics, expressing the tension between ideas about authenticity and technicity surrounding the many definitions of CM.

Tempering Expectations: After the Hype

This section will focus on two specific controversial reports that put into question the viability and the future of the cultivated meat industry, exposing some of the more problematic aspects of the meat, as well as broader concerns linked to process making. The first one relates to a controversial report, a Techno-economic Analysis (TEA) commissioned by Open Philanthropy, an organization involved in "effective altruism," written by David Humbird, a chemical engineer, regarding the scaling-up economics of cultured meat, published in 2020. The other controversy concerns writer and journalist Joel Fassler's reports in two articles, published in the Counter (2021) and the New York Times (2024), about the significant challenges faced by the cultured meat industry. Of interest here are the different stories these two controversies tell and the type of pushback from the CM industry.

Before detailing the issues identified in Humbird's TEA, a short detour about "effective altruism" and Open Philanthropy's link to alternative meats including its donations to the Good Food Institute (GFI) are warranted. As we mentioned before, the GFI is a non-profit organization founded by Bruce Friedrich, a former vice-president of PETA (People for the Ethical Treatment of Animals), in 2016. Friedrich is quoted as saying that he wasn't convinced that the activist campaigns he led against fast-food chains were effective at changing people's mind about eating animals. Instead, he thought that a market-focused approach would be more effective for changing consumer behavior (Stiffman, 2023). Basically, by providing a credible alternative to meat, he believed that people would switch their diets. The Washington-based organization defines its mission as:

> The Good Food Institute is a nonprofit think tank working to make the global food system better for the planet, people, and animals. Alongside scientists, businesses, and policymakers, GFI's teams focus on making plant-based and cultivated meat delicious, affordable, and accessible. Powered by philanthropy, GFI is an international network of organizations advancing alternative proteins as an essential solution needed to meet the world's climate, global health, food security, and biodiversity goals. (GFI, 2024)

Today, the organization has over 200 employees with affiliate offices in Brazil, India, Israel, Singapore, and all over Europe, helping to promote the industry. With the explicit goal of accelerating the transition to plant-based and cellular meats, Friedrich explains how the GFI operates:

> Our Corporate Engagement Team builds collaborative relationships with the largest food processing companies, chain restaurants, and grocery stores to maximize the availability, quality, quantity, and promotion of plant-based

> and cultivated meat. Our Science and Technology Team are the leading experts on the cutting-edge science of alternative proteins. They analyze the state of the industry, identify top research opportunities, engage scientists and engineers from academia and industry, and mobilize funding to accelerate the industry. Along with Corporate Engagement, they also bring people together to form new companies and connect these new companies to expertise and funding. Our Policy Team works closely with legislators and regulators to ensure a clear and efficient path to market for cultivated meat. They also work on the federal and state levels to ensure that plant-based meat can compete on a level playing field. Finally, they work to secure R&D funding for alternative proteins from governments and funding institutions. (Sentient Media, 2019)

The group's revenues have been steadily increasing, as the industry has expanded, going from \$2.9 million in 2016 to \$43 million in 2021. Foundations make up 26% of that support and individual donations provide 61% of funds (Stiffman, 2023). The Good Food Institute has played a major role in providing information about cellular agriculture; it has become a source of important data about the industry quoted by international institutions like the World Health Organization and the Food and Agriculture Organization of the U.N. Environmental Program (UNEP). GFI was the largest funder of academic research until 2023 before the U.S. federal government started financing cultivated meat (Zaleski, 2024). Recently, the Bezos Earth Fund, through its Future of Food program, donated \$5.5 million to the GFI. In a statement, the director of that fund stated that alternative proteins (including plant-based) are essentially "climate tech." Echoing Friedrich's idea that alt-proteins are the "electric vehicles of food," as well as explaining how moonshot thinking worked for the development of electric cars and solar panels, he claims that without policy changes that accelerated their development, these technologies would have remained expensive and inefficient (Stiffman, 2023). Friedrich's lobbying on behalf of the industry has been instrumental in bringing about regulations as well as investments in support of the alternative protein business. He is credited for putting alternative meat on the map (Ibid). The GFI is invested in the future of CM, encouraging governments to increase spending on alt-meats as it brings together the community of food tech start-ups.

GFI embodies a vision of the effective altruism movement and the founder of Open Philanthropy, Dustin Moscovitz, co-founder of Facebook, agrees as his investments in the alt-protein business show. He and his wife have invested in the area of $17 million in funding for alternatives to animal meat (Stiffman, 2023).

Back to David Humbird's report. In 2018, he was commissioned to study the potential of producing cultivated meat at scale. His study was published in the peer-reviewed journal, *Biotechnology and Bioengineering*, in 2021 (Humbird, 2021). His main findings, without going into the technical details of his calculations, were that it was unlikely that cell meat would displace conventional meat because the technology was not financially or technologically capable of producing safe, competitively priced product. He highlights the details of his finding in a 2020 report:

> The analysis identifies a number of significant barriers to the scale-up of animal cell culture. Bioreactor design principles indicate a variety of issues associated with bulk cell growth in culture: Low growth rate, metabolic inefficiency, catabolite and CO_2 inhibition, and bubble-induced cell damage will all limit practical bioreactor volume and attainable cell density. With existing bioreactor designs and animal cell lines, a significant engineering effort would be required to address even one of these issues. Economic challenges are further examined. Equipment and facilities with adequate microbial contamination safeguards are expected to have high capital costs. Suitable formulations of amino acids and protein growth factors are not currently produced at scales consistent with food production, and their projected costs at scale are likewise high. (Humbird, 2020, p. ii)

He argues that in order to get cultured meat to be produced for less than $25/kg (the price for the most expensive plant-based meat analogue) very complex scientific and engineering difficulties have to be overcome. Additionally, he finds the costs of high-quality media components to be limiting factors. In fact, the cell medium is the biggest cost driver for cultivated meat. Of course, some companies have solved the issue of animal-free cell media, no longer requiring fetal calf serum, but the GFI explains that:

> The development of cell culture media formulations optimized for cultivated meat-relevant cell lines can utilize existing techniques that have been pioneered within the biomedical and research fields to develop media tailored for specific cell types and cultivation processes. However, achieving a cost that makes cultivated meat economically feasible is a different task than technical optimization of media for specific cultivated meat-relevant cell lines. (Specht, 2020, p. 7)

Growth media is extremely expensive and complex to produce in high quantities. Sourcing the materials, finding a secure raw material supply presents its challenges. His predictions are that the cell culture process would be plagued by "extreme, intractable technical challenges at food scale" (Fassler, 2021). Basically, the price at the end of the process for ground meat or chicken nuggets isn't worth the trouble. Experts also questioned the potential of expensive biomanufacturing techniques to produce cheap and plentiful human food (Ibid). Besides, it's interesting to think about the "invention" of chicken nuggets as a product of scientific research, poultry scientist Robert C. Baker at Cornell University in 1963. The story is that after World War II, Americans were tired of eating chicken as other meats had been severely rationed as part of the war effort. So, Baker wanted to create new versions of old products, making them look like different foods (McKenna, 2012). Michael Pollan describes chicken nuggets as a cheap bulk commodity, broiler chickens, turned into a high-value-added processed product that put money into the pockets of processors rather than farmers (Pollan, 2016). Nuggets, he suggests, are the abstraction of a chicken, the antithesis of a wholefood. They are emblematic of industrialized agriculture.

A recent article in AgFunderNews quotes an early-stage investor in CM who predicts that 70–90% of companies will fail over the next year, but who also believes that the ones who make it will be able to scale their technologies (Watson, 2023). According to Dr. Ali Khademhosseini from the company Omeat, price parity with conventional meat can be achieved today "if we operate at scale, without relying on future innovations. And we can do that with the 10,000-liter bioreactors we're planning in our pilot plant, whereas others require really large bioreactors that haven't been validated" (Ibid). However, others remain skeptical especially when the costs of materials, and in particular the cellular components used in growth medium at scale, are concerned. Basically, the whole manufacturing model for CM being commercially viable is premised on the market volume. According to Humbird, "None of this stuff makes any commercial sense until everyone's eating it." He adds: "The emperor has no clothes" pointing to the fact that the price of cell media components will come down as a factor of economies of scale, which creates a "chicken-and-egg situation" (Ibid). As already mentioned, industry insiders believe that in order to be both commercially and technically viable, unstructured cultured cells, rather than cultivated whole-cuts of meat, will be added as ingredients to plant-based meats. Blended or hybrid meats seem to be the most logical and affordable way to go at the moment for the industry.

Humbird's report was not received well by the GFI, and they produced a series of revisions before its publication. They commissioned their own TEA, this time by the research consulting firm CE Delft, using proprietary data provided by 15 private companies, and this time the study showed "how addressing a series of technical and economic barriers could lower the production price from over $10,000 per pound today to about $2.50 per pound over the next nine years—an astonishing 4,000-fold reduction" (Ibid). Armed with the results of this TEA, the GFI advocated for more public funds into cultivated meat technology R&D. However, some like Fassler questioned the GFI's interpretation of the second report, which was that cheap, accessible cultivated protein was on the horizon.

A few weeks after the second TEA was released, writer and vegan Ezra Klein published an opinion piece in the New York Times entitled *Let's Launch a Moonshot for Meatless Meat* where he called for the U.S. government to spend billions of dollars in support of scaling up both plant-based and cultivated meat. Picking up on Friedrich's claims that the need for public funding was urgent. Klein, optimistic about the technological advances made in the cell ag industry, writes:

> But so far, most of these advances, most of these investments, are through private dollars, with the findings locked up in patents, by companies competing with one another for market share. We're going to need to move faster than that. "If we leave this endeavor to the tender mercies of the market there will be vanishingly few products to choose from and it'll take a very long time," Bruce Friedrich, co-founder and executive director of the Good Food Institute, told me.
>
> This is where policymakers can, and should, come in. At its heart, the American Jobs Plan is a climate bill. But there isn't a dollar for alternative proteins, despite animal agriculture's huge contributions to both climate and pandemic risk. That's worse than a mistake. It's a failure of policy design. Luckily, it's easily fixed. (Klein, 2021)

Saving the planet, according to Fassler, came down to "doubling down on cultured meat" (2021). Their TEA reported that biomanufacturing facilities would produce 22 million pounds of cultured meat per year, which represents approximately 10% of the whole of the domestic market for plant-based alternatives but, even more shocking, this would be 0.0002% of the amount of conventional meat produced in the U.S. every year, which stands at 100 billion pounds (Fassler, 2021). So, GFI's imagined meat factory, containing hundreds of pricey bioreactors, is estimated to cost $450 million which would actually be more expensive than a conventional slaughterhouse and it would produce less meat (ibid). This means that bioreactors capacity would require about a third of the total capacity of the entire biopharmaceutical industry in order to supply "a sliver of the nation's meat" (ibid).

Moreover, the biomanufacturing process produces a cell slurry or a viscous mass of cells collected from bioreactors that require further downstream processing, in order to turn it into something edible. For cultivated meat, the animal cells need to be formed into muscle-like tissue and this constitutes added costs.

Furthermore, skeptics felt that very little evidence supported the claim that cheap and abundant meat could be manufactured in the near future. Forever expanding timelines by start-ups about the commercial availability of their products show how indeterminate and how uncertain the science of cultivating cells for food remains (Philpott, 2021). One expert,

Huw Hughes, a biomanufacturing expert and consultant and former Pfizer employee, tasked to review GFI's analysis of the second report, found that its interpretation of results was unrealistic and that projected cost decreases underestimated the reality of building a scale-up facility (Ibid). Basically, the GFI's credibility was put in question and Hughes echoed Humbird's sentiment about the feasibility of growing meat in bioreactors cheaply and efficiently. The Open Philanthropy report concludes that the production of cultivated meat would make eating it less efficient in terms of scale and cost than eating plants themselves (Fassler, 2021). For some scientists, cultivated meat remains wishful thinking (Stiffman, 2023). Open Philanthropy announced it would focus its work on convincing governments to fund R&D for cultivated meats and alternative proteins, in line with the Moonshot approach to supporting innovation.

At the heart of this "moonshot" ideal lies the question of whether cultivated meat can actually achieve significant impact in order to feed people, save animals, and improve the climate. Based on the numbers above, it seems unlikely that this meat will benefit a great number of people and be affordable. Moreover, the modalities of government and public funding involvement in the industry remain unclear.

Revisiting the concept of moonshot capitalism, the funding directed toward this industry is based on the logic that it aims to address significant societal challenges, such as agriculture's contribution to climate change and concerns about animal welfare. The problems that GFI wants governments to address are engineering challenges but the issue of how the value created from public investments will be reaped remains open. Whether this will contribute to solving broader societal issues like the farming crisis is questionable. Besides, some question the Effective Altruism community's support of cultivated meat because investments are diverted from funding vegan advocacy and outreach as well as broader dietary activism (Abrell, 2023). In fact, Abrell estimates that animal charities donated over $3 million in 2020 to GFI, considered among its top ranking for grantees. The issue identified by Abrell is that GFI's advocacy work, on behalf of start-ups, represents a "market-based approach to activism" and thus warns against cultivated meat dominating the funding structure of an entire movement. Furthermore, Abrell questions the solutions promoted by GFI, even in the best TEA scenarios, as being able to solve the animal agriculture problem, in terms of reducing CO_2 emissions to "avoid the impending catastrophes of a post-1.5 °C future" (p. 154). If the goal is to reduce the number of animals in industrial agriculture and occupying land, both TEAs confirm that conventional meat is not likely to be supplanted by cultivated meat because of scale, cost, and time challenges that make it impossible to have any real impact on climate change emissions. Furthermore, if, as Abrell contends, the whole strategy is based entirely on market substitution and supporting the animal advocacy work of start-ups instead of other forms of activism to end factory farming, then we should seriously consider the relevance of venture capital for solving the systemic issues plaguing the meat industry.

In his latest opinion piece, Fassler noted that between 2016 and 2022, almost $3 billion were invested in cultivated meat and seafood companies. Venture capital and sovereign wealth funds as major investors, SoftBank, Temasek, the Qatar Investment Authority, for instance, including meatpacking companies like Tyson, Cargill, and JBS were major contributors (Fassler, 2024). His piece is highly critical of the industry's past, current, and future achievements and potential and he writes:

> Interviews with almost 60 industry investors and insiders, including many who have been employed by or been part of the leadership teams of these companies, reveal a litany of squandered resources, broken promises and unproven science. Founders, hemmed in by their own unrealistic proclamations, cut corners, such as using ingredients derived from slaughtered animals. Investors, swept up in the excitement of the moment, wrote check after check despite significant technological obstacles. Costs refused to enter the realm of plausible as launch targets came and went. All the while, nobody could achieve anything close to meaningful scale. And yet companies rushed to build expensive facilities and pushed scientists to exceed what was possible, creating the illusion of a thrilling race to market. (Fassler, 2024)

It seems that the industry is set for some major restructuring as venture capital funds decrease across industries. Several cultivated meat start-ups have either consolidated like the merger between Shiok Meats and Umami Bioworks, in the production of seafood and cellular aquaculture, or they have partnered with other businesses to target specific areas of the supply chain for cellular agriculture, like companies that produce lab-grown ingredients or components used in cell meat production, like collagen, for instance, instead of focusing on a particular end-product like meat. There have also been problems in the plant-based meat sector, with Impossible Foods laying off 20% of its staff in 2023, due to poor sales. In addition, even the largest companies like Upside Foods, valued at a billion dollars, has struggled to deliver on their promise of large-scale production of cultivated chicken. Similarly, as mentioned previously, the trend is toward the commercialization of blended meats due to supply issues. For instance, Eat Just's GOOD Meat division announced the construction of its new manufacturing facility but encountered problems:

> Eat Just's cultivated meat division had the capacity to produce only a small amount of chicken, and that at a steep financial loss. The process still relied on fetal bovine serum, a product of the brutal animal supply chain that cultured meat was supposed to make obsolete. The product, the company said, was about 30 percent plant-based ingredients, a cross between a chicken nugget and a veggie burger. Regardless, the approval was treated as a historic event. (Fassler, 2024, p. 4)

Added to the fact that the battle to win over consumers hasn't yet been won, it seems like the promises and expectations of the cultivated meat industry may not be realistic. Surveys about people's willingness to consume these products show a lot of variability. Nevertheless, the efforts made by GFI on behalf of start-ups to establish cultivated meat as a mission continue to attract public funding. This is because:

> Once a mission is established, societal resource allocation becomes increasingly political, which means that vested interest groups will entrench their connections and abilities to influence government. (Henrekson et al., 2024, p. 11)

GFI's work as a lobby organization has been impactful. Governments are investing in teaching and research infrastructure to support the industry's expansion. The U.S. Department of Agriculture granted $10 million over five years to the Tufts University Center for Cellular Agriculture (Nicholas & Silver, 2021). They have approved the sale of cellular agriculture products in the U.S. and in other countries like Singapore and Israel. However, the way cultivated meat has been received in political circles has been highly problematic, namely, the banning of cultivated products in certain American states, as well as in countries like Italy as we shall discuss later.

The upshot of this whole discussion about the reality of cultivated meat boils down to the fact that there are no guarantees that cultured meat technology will deliver on its promises, and the question is what are we missing about radically changing our food system if we place all our eggs in that basket? Again, if ending animal farming is the goal, then it is clear that the Effective Altruist model of funding start-ups plays into the same capitalist system at the root of the current crisis, by channeling wealth into market-based solutions that do nothing to avert our current ecological crisis. I agree with Abrell's view that cultured meat distracts us from the real work of reforming our food system:

> Perhaps the obsessive desire to have our meat and eat it too as reflected in the cultured meat replacement strategy is a sign that we have transitioned from the Climate Denial stage of eco-grief to the Climate Bargaining stage, but this misguided meat triumphalism only serves to delay the necessary recognition of the pressing need for the radical and immediate transformation of our food system. (Abrell, 2023, p. 157)

Bibliography

Abrell, E. (2023). The Empty Promises of Cultured Meat. In C. J. Adams, A. Crary, & L. Gruen (Eds.), *The Good it Promises, the Harm it Does: Critical Essays on Effective Altruism* (pp. 149–159). Oxford University Press.

Atwood, M. (2013). *Oryx and Crake*. Virago Press.

Awati, R., Bernstein, C., & Wigmore, I. (2022). *What is a Moonshot in Technology?* WhatIs. Retrieved from https://www.techtarget.com/whatis/definition/moonshot

Benjaminson, M. A., Gilchriest, J. A., & Lorenz, M. (2002). In Vitro Edible Muscle Protein Production System (mpps): Stage 1, Fish. *Acta Astronautica, 51*(12), 879–889. https://doi.org/10.1016/S0094-5765(02)00033-4

Bhat, Z. F., Kumar, S., & Fayaz, H. (2015). In Vitro Meat Production: Challenges and Benefits over Conventional Meat Production. *Journal of Integrative Agriculture, 14*(2), 241–248. https://doi.org/10.1016/S2095-3119(14)60887-X

Boyle, A. (2013, August 6). *Google's Sergey Brin Explains Why He Paid $330,000 for Lab Burger*. NBC News. Retrieved from http://www.nbcnews.com/technolog/googles-sergey-brin-explains-why-he-paid-330-000-lab-6C10853442

Castle, N. (2022). In Vitro Meat and Science Fiction. *Extrapolation, 63*(2), 149–179. https://doi.org/10.3828/extr.2022.11

Datar, I., Kim, E., & d'Origny, G. (2016). New Harvest: Building the Cellular Agriculture Economy. In B. Donaldson & C. Carter (Eds.), *The Future of Meat without Animals* (pp. 121–131). Roman and Littlefield International.

Edelman, P. D., McFarland, D. C., Mironov, V. A., & Matheny, J. G. (2005). Commentary: In vitro-cultured meat production. *Tissue Engineering, 11*(5–6), 659–662. https://doi.org/10.1089/ten.2005.11.659

Edible Planet Ventures, I. (2023, May 9). *Mosa Meat Opens the World's Largest Cultivated Meat Campus*. Edible Planet. Retrieved from https://www.edibleplanetventures.com/post/mosa-meat-opens-the-world-s-largest-cultivated-meat-campus

FAO, & WHO. (2023). *Food Safety Aspects of Cell-Based Food*. FAO; WHO. Retrieved from https://openknowledge.fao.org/handle/20.500.14283/cc4855en

FAO (Food and Agriculture Organization of the United Nations). (2006). *Livestock's long shadow: Environmental issues and options*. FAO. https://www.fao.org/3/a0701e/a0701e.pdf

FAO (Food and Agriculture Organization of the United Nations). (2009). *The state of food and agriculture: Livestock in the balance*. FAO. https://doi.org/10.18356/1c04f5eb-en

FAO (Food and Agriculture Organization of the United Nations). (2011). *World livestock 2011: Livestock in food security*. FAO. https://doi.org/10.18356/3d6e8c5d-en

Fassler, J. (2021, September 22). *Lab-grown Meat is Supposed to be Inevitable. The Science Tells a Different Story.* The Counter. Retrieved from https://thecounter.org/lab-grown-cultivated-meat-cost-at-scale/

Fassler, J. (2024, June 29). The Revolution That Died on Its Way to Dinner. *New York Times*.

Fernandes, A. M., Teixeira, O. D. S., Fantinel, A. L., Revillion, J. P. P., & Souza, Â. R. L. D. (2022). Technological Prospecting: The Case of Cultured Meat. *Future Foods, 6*, 100156. https://doi.org/10.1016/j.fufo.2022.100156

Flood, E. (2023, July 31). *From Prediction to Production: A Timeline of Cultivated Meat*.

From Fauna. (2024). *What We Do*. Retrieved from https://fromfauna.org/work/

Gasteratos, K. (2017). *Nature and the Neomnivore*. Retrieved from http://nrs.harvard.edu/urn-3:HUL.InstRepos:33839952

GFI, G. F. I. (2023). *New Research Supports "Cultivated Meat" Term*. Retrieved from https://gfi.org/wp-content/uploads/2023/01/COR23008_CM-nomenclature-fact-sheet_Final.pdf

Good Food Institute. (2024). *Alternative Protein Company Database* [Dataset]. Retrieved from https://gfi.org/resource/alternative-protein-company-database/

Haldane, J. B. S., & Price, C. A. (2001). The Future of Biology. In *Possible Worlds*. Routledge.

Haraway, D. (2016). *Staying with the Trouble: Making Kin in the Chthulucene*. Duke University Press.

Henrekson, M., Sandström, C., & Stenkula, M. (Eds.). (2024). *Moonshots and the New Industrial Policy: Questioning the Mission Economy* (Vol. 56). Springer Nature Switzerland. https://doi.org/10.1007/978-3-031-49196-2

Hickman, L. (2009, December 1). Why I'd Happily Eat Lab-grown Meat. *The Guardian*. Retrieved from https://www.theguardian.com/science/2009/dec/01/lab-grown-meat

Hocquette, A., Lambert, C., Sinquin, C., Peterolff, L., Wagner, Z., Bonny, S. P. F., Lebert, A., & Hocquette, J.-F. (2015). Educated Consumers Don't Believe Artificial Meat is the Solution to the Problems with the Meat Industry. *Journal of Integrative Agriculture, 14*(2), 273–284. https://doi.org/10.1016/S2095-3119(14)60886-8

Humbird, D. (2020). *Scale-Up Economics for Cultured Meat: Techno-Economic Analysis and Due Diligence*. Engineering Archive. https://doi.org/10.31224/osf.io/795su

Humbird, D. (2021). Scale-up Economics for Cultured Meat. *Biotechnology and Bioengineering, 118*(8), 3239–3250. https://doi.org/10.1002/bit.27848

Jönsson, E. (2016). Benevolent Technotopias and Hitherto Unimaginable Meats: Tracing the Promises of In Vitro Meat. *Social Studies of Science, 46*(5), 725–748. https://doi.org/10.1177/0306312716658561

Klein, E. (2019, January 29). *Ending the Age of Animal Cruelty*. Vox. Retrieved from https://www.vox.com/future-perfect/2019/1/29/18197907/clean-meat-cell-plant-impossible-beyond-animal-cruelty

Klein, E. (2021, April 2024). Let's Launch a Moonshot for Meatless Meat. *The New York Times*. https://nytimes.com/2021/04/24/opinion/climate-change-meatless-meat.html

Landecker, H. (2007). *Culturing Life: How Cells Became Technologies* (First paperback ed). Harvard Univ. Press.

Lemkow, L. (2015). Science Fiction. *Food Phreaking, 2*, 42–47.

Listek, V. (2019, May 6). *Organovo: Bioprinting Could be the New Solution to Organ Transplantation*. 3DPrint.Com | The Voice of 3D Printing / Additive Manufacturing. Retrieved from https://3dprint.com/243160/organovo-bioprinting-could-be-the-new-solution-to-organ-transplantation/

McHugh, S. (2010). Real Artificial: Tissue-cultured Meat, Genetically Modified Farm Animals, and Fictions. *Configurations, 18*(1–2), 181–197. https://doi.org/10.1353/con.2010.0006

McKenna, M. (2012, December 28). The Father of the Chicken Nugget. *Slate*. Retrieved from https://slate.com/human-interest/2012/12/robert-c-baker-the-man-who-invented-chicken-nuggets.html

Montgomery, R. A., Stern, J. M., Lonze, B. E., Tatapudi, V. S., Mangiola, M., Wu, M., Weldon, E., Lawson, N., Deterville, C., Dieter, R. A., Sullivan, B., Boulton, G., Parent, B., Piper, G., Sommer, P., Cawthon, S., Duggan, E., Ayares, D., Dandro, A., et al. (2022). Results of Two Cases of Pig-to-Human Kidney Xenotransplantation. *New England Journal of Medicine, 386*(20), 1889–1898. https://doi.org/10.1056/NEJMoa2120238

Niamh, M. (2018). *Aleph Farms CEO on its 3D Cultured Beef: "Unlike Other Companies, Our Meat Grows Together Like Real Meat."* Foodnavigator.Com. Retrieved from https://www.foodnavigator.com/Article/2018/05/02/Aleph-Farms-CEO-on-its-3D-cultured-beef-Unlike-other-companies-our-meat-grows-together-like-real-meat

Nicholas, I., & Silver, M. (2021, October 15). *Tufts Receives $10 Million Grant to Help Develop Cultivated Meat | Tufts Now*. Retrieved from https://now.tufts.edu/2021/10/15/tufts-receives-10-million-grant-help-develop-cultivated-meat

Nie, X., Liang, Z., Li, K., Yu, H., Huang, Y., Ye, L., & Yang, Y. (2021). Novel Organoid Model in Drug Screening: Past, Present, and Future. *Liver Research, 5*(2), 72–78. https://doi.org/10.1016/j.livres.2021.05.003

Philpott, T. (2021, February 8). Is Lab Meat About to Hit Your Dinner Plate? *Mother Jones*. Retrieved from https://www.motherjones.com/food/2021/08/is-lab-meat-about-to-hit-your-dinner-plate/

Pollan, M. (2016). *The Omnivore's Dilemma: A Natural History of Four Meals* (Edition with a new afterword). Penguin Books.

Rajan, K. S. (2006). *Biocapital: The Constitution of Postgenomic Life*. Duke University Press.

Rasko, J., & Power, C. (2015, February 18). What Pushes Scientists to Lie? The Disturbing but Familiar Story of Haruko Obokata. *The Guardian*. Retrieved from https://www.theguardian.com/science/2015/feb/18/haruko-obokata-stap-cells-controversy-scientists-lie

Rose, N. S. (2007). *Politics of Life Itself: Biomedicine, Power, and Subjectivity in the Twenty-First Century*. Princeton University Press.

Seamons, K. (2017, September 30). *A Hungry WWII POW Had Ideas. Some 60 Years Later, Silicon Valley Wanted Them*. Newser. Retrieved from http://www.newser.com/story/249128/hungry-wwii-pow-had-ideas-silicon-valley-now-wants.html

Sentient Media. (2019, December 24). *Good Food Institute: Interview with Bruce Friedrich*. Retrieved from https://sentientmedia.org/good-food-institute-interview-with-bruce-friedrich/

Shay, J. W., & Wright, W. E. (2000). Hayflick, his Limit, and Cellular Ageing. *Nature Reviews Molecular Cell Biology, 1*(1), 72–76. https://doi.org/10.1038/35036093

Specter, M. (2011, May 16). *Test-Tube Burgers*. Retrieved from https://www.newyorker.com/magazine/2011/05/23/test-tube-burgers

Specht, L. (2020). *An analysis of culture medium costs and production volumes for cultivated meat*. Good Food Institute. https://gfi.org/wp-content/uploads/2021/01/clean-meat-production-volume-and-medium-cost.pdf.

Stephens, N., Sexton, A. E., & Driessen, C. (2019). Making Sense of Making Meat: Key Moments in the First 20 Years of Tissue Engineering Muscle to Make Food. *Frontiers in Sustainable Food Systems, 3*. https://doi.org/10.3389/fsufs.2019.00045

Stiffman, E. (2023, October 16). *Does Alternative Meat Need Philanthropy to Take It Beyond?* The Chronicle of Philanthropy. Retrieved from https://www.philanthropy.com/article/does-alternative-meat-need-philanthropy-to-take-it-beyond

The White House. (2022, September 12). *Background Press Call on President Biden's Executive Order to Launch a National Biotechnology and Biomanufacturing Initiative*. The White House. Retrieved from https://www.whitehouse.gov/briefing-room/press-briefings/2022/09/12/background-press-call-on-president-bidens-executive-order-to-launch-a-national-biotechnology-and-biomanufacturing-initiative/

Tsing, A. L. (2017). *The Mushroom at the End of the World: On the Possibility of Life in Capitalist Ruins* (First paperback printing). Princeton University Press.
Van Der Weele, C., & Tramper, J. (2014). Cultured Meat: Every Village its Own Factory? *Trends in Biotechnology, 32*(6), 294–296. https://doi.org/10.1016/j.tibtech.2014.04.009
van Mensvoort, K. (2015, March 3). *In Vitro Meat Godfather Dies at Age of 91.* Next Nature Network. Retrieved from https://nextnature.net/magazine/story/2015/in-vitro-meat-godfather-dies-at-age-of-91
Verbeke, W., Marcu, A., Rutsaert, P., Prada, M., Gaspar, R., Seibt, B., Fletcher, D., & Barnett, J. (2015). "Would You Eat Cultured Meat?": Consumers' Reactions and Attitude Formation in Belgium, Portugal and the United Kingdom. *Meat Science.* https://doi.org/10.1016/j.meatsci.2014.11.013
Waldby, C., & Mitchell, R. (2006). *Tissue Economies: Blood, Organs, and Cell Lines in Late Capitalism.* Duke University Press.
Watson, E. (2023, November 20). *Crunch Time for Cultivated Meat: 'Probably 70–90% of Players will Fail in the Next Year'.* AgFunderNews. Retrieved from https://agfundernews.com/crunch-time-for-cultivated-meat-probably-70-90-of-players-will-fail-in-the-next-year
WEF, W. E. F. (2018). *Meat the Future: Time for a Protein Portfolio to Meet Tomorrow's Demand.* World Economic Forum. Retrieved from https://www.weforum.org/publications/meat-the-future-time-for-a-protein-portfolio-to-meet-tomorrow-s-demand/
Wurgaft, B. A. (2019). *Meat Planet: Artificial Flesh and the Future of Food.* University of California Press.
Yao, T., & Asayama, Y. (2017). Animal-Cell Culture Media: History, Characteristics, and Current Issues. *Reproductive Medicine and Biology, 16*(2), 99–117. https://doi.org/10.1002/rmb2.12024
Zaleski. (2024, March 14). *Meet the DC Activist Behind the Alt-Meat Revolution.* Retrieved from https://www.washingtonian.com/2024/03/14/meet-the-dc-activist-behind-the-alt-meat-revolution/
Zassenhaus, M. (2021, March 27). *How the New Harvest Community Coined the Term Cellular Agriculture.* New Harvest. Retrieved from https://new-harvest.org/community-coined-cellular-agriculture/

6

Mapping Meat Technologies: Biosocialities and Geosocialities of Cellular Meat

The history of cultured meat is interesting and its aims are laudable, but the business model of disruptive innovation it is based on deserves careful consideration, as it sets the stage for major social and technological transformations. This is because behind the language of disruption lies a series of creative, technical, and financial assemblages couched in futurist ideals and profound anxiety about the present (Lepore, 2014). A critical perspective on the rhetoric and significance of cultured meat, at this historical juncture, reveals as much about Western society's enduring beliefs and faith about progress and the potential of technology to save the world as it does about the unlearned lessons from the past, in other words about how we got here in the first place.

The idea of cellular agriculture depicts a future where farming cells rather than plants or animals is the way forward. Not only does cellular agriculture promise to lessen the ecological destructiveness of our global food system, the elimination of animals in food production also promises to resolve many contradictions linked to modern agriculture. It offers a radical re-imagining of how food is produced premised on reducing our reliance on land-based agriculture, opening the possibility to produce food in the laboratory here on Earth or in outer space. While promoters of cell meat may have different ambitions for its future, with some

E. Abergel, *Dead Meat*, https://doi.org/10.1007/978-981-97-9049-4_6

viewing it as a kind of artisanal food while others imagine industrializing its production in order to feed the world, it needs to be said that much of the hype surrounding this product remains highly speculative and that even though the reality of cellular meat may be potentially far off or non-existent, it is an object worthy of critical analysis, in the sense that it is a real phenomenon and that its promise has given rise to concrete practices and situations as well as organizations. As such, it has materialized into an activity and a set of technologies that invites other activities and technologies thus creating the conditions for particular modes of thinking and doing to emerge. It has spawned a network of social actors in Latourian words creating webs of relationships between a variety of human and non-human actors.

Part of our inquiry delves into understanding the transformations that animal economies have undergone in the age of the Anthropocene. Not only do I question the notion of meat through its various material and symbolic meanings, but I also want to underline the various ways in which animals themselves have evolved into machinic organisms via technoscientific interventions and genetic modification. As a result, meat has progressively undergone a series of semantic and material transformations and has historically become disembodied from its source animal.[1] To a certain extent, the same could be said of milk and eggs. Underlying the language developed in butchering and the slaughtering practices used to categorize animal parts (carcass, offals, waste), Vialles suggests that the industrial meat system and, at its base, meat-eating as well as vegetarianism follow a logic of negation of the animal. This negation operates on two opposite levels: on the one hand, by excess, which means that the identification with the animal leads to a refusal to consume its flesh (vegetarianism). On the other hand, by default, meaning meat-eaters who identify the animal as food see its fate to be food, making it possible to

[1] The U.S. Cattlemen's Beef Association wants the U.S. Department of Agriculture to oversee inspection and marketing of lab-grown and plant-based meat. Only animal products from actual living cattle or pigs should be allowed to be labeled as "beef" or "pork." France has banned the use of terms like 'vegetarian sausage,' 'vegan bacon,' and soya 'steaks.' The prohibition, which will also apply to vegan and vegetarian dairy alternatives, will be implemented by a change in French agricultural law. In June last year, the European Court of Justice in Luxembourg (ECJ) decided that marketing products of purely plant origin by using words such as 'milk' or 'cheese' is no longer permitted.

consume its flesh (Vialles, 1988, 1995). Hence, our relationship to animals as meat is complex and cell-based meat further complicates it. The practices associated with deanimalization pose serious ontological questions and place into focus how life is conceptualized when it is defined by technoscientific possibilities. It shows us what it becomes when it is subjected to extreme forms of industrialization and radical technological de/re-materialization. As agriculture and food production became more industrialized, our relationship with nature became more ambiguous, as the dual goal of growing large quantities of food while sustaining the environment proved less and less attainable.

In fact, this ambiguity has led us to imagine a future in which meat could still be produced and consumed without all the negative ethical and ecologically destructive side-effects of factory farming. While the farm animals of today have been technologically optimized genetically and morphologically to produce ever-increasing quantities of meat and other animal-derived commodities, such as eggs and milk, industrial agriculture has not overcome/absorbed the costs associated with the global scale of industrial slaughter and its economic, sanitary, and ecological repercussions. Consequently, industrial animal slaughter has been so technologized that the disassembly of animals into various edible and non-edible corporeal renderings has contributed to what the anthropologist Mondher Kilani calls "the construction of a fictive 'hygienic' death, without spilled blood, or victim, or sacrificer: the modern putting to death of animals no longer possesses a sacrificial dimension" (Reinert, 2007). CM extends this process by bioengineering the animal out of its body, whereby it is no longer death that is fictively hygienic but life itself. In fact, lab-grown meat is an extension of this process as it does not alter the instrumental nature of our relationship to farm animals.

Accordingly, the historical trajectories of industrial agriculture have resulted in cell-based meat which is inextricably linked to new forms of industrial appropriation and the substitution of biological living materials within tissue economies. In turn, these are constitutive of biosocialities which view nature as "modeled on culture understood as practice. Nature will be known and remade through technique and will finally become artificial, just as culture becomes natural" (Rabinow, 1996, p. 99). Writing about developments in the life sciences, Rabinow suggests that the level of transformation and intervention technosciences hold at the biological level open up the potential for profound transformations of collective identities, subjectivities, and social life. As laboratory-produced animal tissues transform into edible meat, they co-construct groups of individuals and eaters with a collective identity, informed by shifting ideas about food and how it might be reconceptualized around moral and ethical imperatives. A new kind of eating biopolitics takes shape, however, dietary choices are also inextricably linked with geosocialities—that is the point of contact between geology and social-cultural processes (Palsson & Swanson, 2016). It has become apparent in light of the Anthropocene that historical encounters with Earth processes have had profound effects on social life and that political, social, and economic agency have been both enabled and constrained by geological forces (Clark & Yusoff, 2017). What Chakrabarty describes as human geological agency and what others describe as the Anthropocene now articulates new forms of geo-politics defined by human entanglements with stratigraphic data and geological timescales.

This logic means that technologies and the Anthropocene have become intertwined, whereby the Anthropocene has been technologized and, similarly, technologies have been 'anthropocenized.' As Swyngedouw and Ernstson (2018) remark, the various narratives around the notion of the AnthropoScene tend to depoliticize it, opening the way for hyper-accelerationist and ecomodernist visions that rely on big science and technology to save the Earth (p. 3). Within their critical approach of 'Anthropocene' narratives, the authors call for renewed thinking about the horizontality of human-non-human relations as conceptualized in post-human ontologies because these are depoliticizing and because they lead to the enfolding of the more-than-human nature within capitalism, thus securing "capital's hold over life" (p. 11). What is at issue for these authors and others are the 'emancipatory' politics of the inseparability of human and non-human agents, which are by no means guaranteed through new non-dualist materialist ontologies. In that sense, by mobilizing the seamlessness of socionatures, these ontologies promise a

> renewed and ecologically sensitive 'hyper-reflexive' capitalism that takes seriously both humans' geo-physical force and the material acting of the non-human, while redeeming the sins of the past. (p. 11)

We will return to this point when we discuss Anthropocene narratives. Nevertheless, the point I wish to make is that geosocialities or "the entangled relations of the Earth and biologic beings" expand and twist "the conceptions of agency, intimacy, and politics that have emerged from biosociality and the biosocial" (Palsson & Swanson, 2016). Biosocialities and geosocialities become linked, in terms of the entangled histories and dynamic processes that enable living organisms and vital materials to engage with non-humans, whereby inert and lively materials come into contact with different geological scales as sites of biocultural entanglements. The Anthropocene, in its multiple accounts described later, has prompted a 'geological turn' implying that humans have become cognizant of Earth systems, highlighting a need for collective action in the face of impending catastrophes brought about by climactic breakdown and mass extinctions. As such, our responses to Anthropocene futures are what this book is partly about. It explains how amid existential risk and apocalyptic events, strategies such as cellular agriculture, which would have seemed implausible or improbable in the past, become implemented, providing new opportunities for investment and for capital circulation (Harper & Specht, 2022).

The production of animal protein from stem cells depends on the highly regulated molecular mechanisms that promote cellular growth and function. These include the regulated transport of nutrients, ions, and signaling molecules from the culture medium into and across cellular membranes, via specialized transport proteins, ion channels, and receptor-mediated pathways. These events ensure the intracellular environment is conducive to essential activities, such as energy production, protein synthesis, and cell differentiation. Additionally, electrical impulses and chemical gradients are critical in coordinating cellular communication, maintaining membrane potential and driving the metabolic activities required for cell replication and maturation into tissue structures. Such intricate molecular interactions are fundamental to the viability of cellular agriculture, while at the same time highlighting the complex interplay of biological, technological, and environmental factors necessary for scaling this process to address global protein demands. While this approach reconfigures "life" as a manipulable substrate that can be optimized by highly controlled biotechnological systems, meat gets defined as "…a collection of cells of various key types […] arranged in a matrix of connective tissue which contributes to the texture and structure of meat" (Specht, 2018). As a result, Specht claims that rethinking meat requires approaching its creation as a contingent outcome of biological processes rather than as a fixed product of naturalized systems (i.e. living animals). The assembly process, starting from the cell biopsies taken from live animals to the stabilization of cell lines, their proliferation in vitro, adhesion and structural formation onto scaffolds, growth and cellular organization to form tissues, is constituted by a series of interconnected networks and flows of information and materials, in a series of stepwise in vivo interventions which operate outside of the body. The disembodied character of these biological and chemical processes puts into relief a series of dualities such as the natural versus synthetic, organic versus inorganic, natural versus manipulated, and experimental versus theoretical. These dualities underscore the potentialities of the technoscientific project in the context of planetary limits as the distinction between biological and geological processes becomes increasingly blurred. Human activities act as catalysts, integrating human biological activities into biogeochemical cycles which produce a distinctive fossil record that includes the bones of domesticated animals and other bio-artifacts. Geological changes resulting from anthropogenic activity, such as increased greenhouse gases, profoundly affect biological systems such as species migration, extinction, and other biological responses in altered ecosystems. Soil degradation affects plant life and ecosystem health, and ocean acidification severely impacts marine life. As a result, the biological collapses into the geologic, and the geologic collapses into the biological.

While the techniques for growing animal tissues are not new, what is new is the fact that these are produced for food and not biomedical purposes. As a result, the novelty of these meats means that we will discuss the notion of edibility as these laboratory foods are new to the human diet and authors like Sexton qualify them as post-Anthropocene foods because they herald the consumption of novel protein in the human diet

(Sexton, 2018). Animal laboratory-grown tissues are rendered edible via a series of biofabrication steps, but developers have had to face a series of technical and financial difficulties, and despite optimistic visions, these products have not yet reached commercialization, at least in terms of industrial manufacturing. At the time of writing this book, only a handful of companies are able to produce small quantities of products, which are allowed to be sold in a few countries and at specific locations—such as high-end restaurants or niche charcuteries.

The Biopolitics of Eating Cell Meat and Virtuous Eating

What biopolitical implications for the body of eaters does eating cultivated meat hold? How does animal body erasure, on the one hand, and individual bodily self-improvement, on the other, become intertwined when this product is consumed? Can we make certain observations or assumptions about the laboratory or informational, molecularized environment used to grow cellular meat and potential interactions with eating bodies? These are some of the questions that this section will try to discuss via the idea of biopolitics.

The driving force behind the creation of in vitro meat rests on two interconnected themes dealing with biosocialities and geosocialities, as mentioned before. The first one, which I call the biopolitics of eating, invokes health claims via a discourse of self-empowerment, purity, and personal discipline over one's body via good consumerist attitudes linked to biosocial ideals of health, wellness, and self-optimization. The second theme involves the idea of *virtuous eating* as a direct response to the environmental devastation and animal suffering caused by industrial agriculture, and in particular intensive animal agriculture, which mediate the affective and material boundaries between human bodies and non-human others, at the molecular and geological scale. The notion of virtuous eating has to do with the moral imperatives that compel individuals to consume products, as moral agents, recognizing and responding to animal suffering in compassionate ways. This may mean eliminating meat from the diet or becoming vegan, which avoids all animal products or it may mean choosing lab-grown meat as an option. Of course, I use the term virtuous eating in both ways, as a moral choice and to signify the virtuality of meat, that is, its status as real animal meat and the illusory thinking that comes along with consuming cultivated meat, that by eating it, eaters are not engaging in any kind of animal abuse or exploitation and environmental destruction, abstracting away the process of obtaining the final product. The injunction that eating CM is better for people, the animals, and the planet opens up debates about the "myth of green and healthy meat," vegetable, animal, or both. John Miller questions whether CM represents a "real choice" for vegans and vegetarians; he writes:

> I argue, in vitro meat appears not as a radical solution to the violent subjection of nonhuman animals within industrial capitalist cultures, but rather as a further symptom of the remarkable extent of this violence. (Miller, 2012, p. 45)

As such, CM remains an unstabilized entity because it requires simultaneous distancing from and connection to the animal (Castle, 2022). This complicates matters for eating subjects, as they attempt to make sense of CM's ontological status.

These issues also engage us to think in terms of a post-animal bio-economy, a market model in which the production of meat (including fish), dairy, eggs, and leather will no longer require the use of live animals (Datar et al., 2016). This relates to specific visions of the future of agriculture and of the role of animals and farmers within them in ways that frame meat as a product of a 'cellular agricultural meat factory.'

Clearly, *the biopolitics of eating* and *virtuous eating* are deeply interconnected; both are linked to the commodification of edible tissues, pointing to the failures of productivist agriculture as potentially risky for human health and the environment, as well as institutionalizing the mass slaughter of billions of animals. Rather than exposing the deep inequalities built into the global food system, these two interweaving narratives provide both a moral and an affective legitimation and cognitive validity to cultured meat while relying on meat-eating as a key feature of Western diets, thus avoiding the need for consumers to radically change their behavior

(Galusky, 2014). Food biopolitics occurs when eating becomes a way of doing and enacting food politics. Considering the unique form of consumption that eating represents, we literally incorporate the outside world into our bodies, ourselves, and our identities; it is interesting to consider the human-nature relationships as well as human-non-human relationships, enacted through the practices of eating. Moreover, Pollan observes, while experimenting with hunting and gathering, that modern industrialized food has kept hidden the biological realities of eating, including our dependency on the complexities and webs of relations with other species. Food is the ultimate socio-nature (Goodman & Sage, 2016). However, food biopolitics also engages us to think about how individuals enact food politics on a daily basis, where eating choices (or lack of choices) create and discipline subjects as bodies (Rose, 2007). Issues relating food to health and medicine or to body image and physical performance make eating increasingly contentious and complex. It has to be said that lab-grown edible animal tissues have never been consumed as food and remain mostly a product of biomedical research; hence, their edibility has to be socially and culturally negotiated (Roe, 2006; Sexton, 2018).

Virtuous eating, on the other hand, relates to many of the issues raised by the anthropologist Mary Douglas in her book *Purity and Danger*, where food avoidance or impure food is connected to a complex social and symbolic system of purity. Dietary rules dictate how we might achieve "the good life," in terms of culture and religion, but by extension they signify a search for moral and physical fulfillment. Claiming something as dangerous, unhealthy, or impure essentially amounts to enforcing conformity within a particular social order, thus what is considered inedible is essentially "matter out of place" (Douglas, 2001). How then does laboratory meat fit into the social classification of food? According to Douglas' ideas, if edible cultured animal tissues are to be accepted as meat, the structure of assumptions about their edibility has to be modified. In order not to be rejected, they must fit in what she calls secular rites and symbolic enactments (p. 64) which link the present with the past. Presenting CM as "meat" represents such a process. Hence, social framing of lab-grown meat is of crucial importance. A recent study has shown that "those who encounter cultured meat through the 'high tech' frame have significantly more negative attitudes toward the concept, and are significantly less likely to consume it" (Bryant & Dillard, 2019). Science-themed media and other depictions of lab-grown edible animal tissues might put off consumers, but the product name is also an issue. According to Douglas, how we name this product affects the ways it is perceived the next time one encounters it. Calling and portraying it as "high-tech" or "laboratory-grown" or "in vitro" conjures up images of food in opposition to "wholesome," "organic," or "natural." Debates among proponents about how to name lab-grown edible animal tissues have been contentious. Should it be marketed as "clean meat," "cell-based," "laboratory-grown," "cultured," or "craft" meat? At the time of writing this book, the consensus among industry players was to call lab-grown edible animal tissues "cultivated meat" revealing much about the process of meaning-making around these products.

This book is as much about bodies, those of eaters and those that are eaten, as it is about the material absence, alteration, and distancing of bodies that become vital fragments and edible tissues. More specifically, it is about environmental and other narratives that drive the need for such developments, taking at its core the ways in which our reliance on technoscientific progress, in the life sciences in this particular case, frames the future of our species in the most intimate and entangled ways.

Meat as Identity Politics: Consuming Cellular Meat

In this section, I will pose the question: what are we eating when we are eating cell meat? And how do our food choices construct who we are? Taking as a point of departure the work of Mary Douglas, *Purity and Danger*, where she interrogates, among other things, the meaning and purpose of dietary rules and how food rituals relate to social order and culture, we will try to think through the ethics of eating ethical meat. More broadly, using the concept of biosociality as a line of reasoning, our aim is to untangle the morality of eating, the process by which cellular meat and cultured animal products become morally justifiable and how future consumers and eaters of these products are produced. The concept of biosociality stems from the idea that the life sciences have altered our understanding of ourselves and our individuality in biological terms, also known as somatic individuality (Brekke & Sirnes, 2011; Rose, 2007). This implies a kind of re-socialization of the body characterized by new forms of collective and individual identification, linked to a neoliberal ethic which promotes a maximization of the self via lifestyle, health, and quality of life (Rose, 2007). Rose describes a kind of biological citizenship defined by somatic individuality whereby the flesh, corporeal existence and individuals experience and reflect upon their life experience "in part in the language of biomedicine" (p. 26). The new possibilities offered by the biosciences become opportunities for individuals to empower themselves in the genomic era, and as such, these biological citizens are active actors as creators and consumers of technoscientific developments and moreover in the capitalization of vitality. In this new ethic, somatic selves navigate between "choice, prudence and responsibility" where biology is no longer destiny and where vitality is constantly negotiated via bioengineering as well as technological promises and hopes. Rose explores how biomedicine and biosocial collectives affect the course of bioscientific research and medical knowledge, demanding access to knowledge and treatments, and how biomedical enterprises create new regimes of hope which alter how individuals see themselves in a biopolitical manner, as self-entrepreneurs who redraw the lines of democratic participation via biological citizenship. In this context, it is clear that food production and consumption have been determined by the limits of technoscientific intervention since the beginnings of the industrialization of agriculture. Through the idea of biosociality, we can understand how individuals increasingly make food choices based on health and well-being.

As cell meat remains "morally ill defined" and "ontologically ambiguous" (Stephens & Ruivenkamp, 2016), it becomes important to understand the ways in which animal replacement technologies used for food impose the creation of new forms of social reordering and how eaters might engage in such reordering. One of the main arguments in support of cultivated meat is its claimed purity—the ability to grow cells in sterile, virus-free environments without the need for antibiotics or chemicals.

This contrasts sharply with the potential dangers associated with consuming meat from animals raised under factory farming conditions, where industrial practices and pharmaceutical interventions are required to keep animals disease-free and ensure rapid growth throughout their existence. As a result, the purity of cell meat is pitted against the potential dangers and 'pollution risks' associated with meat produced from intensive farming operations. This takes on new significance as we live through the worse pandemic since the Spanish flu, and the probable original source of the COVID virus, most likely to have come from an animal, is yet to be identified at the time of writing this book. Hence, new categories of meats are created which provoke profound questioning about the nature of purity and impurity, as live animals are construed as being as morally ambiguous as edible cultured tissues. Therefore, there is a direct correlation between the pollution of animal bodies and the pollution of the planet, signaling potential dangers for eaters. Hence a fear of animal meat and edible animal products is produced, providing the moral justification for the production and consumption of cellular products. If pollution as Douglas states is matter out of place, what then are the social and biosocial processes that order or put in place certain food and eating orders and, by extension, insinuate the restructuring and eventual elimination of animal agriculture?

This is further complicated by the fact that in vitro meat is identified according to the origin of the cells used to produce it. Hence, the identification with the food animal: beef, chicken, pork, etc. creates a sort of animal taxonomy which harks back to the idea of eating the whole animal, without the pollution risks associated with it.

I am interested in what the foods we eat say about who we are and how our identities beyond our diet are defined. As anthropologists such as Abbots have shown, eating is "an embodied experience [that] mediates and is mediated by the relationship between food's matter and meaning" (Abbots, 2017). More broadly, this implies that our relationship to food is intimately connected with our lived experiences and our social realities and how we relate to the world around us. Sensory experiences, symbolic rituals, and identity formation are linked in complex ways. How does cultivated meat fit into this scheme?

Bibliography

Abbots, E. J. (2017). *The Agency of Eating. Mediation, Food and the Body*. Bloomsbury Academic.

Brekke, O. A., & Sirnes, T. (2011). Biosociality, Biocitizenship and the New Regime of Hope and Despair: Interpreting "Portraits of Hope" and the "Mehmet Case". *New Genetics and Society, 30*(4), 347–374. https://doi.org/10.1080/14636778.2011.592012

Bryant, C., & Dillard, C. (2019). The Impact of Framing on Acceptance of Cultured Meat. *Frontiers in Nutrition, 6*. https://doi.org/10.3389/fnut.2019.00103

Castle, N. (2022). In Vitro Meat and Science Fiction. *Extrapolation, 63*(2), 149–179. https://doi.org/10.3828/extr.2022.11

Clark, N., & Yusoff, K. (2017). Geosocial Formations and the Anthropocene. *Theory, Culture and Society, 34*(2–3), 3–23. https://doi.org/10.1177/0263276416688946

Datar, I., Kim, E., & d'Origny, G. (2016). New Harvest: Building the Cellular Agriculture Economy. In B. Donaldson & C. Carter (Eds.), *The Future of Meat without Animals* (pp. 121–131). Roman and Littlefield International.

Douglas, M. (2001). *Purity and Danger: An Analysis of the Concepts of Pollution and Taboo* (Repr). Routledge.

Galusky, W. (2014). Technology as Responsibility: Failure, Food Animals, and Lab-grown Meat. *Journal of Agricultural & Environmental Ethics, 27*(6), 931–948. https://doi.org/10.1007/s10806-014-9508-9

Goodman, M. K., & Sage, C. (Eds.). (2016). *Food Transgressions: Making Sense of Contemporary Food Politics* (1st ed.). Routledge. https://doi.org/10.4324/9781315582702

Harper, E. T., & Specht, D. (Eds.). (2022). *Imagining Apocalyptic Politics in the Anthropocene*. Routledge.

Lepore, J. (2014, June 16). *What the Gospel of Innovation Gets Wrong*. Retrieved from https://www.newyorker.com/magazine/2014/06/23/the-disruption-machine

Miller, J. (2012). In Vitro Meat: Power, Authenticity and Vegetarianism. *Journal of Critical Animal Studies, 10*(4), 41–63.

Palsson, G., & Swanson, H. A. (2016). Down to Earth: Geosocialities and Geopolitics. *Environmental Humanities, 8*(2), 149–171. Retrieved from http://dx.doi.org.proxy.bibliotheques.uqam.ca/10.1215/22011919-3664202

Rabinow, P. (1996). *Essays on the Anthropology of Reason*. Princeton University Press.
Reinert, H. (2007). The Pertinence of Sacrifice—Some Notes on Larry the Luckiest Lamb. *Borderlandes E-Journal, 6*(3). Retrieved from http://www.borderlands.net.au/vol6no3_2007/reinert_larry.htm
Roe, E. J. (2006). Material Connectivity, the Immaterial and the Aesthetic of Eating Practices: An Argument for How Genetically Modified Foodstuff Becomes Inedible. *Environment and Planning A: Economy and Space, 38*(3), 465–481. https://doi.org/10.1068/a3835
Rose, N. S. (2007). *Politics of Life Itself: Biomedicine, Power, and Subjectivity in the Twenty-First Century*. Princeton University Press.
Sexton, A. E. (2018). Eating for the Post-Anthropocene: Alternative Proteins and the Biopolitics of Edibility. *Transactions of the Institute of British Geographers, 43*(4), 586–600. https://doi.org/10.1111/tran.12253
Stephens, N., & Ruivenkamp, M. (2016). Promise and Ontological Ambiguity in the *In vitro* Meat Imagescape: From Laboratory Myotubes to the Cultured Burger. *Science as Culture, 25*(3), 327–355. https://doi.org/10.1080/09505431.2016.1171836
Specht, L. (2018). Meat by the Molecule: Making meat with plants and cells. *The Biochemist, 40*(4), 18–21.
Swyngedouw, E., & Ernstson, H. (2018). Interrupting the Anthropo-obScene: Immuno-biopolitics and Depoliticizing Ontologies in the Anthropocene. *Theory, Culture & Society, 35*(6), 3–30. https://doi.org/10.1177/0263276418757314
Vialles, N. (1988). La viande ou la bête. *Terrain. Anthropologie & Sciences Humaines, 10*, 86–96. https://doi.org/10.4000/terrain.2932
Vialles, N. (1995). De l'animal à la viande: Une mort sans cadavre. *French Cultural Studies, 6*(18), 335–350. https://doi.org/10.1177/095715589500601805

7

The Biopolitics of Cr/Edibility: The Construction of Taste

A few years prior to scientists getting involved in the production of in vitro meat, bioartists Oron Catts and Ionat Zurr, part of the bioart collectives SymbioticA and the Tissue Culture & Art (TC&A), were also experimenting with growing animal cells outside of animals' bodies in an exhibition entitled *Disembodied Cuisine* shown in 2003. As part of a longer project entitled Semi-Living Steak, they explored "the implications of a near-future scenario in which lab-grown tissues" become a source of food (TC&A Project, 2000). Semi-Living Steak came about during their residency at the Tissue Engineering and Organ Fabrication Laboratory at Harvard Medical School in 2000. The proposition of growing food in the laboratory had not yet been explored even though the idea, as we saw, had been around for a long time. In this experimental work, cells were collected from an unborn fetus, exploring in utero tissue culture (Hauser, 2003). The original idea of growing meat came about in the laboratory when:

> One of the research fellows in the laboratory was working on an in-utero muscle tissue engineering project. She harvested skeletal muscle progenitor cells, known as satellite cells, from a fetal lamb that was yet to be born. These cells grow very fast, and in a matter of days the incubator in the lab

E. Abergel, *Dead Meat*, https://doi.org/10.1007/978-981-97-9049-4_7

> has stacks and stacks of tissue culture flasks brimming with them. For whatever reason, this scientist could not bring herself to dispose of the excess cells, so we offered to take some to try and grow meat. (Zurr & Catts, 2023, p. 78)

According to the TC&A, cell culture technology at that moment was used purely for scientific research, not yet for what they called "cultural" purposes. Their goal was to demonstrate the possibility of growing what they called "victimless meat" using cells obtained from the biopsy of a live animal and growing them in vitro, using tissue culture techniques. This is how they describe their work:

> In 2000, the first "steaks," measuring 1 cm in diameter, were composed of pre-natal skeletal muscles cells—cells taken from an unborn sheep—which were seeded on a degradable matrix. The "steaks" were incubated in a 3D tissue culture bioreactor and fed with fetal calf serum, a necessary nutrient for tissues grown in vitro. Considering the amount—and type—of materials used, specifically plastic and animal products, Semi-living Steak investigates, not only the use of lab-grown tissues for food, but also the ethical dilemmas that accompany this use. (TC&A, 2000)

It took three months to grow small pieces of what looked like meat. Ironically, the Institute's commercialization office did not see the commercial potential for in vitro meat and the intellectual property rights linked to the project reverted to the artists (Zurr & Catts, 2023). By contrast, it took five years of work by Dr. Mark Post and his laboratory to arrive at the process used to grow the tissue that was destined to be used in a televised taste testing. The cultivation of 20,000 strips of tissues used to assemble one five-ounce burger took over three months (Laskawy, 2013).

"Victimless meat" was the subject of the *Disembodied Cuisine* installation at the L'Art Biotech exhibition in Nantes, which problematized the idea of growing meat in vitro one step further, by demonstrating that the use of animal-derived products in the cellular growth medium contributed to abstracting the animals whose serum was used in the process of growing "edible" tissues and as a result obscured them as the "very real victims." Recreating a biosafety level 2 laboratory—a level of containment used for moderate-risk biological agents that is applied to teaching and demonstrations—as a kind of theatrical stage within the museum, frog skeletal muscle was produced from cells taken from a live Xenopus frog,[1] seeded on a food grade biodegradable biopolymer structure and then grown in a bioreactor. Next to this setup, an aquarium with the healthy French 'food grade' frogs, watched over a dinner table, set for the final feast planned on the last day of the exhibition. Plant tissue was grown in glass containers as garnish to accompany the meal and was placed on top of the aquarium. The steaks were cooked and eaten in the final feast and the frogs were released in a local pond to hopefully live out the rest of their existence. According to Jen Hauser, the idea was to show that at the same time as the cells taken from the source animal proliferated and were able to produce a "steak" under the right conditions, the healing of the animal also took place (2003). The notion of "victimless meat," they claimed, could comfort people concerned with factory farming and animal ethics, but the artists suggested that producing laboratory foods ushers in a new form of exploitation: that of the "semi-living." From an artistic point of view, Catts and Zurr define the term "semi-living" as "the living tissue constructs that are grown/constructed out of tissues taken from complex organisms and maintained alive with the aid of technological intervention" (Catts et al., 2021). In this case, these tissues, described as living fragments of bodies, have no biomedical purpose other than an artistic one. They are "conceptual prototypes of a new kind of 'life' that is neither living nor non-living" (Zurr, 2008). A form of vitality that borders between biological life and technological construct. What the Semi-Livings represent are objects/subjects that exist in various places such as laboratories, tissue banks, factories, museums, zoos, and art galleries (Ibid). By linking the medium of art with biotechnology, the subject of life and the living is subverted from its technoscientific ontology and becomes a biocultural entity. Thus, Semi-Livings as distinctive living structures interrogate the nature and treatment of 'life,' extending

[1] A type of African clawed frog commonly used in laboratory research as a biomedical model due to its genetic similarity to humans but also as an accessible source of eggs and embryos which can be easily manipulated and studied in embryology, developmental and cell biology. Pluripotent stem cells derived from its embryos can be differentiated into functional organs (Blum & Ott, 2018; Borchers & Pieler, 2010).

this to questions about the uncertain boundaries between human and non-human life.

The final feast culminated in cooking and consuming the frog meat grown in the bioreactors, an act that necessitated overcoming one's apprehensions it seems, as three people were not able to swallow it. Authors, such as Joanna Zylinska in her book *Bioethics in the Age of New Media* (2009), view "performativity at the ontological level" as the true force of bioart, questioning "who is the other" and what kinds of bioethics do extreme forms of alterity entail? For Zurr and Catts (2023), there is a clear connection between their experimental process in the lab, creating *Disembodied Cuisine* and the production of in vitro meat as a commercial venture, as they were contacted after the exhibition by a volunteer, Jason Matheny, at People for the Ethical Treatment for Animals (PETA) to inquire about how the meat was produced as well as its costs. This apparently led to PETA launching its one million dollar U.S. prize "for the first laboratory to use chicken cells to create commercially viable in vitro (test tube) meat" (PETA, 2008). To claim the prize money, researchers had to reproduce the taste and texture of real chicken flesh and be able to manufacture and successfully commercialize the product at a competitive cost based on the average price of boneless chicken breast. The contest, announced in 2008, was extended until March 4, 2014, coinciding with President Hoover's "chicken in every pot" inaugural statement (Ibid). According to one of their online statements released in 2022, although the prize wasn't claimed, PETA patented the findings of one scientist who eventually joined Upside Foods (a cell meat company founded in 2015 formerly known as Memphis Meats) and sold the patent to the start-up at a very low cost (Perle, 2022). PETA declared the contest a success: although it didn't get any entries, it claims to have sparked the debate surrounding cell meat, attracted the attention of investors, and created a research fellowship which led to some patents (Travis, 2014). No mention of the connections between bioart and the nascent cell meat industry was found apart from Zurr and Catts' recollections. Nevertheless, Zurr and Catts recount that Jason Matheny went on to found New Harvest[2]

[2] New Harvest is a non-profit organization dedicated to advancing cell meat research. Their mission is to "maximize the positive impact of cellular agriculture" by supporting researchers and bringing together various sectors (academia, government, philanthropy, and industry) (New Havest, 2022).

in 2004, one of the first research entities dedicated to promoting and supporting the development of in vitro meat as it was called at the time. As it happens, Jason Matheny went on to become C.E.O. of the RAND Corporation in 2022.

Today, the most active research laboratories and start-ups in this field can be found in the U.S., Holland, the U.K., Japan, and Israel. It is the Dutch team, headed by Dr. Mark Post, an angiogenesis expert and cardiologist, that first made headlines in 2013, when it organized an international media event, in London, where the first lab-grown hamburger was cooked and eaten by two food critics and a food writer. It was revealed that in order to make this "hamburger" in the laboratory, cells biopsied from cows were cultured with nutrients and growth-promoting chemicals. Three weeks later, there were more than a million stem cells, which were then put into smaller dishes where they coalesced into small strips of muscle about a centimeter long and a few millimeters thick. The process of producing a small hamburger took three months, was funded by Google co-founder Sergey Brin, and cost approximately $330,000 U.S. (Jha, 2013). These strips were collected into small pellets, which were frozen. When there were enough strips, they were defrosted and compacted into a patty just before being cooked. After this very publicized event, PETA stated: "[Lab-grown meat] will spell the end of lorries full of cows and chickens, abattoirs and factory farming. It will reduce carbon emissions, conserve water and make the food supply safer" ("World's First Lab-Grown Burger Is Eaten in London," 2013). The public relations exercise was to be important as proof of concept, demonstrating the feasibility of cultured meat as a scientific achievement and as a promising economic venture.

These two events are linked in many ways. In their book, Zurr and Catts (2023) describe how they were asked to re-create one of their early experiments, using fetal lamb cells to grow tissue for a commission to do a documentary for the BBC and the Discovery Channel. The documentary aired in 2009, for the COP15 in Copenhagen in order to highlight "near-future" technologies that would revolutionize climate change. Ironically, their work was included in the piece to highlight "extreme technological approaches intended to solve problems created by technology" (p. 90). In 2012, they produced an event called ArtMeatFlesh for the Dutch Electronics Art Festival in Rotterdam, where Dr. Mark Post took part in what they called an "Iron Chef cook-off" that used lab-grown meat. Shortly after this, on August 5, 2013, the scientific press conference took place in London, featuring Dr. Mark Post, then professor at Maastricht University now chief scientist of Mosa Meats, lifting the lid and revealing the raw in vitro hamburger in a petri dish. After cooking it for about ten minutes, celebrity chef Richard McGeown handed a plate to Hanni Rützler, a nutritional researcher for Future Food studio, and Josh Schonwald, an American food writer, and these two were given an opportunity to sample the world's most expensive meat.

Cell-Based Meat Tastings as Spectacle: Performance and Aesthetics in the Construction and Deconstruction of Meat

The study of taste, as a cross-disciplinary area of research, studies how "situated performances and practices of taste and market level processes and outcomes" interrelate (Pomiès et al., 2021). Taste can be understood as a "situated performance shaped by a socio-material world which, in turn, shapes the world" (Ibid, p. 3). While the ways in which taste is performed lies in cultural activities that extend beyond public tastings and focus on people's consumption patterns and their correspondence to what are called 'taste regimes,' our interest lies in juxtaposing two events that place cellular meat at the center of public tastings and relating them to the ways in which they socially, symbolically, and materially produce taste. In doing so, our intent is to establish the contrasting approaches to staging taste. On the one hand, we describe how bioartists stage cellular meat through a reflexive and discursive set of practices that link to the various elements of its *Disembodied Cuisine* installation, establishing a powerful critique of biotechnologies and their (mis)appropriation of life. In contrast, the highly publicized 2013 tasting of the first laboratory-grown hamburger tells a different story and contributes to the official launching of a new unfamiliar food, inaugurating visions of the future largely determined through technoscientific innovation and entrepreneurial science. Each of these performances paints a different picture of the world in which laboratory-grown meat exists. One interrogates the boundaries of life and the limits of human intervention and the logics of industrializing life processes and adopts a critical anti-utopian stance. The other takes the opposite view, placing technology at the center of the taste experience and as a utopian project of transforming palates as well as the way that meat is produced and consumed. The global press conference introducing the first in vitro hamburger was a key moment in rendering this material as a knowable object, by validating it as meat and as a beef hamburger (O'Riordan et al., 2017).

As Hennion (2007) suggests, tastes are not only determined by the properties of objects or foods, but also rely on what he calls the 'techniques and devices' that enable feelings and sensibilities to be perceived. For him, taste is the product of an activity; it is dynamic, meaning that it must be made to appear as it is not inherently present. The making of taste is ultimately a situated activity and although preference for things might be socially determined, as Bourdieu's work demonstrates, Hennion argues that "we rely on others in a reflexive way to constitute our tastes" (p. 103). So, he claims that the presence of the 'object' within taste is problematic, because objects are what we make of them—at once material and symbolic—and "the body that tastes and the taste of the object" are simultaneously co-produced.

These two performances constitute complete inversions that entail making sense for bioartists of 'animal cells as meat' or 'laboratory constructed tissue as meat' versus an understanding of 'meat as animal cells' or 'meat as laboratory constructed tissue' in the case of the launch of cell meat. These subtle differences engage tasters and the viewing public in distinct ways, by ordering the world of cell meat differently and engaging viewers using different modes of representational discourses. In the case of the bioart exhibition, the *process* is emphasized by focusing on the cultivation of cell meat rather than its culinary presentation. The performance piece is constructed as a scene, immersing viewers in the laboratory setting, where they encounter a bioreactor cultivating living tissues intended for consumption. As a result, *Disembodied Cuisine* directs attention on the intricate processes involved in tissue engineering and the manipulation of biological materials and, more widely, leads to an exploration of the ethical dimensions of biotechnological interventions in food production. This entails challenging the notion of food as mere commodity, allowing viewers to consider the interconnectedness of living beings in food systems.

This interconnectedness can be further explored when one considers the idea of consuming tissues produced from animal, microbial, or plant Others. Eating the Other leaves open the possibility that the Other may in fact include humans as a source of edible laboratory-grown flesh, and the interconnectedness and entanglements discussed above make this come into view. It is within the realm of imaginable food as the Ouroboros project has shown. The possibility of producing meat using one's own cells, because mammalian cells are at the base of cellular agriculture, has been called a carnivore's nightmare, a form of self 'cannibalism' which often features in science fiction narratives.

In contrast, at the official launch of the first lab-grown burger, attention is directed toward the *product* itself, making visible the raw cultivated 'meat' tissue and subsequently progressing to the cooked portions. There are few clues about the laboratory setting and no observable indications regarding the methods employed in the creation of the food under evaluation. This ordering of elements significantly shapes our comprehension and perception, a notion central to Michel Foucault's discourse on power and knowledge. Foucault elucidates how power operates through various mechanisms, including the control of discourse and the arrangement of information. Thus, the sequencing of events from the end product (the cooked morsels) back to its origin in the laboratory underscores a deliberate narrative construction. This particular ordering foregrounds the final product, directing attention away from the processes involved in its creation. In this order, the technologies, the material aspects, and complex assembly of living and non-living entities that fit into a wider configuration of elements within the technoscientific system become backgrounded. This is a way of obscuring the complexities and controversies that cell meat might provoke, effectively framing it as a culinary innovation rather than a product of contentious scientific and ethical debates. This manipulation of narrative sequence, as Foucault would argue, not only shapes understanding but also reinforces power dynamics, by controlling the discourse and perceptions surrounding emerging technologies and their implications. This not only structures knowledge about cell meat, but it also carries broader implications about the future of food production, technology, and social attitudes toward innovation. Clearly, the visual performance of *Disembodied Cuisine* offers a counternarrative to the kind of technological triumphalism normally associated with the aspirations of cellular agriculture.

In their study of people's attitudes toward consuming animals as food, Evans and Miele (2012) contend that consumption practices are embodied on both a visceral and symbolic level and that these are deeply interconnected. They describe this process as 'foodsensing,' a way for consumers to "sense and make sense" of the food they purchase, prepare, and eat. They describe foodsensing in the following way:

> we believe that foodsensing is best viewed in a performative fashion. In other words, we believe that foodsensing is a potentially creative process through which novel 'realities' emerge as different food attributes (or better still 'virtualities') are attuned to and 'made to matter'. (p. 4)

Thus, sense making and making sense are best understood in terms of their consubstantiality as they relate to tasting food. The devices and techniques of tasting, that is, the contextual backdrop and intended audiences are important factors to consider as well as reliance on others, as Hennion points out, for establishing reflexivity about eating experiences. In both the art exhibition and the broadcast studio environment, the meaning of food becomes unsettled whether it is represented as bioart or the product of technological mastery. One thing is clear, the performative aspects of public tastings provide an opportunity to explore the ways in which eating—by others through a reflexive lens—involves the encounter between the eater and the eaten, providing knowledge about foods that include the social relations embedded into their meaning as well as their materiality (Abbots, 2017).

Because cell-based meats involve a new kind of food practice, the construction of a consuming body available and willing to incorporate and ingest this material becomes integral to the public tasting experience. In this sense, how the bodies of the test eaters react is not simply an individual experience, their response permeates in the body of others, meaning that during the eating process, corporeal boundaries are being "continuously being made and unmade" and become the "point of intersection between the corporeal body, the stuff of food and knowledge" (Abbots, 2017, p. 19). Hence, these public tastings are highly intersubjective and invite different types of social engagement with lab-grown meat, ones in which "each eating body becomes situated in and constituted through a network of relations with others" (Probyn cited in Abbots, 2017, p. 20). Abbots' work demonstrates the ways in which subjective eating experiences and political dynamics are entangled. Her analysis draws from the philosophical framework of Gilles Deleuze and Félix Gattari on assemblages, essentially defined as a dynamic and heterogeneous configuration of elements that come together in a particular time and space, forming a temporary and contingent unity (Deleuze & Guattari, 1987). These are inherently unstable and are constantly in a process of becoming. Assemblages are composed of multiple elements that are human and non-human, material and immaterial involving relationships between institutions, ideas, technologies, environments, individuals, organized in rhizomatic structures where connections are horizontal and decentralized and non-hierarchical. Abbots' approach challenges traditional notions of agency as located within individual subjects. By situating agency within 'food assemblages,' she interrogates how humans become vectors who define their form, what can be known, believed, and unknown about certain foods, and how these processes in turn participate in "tuning" bodies toward specific tastes (2017). Eating and tasting are thus a form of mediation, she writes:

> eating can be seen as a form of mediation in that it is a transaction between multiple bodies–including human eating bodies, the matter(s) of food and its meaning(s)–and the more self-evident Bourdieusian cultural intermediaries that comprise a food's network. Knowledge, then, both mediates and is mediated by food's matter and the bodies that consume it. In other words, eating bodies, foods and knowledges are all enmeshed in a continuous process of making and remaking, forming and reforming. (p. 25)

The encounter between human and non-human bodies occurs on many levels via ingestion and forms of mediation that play a role in setting the boundaries of a food assemblage. If we consider cell-based meat as a new food assemblage and public food tastings as a form of sociocultural mediation, we can understand the underlying mechanisms by which information and knowledge about this product might be propagated. Both the gallery exhibition and the press conference provide different narratives about the origin story of the food, the process of creating it, who produced it and why, its place within the 'food's network,' the discourse used to name and describe it, including the feelings and sensations it might elicit, providing some clue about what kind of legitimacy, critical and political authority it might acquire.

This starting point will help us show how the lines and boundaries of lab-grown meat have changed and moved over time, creating new configurations of knowledge, bodies, and power, including opening up spaces of resistance. Such an analysis situates lab-grown meat as a contested field, within a series of debates about food futures as well as the powers of technology. Moreover, these two seminal events signal the emergence of cellular meat and as such remain key for understanding the evolution of the field and its instauration as 'food' or 'meat', reaching far beyond the demonstration of their hypothetical edibility.

The performance of eating *Victimless Meat* in an art gallery mirrors the first public tasting of in vitro meat broadcast worldwide. However, there are some marked differences between the two very public experiences.

Food tastings in a museum setting hold very specific purposes, as performance art, when they are performed 'live.' They critically engage visitors and participants to interpret the meaning of taste. For Tasting brings the visceral and material elements, it highlights the deep interconnections between the material and socio-symbolic dimensions of food and the ways in which eaters and the witnessing public engage with the foods being eaten and staged. As we mentioned earlier, the final feast became part of the *Disembodied Cuisine* exhibition, marking the end of the artistic performance. The table setting, the choice of utensils, plates, glasses, etc. were all part of the show. Historians and sociologists have long studied the assemblages of tablewares in different cultural contexts and have linked them to material culture but also to a process linked with dining rituals closely marked by social class, consumption patterns, and power dynamics. In this instance, the dinner was staged as a traditional meal, as a kind of celebratory occasion marking the end of the show, around a long rectangular dinner table. The food was prepared by a French chef, served directly, and eaten by the invited guests.

Eating the Other, in this case frog muscles grown in vitro, means turning a lab experiment into food. This becomes meaningful as an intimate act performed in public as it challenges eaters to make sense of what they are eating beyond the materiality of the 'meat' (Colebrooke & Miele, 2017; Whittall, 2017). According to Colebrooke and Miele, the experience of eating art means that ideas of what food is become unsettled and tasting unfamiliar foods becomes a "dynamic and emergent" performative relation, a practice that holds the potential to make "matter" into food (2017). The *Final Feast* staged to mark the closing of the bioart installation was described by the artists in the following way:

> At the very last day of the exhibition, we invited the curator, Jens Hauser, fellow artist Joe Davies, a philosopher (who will remain nameless) and two volunteers from the public to join us in eating the cultured frog meat. We set up a long table in the clear room, attached to the lab via the constructed corridor; some of the leaves from the plant tissue culture were cut to be used as garnish; and then we got ready to prepare the meat. We removed the small pieces of meat from the bioreactor vessels… As we were concerned about potential bacterial contamination, we marinated the cultured frog meat in Calvados and asked the French chef who was cooking to fry it in garlic and honey, which are considered to have strong antibacterial properties… The meat was cut into tiny pieces, fried in a small pan and served. It was chewy and somewhat slippery (the chewiness came from the polymer scaffold and the tissue itself had a jelly-like texture). The sauce was good. To our delight, three of our guests could not swallow the meat. We picked up the bits that were spat out and kept them in specimen jars. … At the end of the dinner, we took the frogs from the fish tanks and invited the audience to join us for a walk to the local botanical gardens, where we released the frogs into the beautiful ponds. 'We don't need you anymore', we said, 'go, enjoy your newfound freedom'. It was very likely they would have been eaten by a predator that very night. (Zurr & Catts, 2023, pp. 81–82)

While the *Final Feast* is performative and highly critical of the 'meat' produced in the lab, here *Disembodied Cuisine* expresses the engagement people might have with unfamiliar foods. The artists were particularly interested in the fact that three dinner guests spat out the lab-grown tissues. In fact, they felt delight at salvaging the uneaten pieces of frog tissue. In 2004, the rejected pieces of tissue became part of an exhibition entitled *The Remains of Disembodied Cuisine* along with video documentation of the *Disembodied Cuisine* project.

The tasting begins with the removal of the tissue from the bioreactor, it proceeds with the cooking of the meat, and it is followed by the release of the frogs that represented the bodies from which cells originate and were used to produce in the exhibition. Although these frogs were not the donor animals, their mere presence provokes eaters and the public into thinking about what they are actually eating.

The presence of these animals as part of the exhibition works by reminding us of where food comes from and making a direct link to a long-standing history of laboratory science, where Xenopus frogs are typically used as animal models for studying embryology, reproduction, cell biology, and genetics among other things. They are no longer invisible, nor are all the processes leading up to the culturing of frog cells in a bioreactor. Here, we can visualize and contemplate the many aspects of technoscientific and biotechnological labor as well as the practices of isolating living cells in vitro and culturing and transforming them into viable tissues. More than a mere product, that is, meat, the fact that visitors can view the instruments and the many technoscientific manipulations leading up to the final tasting creates a kind of 'lab to table' experience.

By saying "we don't need you anymore" before releasing the frogs and sending them to a certain death, the artists are not only discharging themselves of the burden of having to rehome or dispose of these laboratory animals with no native habitat, but this gesture raises the debate surrounding the need for the original copy. Once the source material for the cells used to produce the tissues has been isolated and immortalized, what does this signify for the animal? Of course, this harks back to the issue of obtaining immortalized cell lines from farm animals as an essential piece for the industrialization of cell-based meat which remains a challenge today. But, in the early days of lab-grown meat and at the time of the exhibition, this was purely hypothetical.

Part of the challenge of the *Disembodied Cuisine* performance was keeping the tissues 'alive' during the length of the exhibition. The artists were also concerned with the safety of consuming the bioengineered tissue, using Calvados as a way to sanitize the morsels in order to kill potential pathogens before proceeding to serve it to their guests.

In the case of the proof-of-concept display and highly mediatized laboratory meat tasting, the purpose of which is to signal the possibility of a potential market for this food, the relationship to taste is very different. This was organized as a global event. The setup for the press conference was a studio, akin to the filming of a television cooking show. The 'burger' was presented before it was cooked as a finished product, it didn't have to be kept alive, it simply had to look like ground beef. The event was livestreamed and the live studio audience was mostly made up of journalists. A description of the event is highlighted in Chase Purdy's book *Billion Dollar Burger: Inside Big Tech's Race to the Future of Food* (2020):

> "Everyone sitting here with bated breath is dying to see what's underneath," says Nina Hossain, a British television personality. It's August 2013, she's on a brightly lit stage with four others. She turns to Dr. Mark Post, on her left, who has flown to London from Central Europe. He's staring down at the big desk before him, at a silver platter covered by a silver dome. "So can you do the honours and lift the lid on your creation?" (prologue).

The premise of the 2013 test-tube hamburger tasting was to demonstrate the promissory economic value and commercial viability of this product, taking attention away from the technological manipulations of biological processes and instead turning to the familiar idioms and visual cues of culinary preparation and cooking shows. Only in the promotional material accompanying the demonstration are the ecological and ethical arguments for this product justified. As O'Riordan et al. suggest, the staging is at once public demonstration and experiment, but it also serves to produce what they call a 'promotional public'—those watching the live event and promotional materials—attesting to the reality of the product and who might eventually advocate for the burger (2017). The staging of the experimental product is not a simple promotional event, but it is seen as creating publicness for in vitro meat, hence signaling its function as a response to "perceived needs or issues that might be" and additionally as a medium to attract investors for the nascent industry (Ibid). The authors describe the media event as follows:

> The burger launch occurred on the morning of 5 August 2013. It consisted of three distinct elements: First, screening of the opening film created by media production company The Department of Expansion that featured Brin [google co-founder and funder of the event], Richard Wrangham (evolutionary biologist), Ken Cook (environmental science/activism) and Post (tissue engineering science); second, cooking, tasting and discussion of the burger; and third, social media and coverage up to, during and immediately after the eating. These three forms together combine promotional materials with demonstration of an experiment and controlled witnessing. (p. 153)

The desire to manage the uncontrollable nature of a live event like a scientific experiment is interesting. The authors identify several tensions present in the staging of in vitro meat, namely, the fact that there is a strange juxtaposition between the highly technological world of synthetic meat and the mundanity of cooking and eating a meal. In particular, they question "the stabilization of the object as meat" as a commodity that might be materialized in the future. However, managing how the food tasters would react, how the "bite and eating" would be received, was crucial to the event. The anticipatory moment of whether this food was palatable is something we will discuss in the following section dealing with edibility.

Annemarie Mol's article about eating subjectivities raises the question of the physical aspects of eating (an apple) and the uncontrollable aspects of how one reacts to specific foods (Mol, 2008). Discussing how one may choose to eat an apple, she states that eaters cannot "train the internal linings of their bowels," and therefore metabolic processes elude control when it comes to how the body responds to certain foods. Hence as Abbots explains it: "agency lies with the matter of food and its ungovernable interactions with the substances of the body" (Abbots, 2017).

Furthermore, food choices extend beyond taste and often "become expressions of value and demands for solutions" (Galusky, 2014, p. 932).

Thus, managing people's reactions to foods is impossible but this lack of control over how the food critics will receive cell-based meat can be negotiated in other ways, via public discourse and the use of particular words and phrases, as we shall see. Furthermore, a positive reaction to cell-based tissue in this tasting does not necessarily lead to a wholesale public endorsement of cell meat. It can be argued, however, that the choice of tasters might have been premised on the fact that they "appeared to be supportive of the technology" and that neither of them nor the presenter viewed the hamburger or eating it as inherently "creepy" (Barclay, 2013; Specter, 2011).

Nevertheless, the unpredictability of the live tasting might be seen in the way clinical trials are conducted, whereby strict records of the experience of ingesting new drugs are kept and where recording the results becomes part of the "proof of concept" for these new products (O'Riordan et al., 2017). In line with earlier discussions about taste and tasting as a dynamic embodied process, O'Riordan et al. suggest that:

> The bite, taste and swallow invite the audience to empathise with the cultured burger as a meaningful food product via the shared corporeal and ritual practice of eating, in this case in witnessing others eat – as prevalent in television cookery programmes – that renders the tasters' embodied experience a key index of reality. (p. 156)

According to Stahl, a paradox emerges in the way fake meats mobilize textuality, texture, and taste at the same time (Stahl, 2017). What he is referring to is the importance of the performative dimensions of the product as part of the 'pitch.' Discussing fake meats more generally, including plant-based burgers, he writes:

> Whether it is the ability to be seared, to melt, or shred, or whether it is the right "bite" and "chew," signifying the right structural mimetics, the success or failure of mock meats also depends in part on rhetorically and visually ascribing to them the ability to mimic meats in all its forms…a

concerted effort to mobilize a multi-mediated realism that extends from package to plate to palate. (p. 16)

The materiality of the final product is thus essential to capture the imagination of future consumers, making clear the relationship between textuality and texturality, that is, both performative and discursive powers come together in the tasting, culminating in the tasters' impressions of the lab-grown hamburger as it relates to its animal flesh Other (Ibid).

So, how did the first lab-grown hamburger taste? How did the tasters describe their tasting experience? One of the tasters at the 2013 press conference, food writer Josh Schonwald, offers a description in a National Public Radio (NPR) interview that took place shortly after the event. He says:

It tastes somewhere between not bad, OK and decent. I mean, it's got a – it's a very - it has a very neutral flavor. I mean, to break that down, I mean, for me, you know, in the waking - in the moments before trying it, of course, I was worried that it was going to be bad. You know, I was going to - worried that it was going to have a - you know, provoke a gagging reflex, you know, since I rarely have a chance to eat in front of, you know, cameras and the whole world. So I was most worried about that.

But what it has is the – there's a, you know - there was a crunchy kind of exterior which I guess you might describe as kind of being like when you're biting into a biscuit. The thing that was most familiar and I thought best about it was that the texture kind of had a familiar feel to real meat. It was very dry, you know? It has – it's like the leanest meat, so it's not very flavorful.

So, you know, kind of the way that I've been describing it to people is it's kind of like the protein equivalent of pasta or rice. You know, it's not something that you would eat by itself; like it needs ketchup, or it needs salt or cracked, ground pepper or something. But, you know, in the end, I was, you know, I was pleased because it feels like a start. You know, they insisted on not allowing us to use any condiments, and.... (Schonwald, 2013)

In the article featuring the event, entitled "First lab-grown hamburger gets full marks for 'mouth feel'," the Guardian newspaper recounts Schonwald's saying that the bite was similar to a conventional burger

(Jha, 2013). He compared the texture as that of an "animal protein cake." Hanni Rützler, the first person to taste the tissue, claims that if her eyes had been closed, she would have recognized with a high degree of certainty the tissue as meat rather than a plant-based analogue. She expected the texture of the meat to be "more soft" and she noted the lack of juiciness at the center of the burger due to the absence of fat (Ibid). Rützler stated that the burger was edible but, as one journalist explains it, she also agreed that it was not delectable (Barclay, 2013). In terms of its appearance, Jha writes that the burger looked ordinary from a distance; however, he describes the fact that the muscle fibers grown in the lab were colored with beet juice and saffron in order to provide authenticity. Schonwald commented on the color of the burger, stating that it was very close to looking like a real hamburger. So, the results of the tasting were neither overwhelmingly positive nor negative, both tasters seemed to endorse the product and its promise.

The term mouthfeel and details about the texture and the bite of the burger featured prominently in the press coverage following the public tasting and this became the focus of media coverage after the event. In other interviews, Schonwald expressed the lack of flavor of the meat, since no ketchup or salt was added but much of the press coverage relayed his comments about texture and mouthfeel. Authors O'Riordan et al. (2017) notice that the word 'bite' is repeated five times by Schonwald and Rützler, in the first two minutes of the tasting, the term "mouth feel" was also repeated several times by Schonwald and the presenter, suggesting that these words were used to articulate the experience. Paradoxically, O'Riordan et al. highlight the fact that:

> The emphasis on mouth feel, texture and bite invokes a fleshy feel to meat eating that connects it to other bodies, extending the intercorporeal encounter not just to other humans but also to other fleshy bodies, including the edible bodies of the current meat industry, offering a tantalising intercorporeality with cows. (p. 159)

This encounter, as they suggest, was not what the promoters of lab-grown meat wanted. Instead, the idea was to distance the burger and the eater from the corporeality of the cow. However, this topic was not taken up during or after the event, instead it was played out in social media, prompting the authors to explain that the tense exchanges that took place on the platform Reddit, between audiences that witnessed the live audience and a representative of New Harvest, were consistent with how the plausibility of the cell-based meat industry might be negotiated in the future. This suggests, on the one hand, that the exclusion of the animal from the process of producing the burger, achieved by invoking the adverse impacts of the meat industry in terms of disease, cruelty, slaughter, etc., is fundamental to the objective of cell-based meat. However, on the other hand, the whole enterprise involves replicating the taste and appearance of conventional meat, with the animal as its reference point. This tendency of cell meat to focus attention on the negative aspects of the animal source leads to what Stahl calls a dualistic stance that is both hegemonic and anti-hegemonic concerning the relationship between meat consumption and animals (Stahl, 2017).

Returning to the sensory aspects of the burger, it seems that the lack of fat and connective tissue, which give meat its complex structure and taste, was problematic for both tasters. It became apparent that cell culture techniques combining these other tissues as well as food technology methods to add taste and texture would have to be perfected to achieve a product that didn't lack the essential components found in conventional meat.

The terms mouthfeel, bite, and texture are used in food sciences to describe the sensory attributes associated with foods. Things like the tactile sensations experienced in the mouth, like texture, consistency, juiciness, tenderness, viscosity, etc. Food scientists use sensory evaluation techniques, obtained from descriptive analysis and consumer testing, to quantify and analyze mouthfeel characteristics in a systematic way. However, it should be noted that the idea of taste goes beyond this understanding and the notion of mouthfeel is used by the industry to convey "a complex appeal to texture" thus reaffirming a "carnivorous hegemony" (Stahl, 2017). While discussing mock meats in his article, Stahl claims that the meaning of mouthfeel goes beyond texture, instead he sees it as sociosemiotic and biosemiotic, because it is associated with the

> denotation and the kinds of connotations associated with various foods and what they evoke, and bio-semiotic, in terms of the visceral experience of texture as an embodied experience of both food as a bolus in the mouth, as well as where that food came from ("terroir," etc.). (Stahl, 2017, p. 18)

The adequacy of fake meats' mouthfeel can therefore be understood in light of prior experiences and expectations of eating meat. Since cellular meat has not been consumed by humans in the past, we may question the reason for emulating the taste and texture of real meat and what the quest for "appropriate" or "authentic" mouthfeel in this type of product might entail (Stahl, 2017). Eating lab-grown meat then might be related to what Stahl describes as the "taste-image," that is, the coming together of the taste *of* something with the taste *for* something. The taste-image in the case of cell meat might describe the biopolitics of eating as described by Sexton (2018), where the producers of this meat seek to "materialize normative notions of what qualifies as meat…by simulating the molecular, physical, visceral and discursive aspects of conventional meat products" (p. 2), and in doing so, they urge consumers to accept CM by responsibilizing them to improve their own health as well as society's. One of the journalists invited to taste CM chicken said that knowing how the meat was produced made it taste even better than she expected. This brings the notion of biopolitics of eating CM into the picture, as discussed in the section dealing with biosociality and geosociality.

In fact, rather than a press conference to launch the first lab-grown burger, the streamed event was almost conducted like a sensory study held in public. Sensory evaluations consist in providing "valid and reliable information to R&D, production, and marketing in order for management to make sound business decisions about the perceived sensory properties of products" (Meilgaard et al., 2014).

After the tasting, Post claimed to be satisfied with the event and with the fact that this was a positive beginning for lab-grown meat. Chase Purdy (2020) describes the aftermath of the London event in the following way:

> Triumphant at the success of his showcase, Mark Post packed up and flew back home to the Netherlands. He returned to his Maastricht University

laboratory where he got back to work, quietly ushering his expensive little burger into its next phase. (p. 20)

Post succeeded in attracting attention to the fact that growing tissue in a laboratory could be achieved. The next challenge would be to step up research in order to make the techniques of cell culture more efficient, less costly, and less time-consuming. Another major challenge would be to demonstrate the economic viability of this venture and its scalability. Anna Lowenthraup Tsing's notion of scalability describes the capacity of capitalist enterprises and systems to expand or contract production and distribution processes in response to market demands (Tsing, 2017). Her work on the Matsutake mushroom demonstrates how scalability shapes the commodification and exploitation of natural resources within global capitalist networks. Furthermore, she describes the ability to scale the production and consumption of goods as being central to capitalist economies because this enables rapid growth, market penetration, and adaptation to changing economic conditions. Some activities, such as the inability to domesticate Shitake mushrooms on an industrial level, defy scalability and the economic viability of this mushroom becomes a function of the communities that collect them and their entanglements with the forests where they grow. In addition, Tsing weaves together the historical events and migration stories within the context of disturbed ecologies where Shitakes thrive, describing how various displaced communities and people survive on the edges of capitalism. In the case of CM, however, it is assumed that industrial scalability is the absolute requirement toward building the meat of the future in order to feed the planet as well as provide nourishment during space colonization. It may be that the scalability of CM may be a difficult, if not impossible, problem to resolve and might resist utopian ideals. We shall discuss this later in the chapter about promises, expectations, and imaginaries.

The livestreamed event created the perfect setting for highlighting the need for research funding and investments as well as for showcasing the promise of the lab-grown meat. Generally, proof of concept serves this purpose. As a validation for the feasibility and potential for new technologies, it plays an important role in guiding research and development efforts, especially for securing funding and attracting investors as well as garnering the interest of industry stakeholders and the scientific community. One could argue that the London event served as an early test to gauge the scalability of cultured meat, by showcasing the capability to produce a viable and functional product through new and innovating approaches. The full-scale implementation or commercialization of the product was presented as a future opportunity deserving funding. Nevertheless, the primary purpose of this product launch was to establish the trajectory for its industrial and capitalist expansion.

Recently, more public tastings of cell meat have been in the news. An event hosted by the Species Unite, an organization described as "a solutions-driven media company for animals" (www.speciesunite.com) featured Elizabeth Novogratz, its founder and co-host Chris Anderson, the current curator of TED conferences, who had an opportunity to taste cultivated chicken produced by UPSIDE Foods. Uma Valeti, the CEO and founder of the start-up, was there to present his product having recently received FDA approval for sale in the U.S. Novogratz, who had lived on a plant-based diet for over 40 years until that moment, seemed a bit overwhelmed by the idea of putting chicken in her mouth. However, she explained that:

> The weirdest thing about it was that it wasn't that weird. I expected to be a little horrified about the actuality of eating meat, but since I knew that there was no animal involved, it was one of the most exciting eating experiences of my lifetime [...] Tasting that chicken was tasting the end of animal agriculture and I don't think anything on Earth tastes better than that. (Jones, 2024)

For her, eating meat was not a question of taste or similarity to conventional chicken, it was more of a thought experiment. The sensations did not come from a corporeal or sensory experience but rather from an ideal of ending animal agriculture. In fact, she could *taste* it. On the other hand, Anderson, commenting on the same event, tweeted the following:

> Last night I was lucky to be among the first group in New York to get to taste chicken grown from a droplet of muscle cells. It was... DELICIOUS. Succulent, moist, the right texture. Chicken as it should be.

> Even though full commercialization is a way off, it made me excited for the future. (Galpin, 2024)

Unlike the previous taster, Anderson considers himself privileged to be served this meat and while he contemplates the way in which it was produced "from a droplet of muscle cells," he does express that he found the meat "delicious" and "succulent" possessing "the right texture," recognizable as actual chicken. Despite his description of the taste, his appraisal of cultivated meat is highly aspirational as he contemplates the future of the product. These statements about the exciting future of the cellular meat industry are not surprising since this event happened to coincide with the publication of a New York Times opinion piece that challenges the promises made by industry, as we previously discussed. In particular, the article criticizes UPSIDE Foods as being overly optimistic about its prospects, which explains the types of response given by the two tasters, Novogratz and Anderson, as they represent the kind of spokespeople who support the industry's efforts and stand behind company CEO Valeti's proclamations about the future of cell meat. Clearly, Novogratz's message appeals to consumers concerned about animal ethics and the environment, while Anderson's organization TED, the global communications company which stands for Technology, Entertainment and Design, views this technology with great faith. In fact, he believes that meat-eating is part of the future but that only through this technology can it happen (Jones, 2024). Both tasters aimed their message at critics of the cell meat industry, beyond the quality and taste experience of the product.

Joe Fassler, the author of the scathing piece in the NYT, also got to taste Eat Just's cultivated chicken, like the one served in Singapore. His reaction was as follows:

> During my recent visit to Eat Just's offices, I finally got my chance to try the product. In the test kitchen, chefs brought out a single large hunk of cultivated chicken on a gleaming white plate. They cut it into six slices, plating four for me with mushrooms and broccolini on top of a lovely, lavender-colored purée, and kept two for themselves and my host, Eat Just's communications director, Carrie Kabat. Getting to taste it was still a special occasion for them, it seemed. "This cost $10,000," one of the chefs

> joked. The chicken was flavorful, with a mild savory umami taste, but the experience was closer to eating tofu or seitan than chicken, which makes sense since it's filled with plant protein. It would never satisfy a hard-core meat eater, which, after all, was the point. So what was all this effort for? (Fassler, 2024, p. 10)

Fassler's reaction clearly doesn't carry the same level of enthusiasm as the previous tasters. Despite the close to two decades or so since the first official tasting, it seems cultivated meat hasn't quite reached the level of perfection it originally aspired to.

The next section discusses how the two events, involving the first instances of lab-grown meat, produced two important artifacts, one became part of an artistic performance piece that signaled the continuation of *Disembodied Cuisine* by Catts and Zurr, while the other displayed a prototype of the first lab-grown burger in a History of Science and Medicine Museum.

Relics of the Future

Looking to the future of cellular meat, it strikes me that there is another interesting parallel between the proof-of-concept tasting and the bioart exhibition—which as it happens is also a kind of proof of concept but with very different intent—having to do with the situatedness of the displays of 'first lab-grown meat' specimens, either in science museums or as part of an art exhibit. Firstly, it should be said that science museums and art galleries serve distinct yet complementary roles in society, each with its own unique mandate, objectives, and methods of engagement. While art galleries promote cultural preservation, aesthetic appreciation, and artistic expression, science museums tend to focus on promoting scientific literacy and experiential learning to introduce the public to scientific concepts and phenomena. Both educate and entertain. Art exhibitions create spaces for reflecting and interpreting artworks; they are thought provoking and challenging explorations of science and its limits. Science museums, and technoscientific exhibits, in particular, seek to show the exciting potential of technology and scientific discovery. They tend to be informative and less conceptual. However, it is unclear in the case of technoscientific displays where the differences might lie. One might consider the type of storytelling told in the history of science and medicine museum as principally depicting modern science within the context of Imperial science, that is, an understanding of the politics of empire characterized by colonial forms of property where the non-human and non-Western other is appropriated and excluded from view in the presentation of technoscientific achievements and artifacts. The entanglements between human and non-human bodies as well as the transnational character of modern science get subverted in order to neutralize and depoliticize the various types of knowledges that get produced as others remain hidden or subjugated. The power of technoscience also works by naturalizing the superiority of Western forms of knowledge, in particular science and technology, but also the philosophical ideas that underpin them. The concept of Imperial science challenges the notion of science as a neutral, universal enterprise, highlighting instead its historical and ongoing complicity in structures of domination, exploitation, and cultural assimilation. Uncritical claims about the achievements of technoscience fail to acknowledge the constructed nature of progress and the unquestioned role played by science and technology in industrial development and economic growth. It also misses discussions about the violence involved in extracting knowledge, material objects, living organisms, or dead specimens from the Global South. The history of science is full of accounts about the sourcing of plants and vegetal matter for botanical gardens as well as the capture of wild animals for zoos during various periods.

In the *Disembodied Cuisine* exhibition, the aim was to keep the materials 'alive' until the end of the event, cooking the tissues as close to the time of the final feast. A subsequent art performance, inspired from the uneaten and chewed up fragments of frog tissue from the final feast, was entitled *Remains of Disembodied Cuisine,* as previously mentioned. In this piece, the dead meat tissues were preserved like relics, thereby extending the series of works by the same artists centered on the concept of *Semi-Living Beings* and *Victimless Meat.*

Derek Mead, a journalist for the publication *Vice*, writes that during the launch of the lab-grown burger in London, neither of the two tasters were eager to eat the rest of the burger after one bite. He hints at the fact that leftovers might have been included as part of the prior-made agreement. In this case the fate of the leftovers is unclear; however, there was another in vitro burger cooked beforehand that had been used for promotional images, which truly was the first burger of its kind (O'Riordan et al., 2017). Donated by Mark Post:

> This uneaten burger became an exhibit in the Boerhaave Museum (Dutch Museum for the History of Science and Medicine), emphasising its public status as proof of concept and summoning a different kind of public, one that relates to education and national heritage. Thus, important aspects of this event are what was not eaten, who did not eat and the way the live eating was staged in a carefully controlled manner. (p. 158)

It became part of the museum's permanent collection. While working on his book, *Meat Planet: Artificial Flesh and the Future of Food*, food anthropologist Benjamin Wurgaft encountered the burger in a 2015 exhibition about the future of food at the Dutch museum. It was preserved on a dinner plate protected under plexiglass. Wurgaft comments on the implications of the display stating that "the burger will be so important that it already belongs in the historical record" (Wurgaft, 2019, p. 45). This was only two years after its official launch as proof of concept and at the time of writing this book, as of 2024, the product has not yet been released commercially. So, as Wurgaft alludes to in an interview shortly after the publication of his book, this is rather strange from a historical perspective. He explains that the cultural significance of the exhibit resides in the narrative it conveys about science and innovation, which might hold greater significance than the product itself (Sklar, 2019). Reflecting on the preserved burger, he says: "It looked like somebody taking the most recent supercomputer and putting it in a museum exhibit saying, 'This is the supercomputer that changed everything,' as if you were already 100 years in the future, looking back" (Ibid).

As it happens in both cases, an uneaten burger and parts of another engineered frog steak, ended up in a museum and a gallery display. Bits of semi-living frog tissues that were spat out by some of the invited dinner guests were preserved in a small petri dish and used for the installation entitled *The Remains of Disembodied Cuisine.* The installation consisted in a video triptych that documented the process of the original exhibition. Three screens were placed along a table and, opposite to them, three table settings with plates containing the half-chewed pieces in a small petri dish. The intent was to show the relics of the original exhibition and highlight how people consume and reject new technologies. The use of the term relics is used deliberately by artists Zurr and Catts (2023) to describe this artwork. In using this term, they portray the sacredness of death. The relics describe the death of a semi-living tissue as uneaten leftovers—the remains—but can also be interpreted as what stays behind once the ephemera of living art and what they call 'messy vitality' vanishes. The uneaten pieces of frog tissue are essentially fossilized, investing them with spiritual significance and power as products of technoscientific intervention, fabrication, and design of life. In her introductory essay to a special journal issue about relics and remains, historian Alexandra Walsham proposes a description of relics as a material object linked to a particular person or an event associated with them. In Christianity, relics are more than mementos; they are thought to have healing power. Outside of their religious significance, relics act as political "instruments of legitimation and as vectors and embodiments of authority" (p. 24). The relic is different than other material objects because it is capable of operating "as a locus and conduit of power" (Walsham, 2010, p. 13), and this power is inherent in the object itself but also where and how it is displayed and curated imbues it with further significance. Outside of the cultural matrix that creates them, cultivated burger and remains of unchewed frog tissue may be considered completely meaningless and worthless. Their symbolic value rests with the importance we place on such items and the context that created them. If anything, whether intentional or not, what the idea of relics points to is a kind of blind faith in science and technofixes. So, what remains are the fossilized bits of once semi-living tissues as a material semiotics. These become material objects imbued with a specific physical presence. However, they also carry symbolic meanings and implications related to food production, ethics, sustainability, and technological advancement. The fact that these preserved pieces of tissue were rejected (spat out) implies a form of resistance to the technoscientific production of food. Their construction of a particular food future that gets framed within the realm of the power of science and technology and their manipulations and interferences with living organisms and processes disconnected from nature. In this case, the physical and metabolic rejection of the 'meat' acts as a critical discursive feature of the exhibit and speaks to the fact that some technologies spontaneously fail to be socially, ethically, and culturally integrated.

Similarly, the burger created by Post and his laboratory ahead of the official tasting in London, now preserved through plastination and included in the permanent collection of the Dutch Museum of History of Science and Medicine, also embodies material-semiotic significance. It simultaneously carries the physical characteristics (as once living tissue that underwent death, cooking, and plastination) and the social, cultural, and ethical meanings associated with in vitro meat production. Its physical presence and symbolic significance in the History of Science and Medicine Museum are deeply connected with intricate networks of relationships involving scientists, consumers, regulatory bodies, and broader societal discourses about food, technology, and environmental sustainability.

Furthermore, the lab-grown prototype represents a web of practices that holds together elements as heterogeneous as people; animals; tools or artifacts; lab equipment; practices; skills; technologies; electricity, etc. (Law, 2008). Hence, the material-semiotic approach enables us to think with and about the effects of these objects beyond their purported use. It serves to remind us about the material realities of our food as well as its relation to bodies and entangled natures in complex ways and not simply as discursive products.

In other words, it's important to attend to the meaning of these objects such as the rejected frog meat or the plastinated burger prototype, because they participate in shaping practices, subjectivities, and social experiences. The nature of relics also lies in their charismatic power, producing a kind of political discourse.

In both situations, the objects on display are essentially time capsules; they embody what Barbara Adam calls the present-future and the future-present. In her article entitled *The History of the Future*, Adam who is a sociologist of time discusses how modernity made way for a vision of the future that replaced ancient sacred myths and supernatural beliefs about the ordering of time and what she calls 'pre-existing futures' also known as fate (Adam, 2010). Instead, moderns now regard the future as an empty space to own, shape, and exploit. Specifically, this shift, toward a naturalized empty future, enabled the future to become commodified, and thus to be incorporated into the present, meaning that we are now experiencing the legacy of past generations' imagined future. Their empty future is our present, with all their dreams, aspirations, "innovations and impositions, their creations" (p. 368). Furthermore, she claims we are now living in a technologically produced, long-term future which makes humans responsible for unknown and unknowable futures. She talks about future-makers versus future-takers as processes of appropriation of time within capitalist spaces.

Taking from the science and technology studies literature (STS), authors such as Collins, Shapin, and Schaffer have noted that the material form, the embodied form of scientific knowledge production, and its symbolic meaning are key to understanding the scientific practice of experimentation. Citing these authors, Anne Lorimer points out, in her study of the Chicago Museum of Art and Industry's 1989 exhibit about the human brain and its representation, that:

> Indeed taking material form is central to the scientific practice of experimentation, and even here, at the heart of science, the production process does not flow smoothly and unilinearly from conceptual intention to material execution to scientific outcome; rather, conceptual intentions become solidified and revised retrospectively when indeterminacies in how materially to set up the experiment are resolved. (Lorimer, 2022, p. 88)

This means that the final materialized form of laboratory-grown meat is bound to change as scientific experimentation evolves. But this also means that the experimental setup, rendered visible in the *Disembodied Cuisine* exhibition, is missing from the History of Science and Medicine display of the first in vitro burger. In the science exhibit, the experimental nature of synthetically produced meat remains hidden, the final materialized product is presented as a fait accompli, obfuscating the uncertainty and the complexity of the "material execution" of this object. It offers a 'sanitized' and purified image of what is actually at stake. This lends the in vitro burger the status of representing specific visions of food futures, leaving out possible scenarios of what these might look like.

Lorimer argues that in order to understand how scientific representations acquire special resonance in the public sphere, we must pay attention to mediating processes, "which involve the labor not only of scientists but also of, for example, medical illustrators and photographers, taxidermists and big game hunters" (Lorimer, 2022). I would add bioartists, media technicians, and museum curators to this non-exhaustive list. These mediating processes are especially relevant in human anatomy exhibits, where displays of the brain, for instance, are interpreted by the public as "the seat of our own subjectivity and the material object of scientific investigation," thus creating what she calls a symbolic distance between subjects and their mediated counterparts. In the case of the prototype of the first lab-grown burger, such distance occurs on a visceral level, between our subjectivity of what this object might taste like and its legibility as a familiar food, between our bodily understanding of this scientific artifact as food and the rational understanding of what the lab-grown burger foreshadows. There is a symbolic distance between the material object of a technoscientific experiment and its status as animal meat, beef, in this case.

This symbolic distance also becomes apparent when we think of the *Disembodied Cuisine* exhibit. This is where the term disembodiment acquires special resonance as it creates the symbolic distance Lorimer was describing. The frogs act as witnesses of the making of edible tissues from cells, as tissues are cultured in the bioreactor, their presence indicates that without the original animal, there would be no cell meat. However, it is also true that our gut reaction is no longer conditioned by the cow or chicken we normally eat, instead the lab becomes the site where the obsolescence of animal life becomes possible, and the bioreactor is where this potentiality occurs via a simulated fictitious in vitro body-reproducing-apparatus.

Both events exhibited many similarities that culminated in showing the possibility of producing edible tissues from animal cells. Whether these were palatable or not, both the scientists and bioartists produced these tissues with very different intents, each pointing to contrasting visions of the future embodied within those creations. Juxtaposing these events shows how in some ways, stories about the possibilities of creating in vitro meat can be informed by very different perspectives and intentionalities. These two events are linked in real ways but take place in very different settings. Reminiscent of Oscar Wilde's famous phrase, it seems to be a case of life imitating art.

The common thread running through these two displays is a contrasted view of the powers of technoscience to create 'living' food in vitro. What can be valued as subversive and ultimately dangerously uncontrollable in one situation becomes emancipatory and transformational in another. In both cases, whether transgressive or disruptive, publics were conveyed to witness the 'reality' of lab-grown meat.

On the one hand, Zurr and Catts showed the aesthetics and practical ways that the semi-living come into being, the many intertwined processes, living species, and inert materials that enter into their creation, bringing into view the laboratory as a kind of stage as the site of production of hybridity and alterity. The presence of the model organisms, the Xenopus frogs in an aquarium, reinforced the continuity of life as well as its disruption. It also served to 'personalize' the process—even if the cells used in the process were not those of the frogs on display—inviting the public to reflect upon the notion of disembodiment and on the fact that lab-grown meat is not totally victimless. The release of the frogs—to their certain death—marking the end of the performance also conveys their fate as lab-bound animal models for biomedical science research, but also as natural beings trapped in an unnatural ecosystem that is the laboratory or the museum exhibition. Hence, even the few moments of freedom in this foreign pond gave these animals a glimpse of life outside the confines of scientific research as mere lab instruments. This also shows the unnatural aspects of the situation since these frogs had probably never experienced fresh air, a pond, the sounds and sights of the outside world. In effect, they had become natural aliens,[3] proxies for our alienation to the natural world. They therefore stand as a metaphor for humans, living in an increasingly hostile and ecologically degraded environment.

[3] My original doctoral supervisor, Dr. Neil Everndern, wrote a book entitled *The Natural Alien* (1999), dealing with how humans respond to the environmental crisis.

By way of concluding this section, I would like to return to the notion of relics of the future. It is important to highlight that the careful preservation of the undigested bits of laboratory-grown tissues holds meaning. Not simply as an oddity or a curious piece of technological prowess, but as signs of a hypothetical future. In the case of the bioart exhibit, the bits were rejected by the first humans to taste this human-made edible muscle tissue. They were carefully preserved in order to tell a story of potential food futures through *Disembodied Cuisine*. In contrast, the Dutch Museum exhibit tells a different story, as the first lab-grown burger marking a landmark moment in the history of science and technology. It represents the ideologized form of the process leading to the production of the preserved burger and of the promise it holds for the future. This artifact leaves no room for critical thinking or for potentializing alternate food futures. The projected future it projects is as plastinated as the burger itself. It remains a fetishized object imbued with excessive value or significance beyond its practical use, often obscuring the labor, social relations, or material realities behind its production. The artifact is thus seen as an object of misplaced reverence or desire, celebrated not for what it truly represents or could offer, but for its symbolic status as a technological novelty. This fetishization masks the complex and potentially problematic realities of its existence, such as the ethical questions surrounding synthetic food production, the environmental impact, or the social consequences of further commodifying life itself.

The question of temporality and technology is one that is recurrent in this book; it appears in many forms and in different chapters for different reasons. Relics speak of the past but present glimpses of the future, one utopian the other dystopian, as well as the temporalities associated with expectations, promises, and imaginaries as we shall discuss later.

Bibliography

Abbots, E. J. (2017). *The Agency of Eating. Mediation, Food and the Body*. Bloomsbury Academic.

Adam, B. (2010). History of the Future: Paradoxes and Challenges. *Rethinking History, 14*(3), 361–378. https://doi.org/10.1080/13642529.2010.482790

Barclay, E. (2013, August 5). *Long Awaited Lab-Grown Burger Is Unveiled In London | KQED*. Retrieved from https://www.kqed.org/bayareabites/67631/long-awaited-lab-grown-burger-is-unveiled-in-london

Blum, M., & Ott, T. (2018). Xenopus: An Undervalued Model Organism to Study and Model Human Genetic Disease. *Cells, Tissues, Organs, 205*(5–6), 303–313. https://doi.org/10.1159/000490898

Borchers, A., & Pieler, T. (2010). Programming Pluripotent Precursor Cells Derived from Xenopus Embryos to Generate Specific Tissues and Organs. *Genes, 1*(3), 413–426. https://doi.org/10.3390/genes1030413

Catts, O., Salter, C., & Zurr, I. (2021). Living and Semi-living Artefacts on Display. In I. L. Muller & C. S. Langill (Eds.), *Curating Lively Objects* (1st ed., pp. 115–131). Routledge. https://doi.org/10.4324/9780429053481-10

Colebrooke, L., & Miele, M. (2017). Eating Art and the Art of Eating: Unsettling the Practices of Taste. *Performance Research, 22*(7), 102–108. https://doi.org/10.1080/13528165.2017.1353203

Deleuze, G., & Guattari, F. (1987). *A Thousand Plateaus: Capitalism and Schizophrenia*. University of Minnesota Press.

Evans, A. B., & Miele, M. (2012). Between Food and Flesh: How Animals are Made to Matter (and Not Matter) within Food Consumption Practices. *Environment and Planning D: Society and Space, 30*(2), 298–314. https://doi.org/10.1068/d12810

Fassler, J. (2024, June 29). The Revolution That Died on Its Way to Dinner. *New York Times*.

Galpin, K. (2024, February 15). *Tech Expert Shares Opinion after Trying Chicken-like Meat Grown in a Lab*. The Focus. Retrieved from https://www.thefocus.news/lifestyle/tech-expert-shares-opinion-after-trying-chicken-like-meat-grown-in-a-lab/

Galusky, W. (2014). Technology as Responsibility: Failure, Food Animals, and Lab-grown Meat. *Journal of Agricultural & Environmental Ethics, 27*(6), 931–948. https://doi.org/10.1007/s10806-014-9508-9

Hauser, J. (with Le Lieu unique). (2003). *L'art biotech': Exposition, Nantes, Lieu unique, 14 mars-4 mai 2003*. Filigranes éd. le Lieu unique.

Hennion, A. (2007). Those Things That Hold Us Together: Taste and Sociology. *Cultural Sociology, 1*(1), 97–114. https://doi.org/10.1177/1749975507073923

Jha, A. (2013, August 6). First Lab-grown Hamburger Gets Full Marks for "Mouth Feel." *The Guardian*. Retrieved from https://www.theguardian.com/science/2013/aug/05/world-first-synthetic-hamburger-mouth-feel

Jones, A. (2024, February 16). *Our Founder Ate Chicken for the First Time in 40 Years (And No Animals Died).* Species Unite. Retrieved from https://www.speciesunite.com/news-stories/our-founder-tried-cultivated-meat

Laskawy, T. (2013, August 6). *Don't have a Cow: Lab-grown Meat Inches Closer to your Plate.* Grist. Retrieved from https://grist.org/food/dont-have-a-cow-lab-grown-meat-inches-closer-to-your-plate/

Law, J. (2008). Material Semiotics. Heterogeneities.Net: John Law's STS Web Page. Retrieved from http://heterogeneities.net/publications/Law2019MaterialSemiotics.pdf

Lorimer, A. (2022). Chapter 4: Embodied Brains: Why Science Studies Needs the Anthropology of Museums. In J. Edwards, P. Harvey, & P. Wade (Eds.), *Technologized Images, Technologized Bodies* (pp. 85–116). Berghahn Books. https://doi.org/10.1515/9781845458300-006

Meilgaard, M. C., Carr, B. T., & Carr, B. T. (2014). *Sensory Evaluation Techniques* (4th ed.). CRC Press. https://doi.org/10.1201/b16452

Mol, A. (2008). I Eat an Apple. On Theorizing Subjectivities. *Subjectivity, 22*(1), 28–37. https://doi.org/10.1057/sub.2008.2

New Havest. (2022). *What We Do.* New Harvest. Retrieved from https://new-harvest.org/what-we-do/

O'Riordan, K., Fotopoulo, A., & Stephens, N. (2017). *The First Bite: Imaginaries, Promotional Publics and the Laboratory Grown Burger.* https://doi.org/10.1177/0963662516639001

Perle, D. (2022, November 16). *PETA Statement re Culture Meat's FDA Milestone.* Retrieved from https://www.peta.org/media/news-releases/peta-statement-re-cultured-meats-fda-milestone/

PETA, P. for the E. T. of A. (2008, October 6). *PETA's "In Vitro" Chicken Contest.* PETA. Retrieved from https://www.peta.org/features/vitro-meat-contest/

Pomiès, A., Arsel, Z., & Bean, J. (2021). Taste. *Consumption Markets & Culture, 24*(1), 1–5. https://doi.org/10.1080/10253866.2020.1803846

Purdy, C. (2020). *Billion Dollar Burger: Inside Big Tech's Race for the Future of Food.* Portfolio/Penguin.

Schonwald, J. (2013, August 9). *Biting Into the First In Vitro Burger* [Interview]. Retrieved from https://www.npr.org/2013/08/09/210508782/biting-into-the-first-in-vitro-burger

Sexton, A. E. (2018). Eating for the Post-Anthropocene: Alternative Proteins and the Biopolitics of Edibility. *Transactions of the Institute of British Geographers, 43*(4), 586–600. https://doi.org/10.1111/tran.12253

Sklar, J. (2019, September 9). *With Lab-Grown Chicken Nuggets, Dumplings, and Burgers, Futuristic Foods Aim to Seem Familiar*. Upworthyscience.Com. Retrieved from https://upworthyscience.com/with-lab-grown-chicken-nuggets-dumplings-and-burgers-futuristic-foods-aim-to-seem-familiar/

Specter, M. (2011, May 16). *Test-Tube Burgers*. Retrieved from https://www.newyorker.com/magazine/2011/05/23/test-tube-burgers

Stahl, G. (2017). Making a Mockery of Meat: Troubling Texture and the Failings of the "Flesh". *Journal of Asia-Pacific Pop Culture*. https://doi.org/10.5325/jasiapacipopcult.2.2.0184

TC&A Project, T. T. C., & A. P. (2000). Semi-living Steak. *The Tissue Culture & Art Project*. Retrieved from https://tcaproject.net/portfolio/semi-living-steak/

Travis, J. (2014, March 3). *PETA Abandons $1 Million Prize for Artificial Chicken*. Retrieved from https://www.science.org/content/article/peta-abandons-1-million-prize-artificial-chicken

Tsing, A. L. (2017). *The Mushroom at the End of the World: On the Possibility of Life in Capitalist Ruins* (First paperback printing). Princeton University Press.

Walsham, A. (2010). Introduction: Relics and Remains. *Past & Present, 206*(Suppl_5), 9–36. https://doi.org/10.1093/pastj/gtq026

Whittall, E. (2017). Taste, Performed: M. Clavel and the Live Feed. *Performance Research, 22*(7), 44–47. https://doi.org/10.1080/13528165.2017.1353191

World's First Lab-Grown Burger is Eaten in London. (2013, August 5). *BBC News*. Retrieved from https://www.bbc.com/news/science-environment-23576143

Wurgaft, B. A. (2019). *Meat Planet: Artificial Flesh and the Future of Food*. University of California Press.

Zurr, I. (2008). *Growing Semi-Living Art* [PhD, University of Western Australia]. Retrieved from https://api.research-repository.uwa.edu.au/ws/portalfiles/portal/3225984/Zurr_Ionat_2008.pdfb

Zurr, I., & Catts, O. (2023). *Tissues, Cultures, Art*. Palgrave Macmillan.

Zylinska, J. (2009). *Bioethics in the Age of New Media*. MIT Press.

8

Erasing the Body

Biotechnologies and animal genomics have enabled the genetic modification of animals, as well as the crossing of species boundaries in order to create new animal varieties. Genetically altered animals, clones, transgenics, knockouts, and chimeras are established in scientific research as model systems for human diseases and in industrial agriculture for improved efficiency as food. However, by presenting new conceptions of humans and animals, these technologies also shift our understanding of what animals are and in doing so, "redefine the human-animal relationship within both science and society" (Ung-Lanki, 2014).

This includes how we view the animals we eat, with the ultimate stage of the industrialization of agriculture consisting in CM, where animal bodies are no longer necessary for meat production. The discourse surrounding cultured meat fluctuates between narratives of preserving nature and animals and those that essentially erase or transform animals into post-animals or create meat markets with no-animals.

In her book, *The Sexual Politics of Meat* Carol J. Adams uses the term unanchoring to describe the disembodied form of the animal that is the 'absent referent.' She links this absence of the subject to women's oppression, where she compares the violence inflicted upon animals to the violence against women under patriarchy. Butchering is what transforms

E. Abergel, *Dead Meat*, https://doi.org/10.1007/978-981-97-9049-4_8

animals into absent referents because for meat to exist, the animal must be made absent (Adams, 2015). She writes:

> Behind every meal of meat is an absence: the death of the animal whose place the meat takes. The "absent referent" is that which separates the meat eater from the animal and the animal from the end product. The function of the absent referent is to keep our "meat" separated from any idea that she or he was once an animal, to keep the "moo" or "cluck" or "baa" away from the meat, to keep something from being seen as having been someone. Once the existence of meat is disconnected from the existence of an animal who was killed to become that "meat," meat becomes unanchored by its original referent (the animal), becoming instead a free-floating image, used often to reflect women's status as well as animals'. (Adams, 2015, p. xxiv)

According to Adams, animals are not only physically absent but erased also through language, when specific terms associated with butchering and gastronomy are used. In her words, "one does not eat meat without the death of an animal" (2015, p. 21). Another form of absence is metaphorical according to her, meaning that the names of animals are used to describe a human experience, for instance, Adam refers to the expression "feeling like a piece of meat"—used to describe the experiences of battered women or victims of rape. The bodies of the animals used as a source of cells become the new absent referent because the origin of the meat tissue is obscured through the process of growing it in the laboratory. The technology rather than the animal becomes the focus of attention. Additionally, when CM proponents describe their products as clean or pure, they further fetishize the object of meat, making the animal either totally invisible or visible because it stands in contrast as impure or unclean. Extracting cellular material from animals raised in industrial agriculture involves utilizing bodies that are already alienated from their natural contexts. Cellular agriculture intensifies this alienation, particularly through the establishment of cell banks, for example, the Frozen Farmyard project. By treating these animals as mere blueprints for cellular agriculture, the notion of the absent referent is profoundly deepened, as the animals are further abstracted from their biological and ethical identities.

Along with the notion of post-animals, we are dealing with the possibility of creating postnatural beings, with the term postnatural implying "after nature" but often used to describe a colonized nature and, specifically:

> When we speak of the "postnatural," we refer to anthropogenic interventions into evolution that are both intentional and heritable, regardless of their subsequent unintentional consequences. The postnatural therefore is not an epoch of Earth's geohistory, but a conceptually inclined adjective used to describe the purposeful and permanent modification of living species by humans through domestication, genetic engineering, and synthetic biology. (Pell & Allen, 2015, p. 33)

This is part of a social project that began with the industrial revolution and which signals the purposeful transformation of the living world, culminating in the anthropogenic activities linked to the Anthropocene. As others have suggested, the long-term impact of the gradual elimination of farm animals goes with the underlying and unquestioned assumption that creating meat without animals inevitably opens up the question of animal agriculture without animals, land without farms, and, eventually, a world without farmers.

By creating the technological conditions to engineer animals out of their bodies (we're still a long way from accomplishing this), cell ag advocates are combining a set of emergent, "disruptive" technologies such as genetic engineering, gene editing, tissue and cell engineering, fermentation technologies and bioreactor systems, bioprocessing, alternative protein sources, synthetic biology, strategies for combining plant and animal proteins, 3D bioprinting, AI, robotization, blockchain, etc. Cell-based farming has been identified as belonging to the so-called fourth agricultural revolution, a term describing a very undefined set of technologies, often framed as precision farming, digital agriculture, smart farming, etc. While there is great hype surrounding this fourth agriculture revolution in media and policy discourse, it seems the term fails to resonate with farmers (Barrett & Rose, 2022). I question this as a metaphor of the consequences of unchecked technological progress, within the broader context of environmental concerns and humanity's impact on the planet. It has also been referred to as the second domestication of animals as we will discuss in the section about disrupting the cow (Tubb & Seba, 2020).

While animal replacement products constitute an important part of this new agricultural paradigm, cellular agriculture as a whole provides the industry with a set of converging technologies and scientific infrastructures that play into the promissory economy as well as narratives about the salvatory powers of technology. Some might say, cell-based meat narratives play into post-human scenarios that consider the end of nature and our decreased reliance upon it as a form of human-directed evolutionary enlightenment. By contributing to discourse about cellular agriculture, the developers and supporters of cell-based meat have insinuated themselves into narratives about the post-animal, anticipating a future characterized as "after" animality.

The Transformation of the Laboring Animal into a Protein Platform

The industrialization of agriculture has intensified the extraction of surplus value from animals. Through practices such as selective breeding, feed optimization, and the use of growth hormones, animals are engineered to produce more in less time, thereby increasing the rate of capital accumulation. These factors affect the lifespan of animals as well as their living conditions. In political economy, animals can be considered part of the broader labor force within industrial agriculture, challenging the anthropocentric notion of labor relations and non-human workers within industrialized food systems. However, the question of animal labor is often neglected in the industrialization process because animals are primarily considered as units of production and commodities. Hribal traces the role farm animals played during the industrial revolution under regimes of expropriation and exploitation and places them in the laboring class, as they were "plowing, drilling, mowing and harvesting...ginning, crushing" but also transporting materials and people while also "supplying their flesh, eggs, and milk" (Hribal, 2003). As animal exploitation reached its peak, resistance movements that linked the fate of humans and animals appeared, identifying their struggle from a class or proletarian perspective (Ibid). Taking a more sociological point of view, Jocelyne Porcher thinks that all animals, be they farm animals or companion species, enter into relations of labor with humans so that keeping company represents a form of labor for a domestic animal. In her view this marks a distinction between the livestock industry and animal husbandry, the first treats animals like machines and relies on zootechnics while the latter is described as a working relationship with animals. As such, working with animals for Porcher means also living with them. Historically, she claims that animals were full participants in the social world, with a status that extended beyond merely being objects of labor (Porcher, 2017).

Clearly, the question of animal labor is complex as Porcher illustrates and the distinction between zootechnics and animal husbandry represents fundamentally different ways of conceptualizing the material and sociological aspects of animal agriculture for meat production. It is interesting to note that the conceptualization of the labor process involving humans and animals is central to a Marxian critique of capitalism, because it appropriates non-human nature and transforms raw material into commodity forms. However, using Fish's analysis of animals being transformed into living factories by harnessing molecular processes through what he calls the 'transgenic labor process,' similarly, cellular agriculture means that life now constitutes the raw material, technology, instrument of production, and product of the cellular labor process, defined *by* and *for* human needs following the logic of industrial production (Fish, 2013).

Porcher views the question of animal and human labor as a significant oversight among advocates of cellular agriculture. She argues that the complex, relational dynamics between farmers and animals are often oversimplified, reduced to a "unilateral operation of exploitation and production of animal matter." This reductionist view, she contends, allows the cultured meat industry to assert that CM is equivalent to meat derived from animals. Farm animals are reduced to their muscular mass, a cow is only its meat outside of all other forms of existence. This view maintains them as anonymous entities. What makes them knowable beyond their muscular mass is labor:

> Labor, not just as a productive activity, but as a set of rationalities, one of which—essential for farmers—is relational rationality. Working with animals means living with them, transforming nature with them, and growing with them. Labor is what constitutes the hyphen in the "human-animal" relationship. There is no domestic human-animal relationship without work, nor are there domestic animals without humans. (Porcher, 2020, p. 306) (my translation)

Moreover, she posits that critically examining the role of human-animal labor is vital for transcending reductionist dualistic frameworks that portray animals merely as 'meat' or as passive 'victims.' This perspective reflects a specific conceptualization of farm animals within the broader structures of political economy, where they are embedded in systems of industrialized food production and environmental degradation. In these systems, animals are positioned not only as resources to be exploited but also as problematic entities that disrupt the efficiency and ethical ideals of modern consumption. This results in a commodification of animals that abstracts them from their living, sentient existence, reducing them to units of production.

In this context, the logic of eliminating animals from agricultural production can be seen as a response to the socioeconomic and ecological contradictions inherent in industrial capitalism. By removing animals from the equation, proponents aim to address the ethical, environmental, and health concerns associated with meat production, thus aligning with a consumer desire for 'clean' and 'guilt-free' products. However, this move also reflects deeper political-economic dynamics, where the restructuring of food systems around synthetic or lab-grown alternatives serves to reinforce the commodification and technologization of animal-derived life. The goal of replacing animal bodies in the food chain as 'animal protein' can also be seen, beyond the desire to minimize costs (the efficiency of input vs. output costs) and risks (disease outbreaks, animal welfare, regulations, etc.) as a shift to exert greater control over production processes managed through technology. As Galusky explains:

> It is important to realize, however, that the body of the animal is not simply removed, but rather it is replaced in a more mechanized form. That is, in

> the in vitro context, the animal body with all of its complexities needs to be replaced by a lab-based "body" that does one primary task—support the production of protein. The animal body that does many things (some desirable and others less so) is replaced by a kind of artificial body that has fewer tasks—a function specific alternative. This move represents a kind of ideal—the very goal of previous animal-based agriculture has been to reduce the function of the body to the exclusive production of protein, but the body has been more recalcitrant to such simplification. Hence, the move from animal to lab-based body. Even in this simplified form, however, the goal of a substitute leads to some profound complexities. (Galusky, 2022, p. 58)

The material contradictions embodied within neoliberal technocapitalism are resolved not just in the sphere of surplus distribution, but in bodies (Guthman, 2009), human and non-human. This refers to the way technocapitalism addresses its inherent contradictions not only through the economic mechanisms of surplus distribution but also through the manipulation and exploitation of both human and non-human bodies.

Concretely, this is achieved through the manipulation of biological entities and non-human bodies which impacts human bodies, in terms of how these might be affected by consuming CM. While the bodies of eaters are empowered by cell-based foods for health, environmental, and ethical reasons, the disposability of animal bodies (and the humans that are in interaction with them, the farmers) is also being negotiated.

Meat imaginaries, and the narratives used to construct them, are constantly alternating between saving nature and animals and rendering them obsolete. Post-animals are emerging figures within food futures as no real scenario for the billions of farm animals currently existing for the sole purpose of feeding us is seriously anticipated. Projects like the *Frozen Farmyard* aim to create a cell line repository for "agriculturally relevant animals" (GFI, 2024). This project, led by Gareth Sullivan at the University of Oslo, plans to create the first centralized biobank to store animal cell lines of interest for the cultivation of animal muscle and fat (Rodríguez Fernández, 2019). Sullivan's expertise is in the conservation of endangered species, he isolated stem cells from a rare species of rhinoceros. By collecting biopsies from farm animals from across Europe, the researchers hope to establish cell lines by inducing pluripotent stem cells (iPSCs) in order to immortalize them and use them as a limitless source for the cultured meat industry. Under optimal growth conditions, these cells can multiply indefinitely. The use of adult stem cells requires getting biopsies from live animals because these cells are multipotent, they can transform into a limited number of cell types. *Frozen Farmyard*, as its name suggests, symbolizes the cryopreservation of animal vitality for the production of meat, but it also conjures up the idea that animal farming may be a thing of the past, frozen in time. Of course, cryopreservation has a way of resurrecting the past and bringing it into the present, unsettling the biological as well as geological temporalities. But as the name *Frozen Farmyards* evokes a paradoxical image: it contrasts a farm landscape with living animals with a static vision of animal cells preserved in liquid nitrogen. It may evoke a world where farm life is no longer necessary or relevant.

The Body Problem in Conventional Agriculture

At first glance, the elimination of animal bodies appears to be a positive advance from an ethical standpoint, no need to slaughter billions of living creatures. But this also means that body materiality and physicality need to be rethought in the process of technologizing life, within the various technical aspects of growing edible muscle tissues. In fact, the absence of the body means that corporeal materiality and tissue physicality must be recreated in the laboratory via tissue culture technology, using scaffolds to structure flesh and to enable cells to proliferate and differentiate, as well as mechanical means to maintain the viability and minimal functionality of the tissue, creating the conditions for muscle tissues to contract during cultivation.

Stripped from bodily integrity, lab-grown edible animal tissues are decorporealized and displaced; they exist within a space of non-belonging (Lee, 2005). This makes eating cultivated meat complicated, for the body of the animal, even in its most industrialized form, makes it possible to imagine it as an embodied subject. At some level, the bodily experience of eaters of meat produced from animal stem cells, having extracted some kind of corporeal essence, abstracts our understanding of what we are doing when we are consuming such a product. Removing the bodily aspects of meat reduces it to a kind of intellectual abstraction, a purified substance from nowhere and no-body. An absent referent to use Adam's language.

Scientists involved in CM explain that their goal is to completely separate the animal from meat-making. This is accomplished by focusing solely on one piece of the animal, one type of tissue, making it simpler to reconstruct while mimicking its functionality. Living tissues are complex by nature but by removing the animal, one can focus on the cellular processes or cellular cycles rather than the reproductive cycle of the animal.

However, some start-ups insist on individual animals when presenting their processes. This form of acknowledgment of an animal represents a form of temporary erasure, where an animal is identified, as a way to obscure the deeper issues of reducing the cow to a source of cells and an abstraction of its corporeal nature as a resource. An example of this is Aleph Farms. The company introduces the public to the cow from which cells were harvested to produce their cell-based steaks. This is how they describe their process and where they source their cells:

> Lucy was a premium Black Angus cow from a breeding farm in California. Ranchers around the world would purchase Lucy's fertilized eggs and use them to improve the quality of their livestock. We needed a source of premium cow cells and needed to look no further than Lucy. She passed away in June 2022.
>
> We have no plans to raise cows because all we need is a one-time collection of Lucy's fertilized eggs. From there, we can grow high-quality, tasty, and nutritious Aleph Cuts (without the rest of the cow).
>
> When at scale, we expect to grow thousands of tons of steak from this single collection. (Aleph, 2024)

Here we meet Lucy, whose cells provide the raw material, the original matrix from which the Petit Steak is made. We learn that Lucy lived and died on a farm in California, that she was used for breeding. A beautiful photo of Lucy shows her as a prize animal. This explanation is given on Aleph's website:

> The first Aleph Cut to be introduced to diners in Israel—the cultivated Petit Steak—is made of non-modified, non-immortalized cells of a premium Black Angus cow named Lucy, as well as a plant protein matrix made of soy and wheat. Aside from starter cells that come from one of Lucy's fertilized eggs, there are zero animal-derived components (i.e., no fetal bovine serum) in the cultivation process and the final product. No antibiotics are used in production, nor are any present in the final product. A controlled and traceable process, including an aseptic production environment, increases transparency and greatly reduces any risk of contamination. (Aleph Farm website)

Apart from the photo and the mention of her name, there is no further information given about her, other than she is the source of the materials from which the Petit Steak is made. This reinforces Porcher's point about the animal being reduced to its biological matter.

In 2018, the Eat Just company released a short video entitled *Cultured Meat: A Vision of the Future* which features lab-grown chicken nuggets. In the course of describing the process of making the nuggets, we are introduced to Ian the chicken in the arms of a farmer. Ian is a special chicken, a fine specimen of industrial agriculture, a product of generations of genetic manipulation and breeding. Ian's feathers were used as source materials and his cells were used to make "clean meat" in the form of chicken nuggets. We then see a chef cooking nuggets outdoors and serving the meal to Eat Just employees sitting around a picnic table. The chef says: "I said to myself, I'm going to make the best chicken nuggets ever" and later,

> it was an *out-of-body experience to sit there and eat a chicken but have the chicken that you're eating running around in front of you*. You don't imagine doing something like that but then you have this realization that *we've figured out how life really works and now we don't need to cause death in order to create food* and we're gonna have to do it if we want to continue living on this planet. (Just Egg, 2018) (video transcript, 2018) (Emphasis added)

In this scenario, Ian's life is spared because the process for making nuggets utilizes cells from his feathers. He is able to participate in the process of 'being consumed,' in what the commentator calls from his perspective "an out of body experience."

It is interesting how these two examples of highlighting individual animals challenge the traditional notion of farm animals as a species rather than as distinct subjects. This is particularly true of industrialized farms, where there is a shift in animals as livestock or cell stock. The individual lives of these animals, and the people who isolate their cells or use their cells, intersect in different ways. By focusing on individual farm animals, this perspective provides a glimpse into their individual life experience as products of industrial farming. There is clearly a distance between Lucy and the Aleph Farm scientists that use her cells and no social interaction with Lucy except through her cryopreserved embryos. Besides, we learn that Lucy died, even though she will live on through her cell stock and as materialized cell-based steaks. By recounting her story, there is a sense of wanting to establish a sense of caring for the animal because Aleph Farms and other cell meat companies want to alleviate animal suffering. In the case of Ian the chicken, he was very much alive during the video and he was the hero, at the center of the story of the making of cell-based chicken nuggets. We witness Ian running around the table where people are enjoying their meal while glancing at him in the process. Castle points out the absurdity of this situation, Ian is "both alive and present and freshly fried into chicken nuggets" (Castle, 2022).

It is clear that the material value of these two individuals lie in their biological and reproductive potential. Even though they are presented as subjects, they contribute to the enframing of cellular meat as animal liberation and as an escape from death. However, it is difficult not to see them as "ghosts and as fungible things" (Weis, 2018). According to Weis, in a discussion about animal obsolescence in animal agriculture:

> The imagery of animals as unseen things in industrial livestock seeks to convey the intensive interspecies violence of these systems, signaling the total domination of animal lives and disregard for their anguish. (Weis, 2018, p. 140)

His observation also applies in the case of cell meat, because rather than being a radical solution to the violence inflicted on animals within industrial capitalism, it is a "further symptom of the remarkable extent of this violence" (Miller, 2012). Both Lucy and Ian embody this violence as constructed subjects of the agricultural system perpetuated by considering them solely as source materials for CM. The visibility of each animal is a reminder of their corporeal being as mere material substance, part of the process of making this meat. They are examples of "capital realizing itself through animal figures and flesh" (Shukin, 2009). In the case of ethical eating as promoted through lab-grown meat, both Lucy and Ian's mise-en-scène aim to provide a humane narrative that sidesteps the reality of their existence as farm animals and exemplars of industrial breeding. The fact that a tangible part of their body was used in producing in vitro tissue, and was reproduced through a kind of mimesis, creates a ghost-like experience in the case of Ian, where the animal watches its flesh being cooked and eaten without fear of being slaughtered. Both these animals become witnesses of the coming post-animal economy. They are the "present-absent-referents" to extend Carol Adams' concept of the absent referent. However, the depictions of Lucy and Ian show them as marketing tools rather than as fully realized beings, thus maintaining the dynamics of the absent referent albeit in a more sophisticated form. In this case however, these animals remain passive actors, representatives of genetic 'improvement' and selective breeding for agricultural purposes, rather than as individuals for their own sake.

Both Lucy and Ian represent what Nicole Shukin calls animal capital. She writes that actual animals no longer exist within the category of "species," their animality essentially always rendered available to humans. She articulates it this way:

> The phrase animal capital points to the paradox of an anthropomorphic order of capitalism whose means and effects can be all too posthuman, that is, one that ideologically grants and materially invests in a world in which species boundaries can be radically crossed (as well as reinscribed) in the genetics and aesthetic pursuit of new markets. (Shukin, 2009, p. 11)

In the process of replacing the animal, the production of cellular meat outside of the bodies of animals via the cellular reconstruction of tissues necessitates a conceptual re/materialization of the animal in order to make these legible as food, on the one hand, but to be identified as 'beef' or 'pork' or 'chicken' or even 'seafood,' on the other hand. Hence, both material deconstruction and imaginary regeneration of animal bodies, for the purpose of identification and edibility, take place simultaneously. The co-occurrence of physical disembodiment and imaginary embodiment of animals, the disappearance of bodies and their conceptual reappearance, the play between materialism and immaterialism raise questions about the status of farmed animals in the in vitro meat project, but more widely in human/animal relations. The process of making/unmaking bodies also carries important implications in terms of the biopolitics of eating as it relates to human subjectivities and food choices. Although the food is completely disconnected from the animal, what is of interest is how the animal is made to 'matter' in the physicality of the food and in the name of the final product. This means that eaters must actively 'make sense' of what they are eating when they consider cell-based meat, they must engage with this food both on a visceral and cognitive level (Evans & Miele, 2012). The designation of cellular meat as 'real animal meat' involves a conceptual reframing of animals and how they are made to matter or not matter through both language and sensory practices of production and consumption. As a result, the processes by which eaters engage in the material connections and disconnections between animal lives and animal foods suggest that "objects come into being—and disappear—with the practices in which they are manipulated" (Mol, 2007, p. 5). Mol suggests that in this way, the animal is made multiple, the farmer, the cell meat scientist, the technologies, the start-up investors, the chef, the butcher and slaughterhouse worker, etc. derive different meanings from the same object as it differs from one site to another, and she suggests these meanings depend on one another.

Human/Animal Entanglements: Post-Animals, Material Vitality, and Bio-objects

What cellular agriculture proposes is to engineer the animal out of the meat production process by creating muscle protein in vitro, in the laboratory. This type of agriculture aims to resolve several contradictions by virtuously invoking ecological and human health problems and animal suffering as major critiques of the current food system. By providing the grounds for individual self-empowerment, laboratory meat developers argue for a new biopolitics of meat-eating and for the replacement of living animals in agriculture and as food, based on what has been called a promissory economy represented by technoscientific innovation, all the while embodying and absorbing the contradictions of industrial farming and technocapital. Rather than disrupting the food system and productivist animal agriculture, I contend that cellular agriculture provides the grounds for a series of industrial realignments within Big Ag along the lines of a "protein-based" production model that includes the novel development and bioengineering of plant, animal, insect, algae, and fungal proteins. Already, our industrial food system conceives of animals as "protein machines with flaws" (Galusky, 2014), with the proviso that cell-based meat will rectify the imperfections, inefficiencies, and unpredictability linked to rearing whole animals. As such, CM proposes to fix the flaws by creating perfect protein machines without the animal (Ibid).

Freeing Cells from the Body and from Nature

Cultured meat invites us to consider ideas about tradition and innovation, continuity and rupture with past agricultural and farming practices as well as foodways. As such, food futures probe our understanding of the materiality of beings like never before, at a time when the biological becomes a major creator of value in a context of deteriorating natural resources. For authors like Galusky, industrial animals such as chickens are bred to be one-dimensional, their shape and existence are conditioned by the imperatives of a system that dictates control, purity, stability, confinement, genotypes, etc., and although these express an ideal of human mastery over nature, their more significant role is to "mediate between the nature that is possible and the nature that is technologically necessary" (Galusky, 2014). As a result, the wholesale elimination of animal bodies from birth to death and the use of stem cells to produce viable muscle tissue with fat and other components presuppose a particular understanding of how whole bodies function.

In her book, *Culturing Life: How Cells Became Technologies* (2007), Landecker retraces the history of tissue culture in order to show how cells have become the central unit of biological thought and practice. Her account is about the plasticity and temporality of living things, the ways in which they can be altered and made to work for human purposes. Most importantly, she describes the ways in which biotechnologies have altered our relationship to living matter through its reconfiguration as technology. Specifically, she analyzes how the significance of the capacity of cells to continually reproduce as "endlessly proliferating cell lines" affects our understanding of immortality, hybridity, and individuality (Landecker, 2007). The productive potential of cells extends beyond their scientific significance for research and clinical or other applications; their economic potential is enormous as productive and patentable entities. She writes:

> Cultured cells are characteristic of the present of the human condition, they function within well-established systems of labor and exchange, they are normalized in and by these systems; yet they also represent profound and recent change to a new state of being, as routine tools, alienable commodities, and sites of production. (Landecker, 2007, p. 16)

Yet, the cultured cell represents a different kind of life form, a technological one which can only exist in the laboratory, that has no real resemblance to "the body plan or lifespan" of the original organism. It is the dis-location of life from the bodies of plants and animals in the laboratory that is at the basis of cultivated meat today. As a result, "this separation of cells and their reproduction outside the body erases distinctions of inside and outside, tissue and organ making them all equally accessible to experimental manipulation of their capacities to grow and to differentiate" (Landecker, 2007, p. 23). Furthermore, the cell now reduced to living matter, living and proliferating outside the individual forms of an organism, has been transformed into a "disembodied, distributed and continuous form."

Cultured cells now are the basis for creating new biotechnological objects, from tissue engineering to reproductive science (p. 6). The importance of stem cells in biotechnological research and applications today can be understood through Landecker's historical account of cell culture. Her analysis of the strategic importance of cells as productive units in biotechnology extends to their scientific, technical, philosophical, and economic transformation. As the plasticity of cells came to be exploited, she describes how our ideas and manipulation of living things have changed. The two main aspects of the technologization of cells, via cell culture, are biological plasticity and temporality which we have discussed throughout the book in various contexts. With regard to cell culture, biological plasticity can be understood in three ways: as a fundamental characteristic of living matter; as what resists human intention and construction; and as a form of practice belonging to contemporary biological science (p. 10). Temporality refers to the potential immortality of cells in culture, the shift from in vivo to in vitro experiments, which gave rise to new ways of shaping living matter, creating technologically mediated life forms not found in nature, capable of existing in space and time within the laboratory. Clearly, biological plasticity and temporality are connected, since the manipulation of cells affects their life cycles and how the timespan of these cycles get altered as cells become immortalized, for instance. The temporality of cells can also be changed by manipulating the cell growth medium in which they are cultured. The various substances, growth factors, temperature, gases, fluids, salts, etc. that enable cells to grow also control biological events such as growth, differentiation, infection, repair, etc. depending on what scientists want to achieve. She writes: "The manipulation of medium was, and is, in turn tied to the manipulation of cellular time" (Landecker, 2007, p. 25). The culture medium is among the most important and costly factors in the production of cellular meat as we saw earlier. Disembodiment and distributed continuity are two other themes detailed in Landecker's book which apply to the processes of making CM.

Regarding our understanding of how the whole bodies of organisms function, Galusky suggests, "when meat was produced in the context of the animal body, one did not have to fully understand how that process worked to achieve a particular result. [...] By removing the body, humans have taken on the responsibility of understanding and functionally replacing the animal body (and the ecology) necessary to produce this form of protein" (Galusky, 2014). The question is how is this responsibility taken up by cellular agriculture scientists? Eric Schultz, a senior scientist at Memphis Meats, the first company to produce cell-based meatballs and chicken strips, declares "The cells don't know they aren't in an animal anymore. We have to train them to realize that. We call it the 'second domestication.'" He goes on to say: "We chose to create both mammal and avian meat because we wanted to show that we can handle the differences between the two." The differences between their muscles are incredibly distinct" (Beckett, 2017). Discourse surrounding domestication and the intervention of humans in plant and animal agriculture have been discussed in the section dealing with forms of biological appropriation and substitution. But the notion of domestication engages us in a deeper way to think about the history of domestication, not as a natural process but as a process characterized by systemic violence, slavery (Hribal, 2003), and as an effort to transform nature as a product of human design. The Anthropocene is a story of domestication of the environment on a global scale over a long historical period:

> Domestication is, from this perspective, a process that presupposes human mastery in the form of modification and manipulation of the behavioral dispositions, morphology, and lifeworld of other beings, as well as of their distribution and whereabouts. (Tønnesen et al., 2016, p. xiii)

Marx stated that in transforming nature, we also transformed ourselves and in so doing, animals became alienated from nature while humans lost their capacity to understand the extent of their impact on nature (Porcher, 2017, p. 32). So, domestication is a relation of exploitation based on the enslavement of animals and the transformation of their bodies strictly for the purposes of labor and as consumable products. However, others believe that domestication is a two-way process, a symbiotic relationship that means that plants and animals have played an important role in domesticating humans (Tønnesen et al., 2016). This view highlights the interdependence between species and their deep entanglements. So, how can we understand this idea of a second domestication of nature through the processes of creating vital tissues in the laboratory? What this entails is the idea that there is a historical continuum between the first and the second domestication, essentially extending in vitro culture as the natural evolution of animal agriculture and focusing attention on the productivity of biological processes at the cellular level. This produces a naturalized account of cellular agriculture which provides a frame for further asserting human domination over nature and living organisms. It also overlooks the fact that the process of domestication engages humans with animals. What is being domesticated then is not animals or existing life processes but the capacity to design and construct new biological parts, devices, and systems or to re-design existing natural biological systems for useful purposes. This includes domesticating the humans who consume the meat. This represents a radical extension of domestication, where the manipulation and control of life forms transcend traditional breeding and cultivation to involve the direct engineering of genetic material. Philosophically, this raises significant questions about the nature of life and the ethical boundaries of human intervention.

The Anthropocene is a story of domestication of the environment. Domestication can best be understood using the concept of the Plantationocene. According to Donna Haraway:

> It's an epidemic friendly way of rearranging species life in the world. It is a system that depends on forced human labor of some kind because if labor can escape, it will escape the plantation. The plantation system requires either genocide or removal or some mode of captivity and replacement of a local labor force by coerced labor from outside, either through various forms of indenture, unequal contract, or out-and-out slavery. The plantation really depends on very intense forms of labor slavery, including also machine labor slavery, a building of machines for exploitation and extraction of Earthlings. I think it is also important to include the forced labor of nonhumans—plants, animals, and microbes—in our thinking. (Haraway, 2015)

For Anna Tsing, the plantation evokes an ecological simplification through monocultures which made scalability possible, as the central tenet of the capitalist agricultural modernization (Guthman, 2022).

The goal of cellular agriculture is to produce meat. This is because agricultural projections suggest that meat consumption will double by 2050 but also because it is so entrenched in Western culture. Carnism creates a specific social order, often associated with masculinity and sexual politics (Adams, 2015). As a result, meat-eating provides a unique way of understanding the often-blurry boundaries between macro-social processes and ordinary life, especially when we consider that cultured meat is made to reflect specific current values (Galusky, 2014). Conventional meat is the product of active human interventions; in vitro meat creates another level of disarticulation from the context of its creation—the animal and its environment. Galusky goes on to say that cultured meat:

> [It] eliminates the environment in which animals would normally live. It eliminates wasteful translation of energy into non-consumable elements. It eliminates moral prohibitions against eating exotic animals, or animals at all, because it eliminates the animal. And, what's more, it eliminates the need for consumers to change. Little else [changes], that is, except a willingness to eat meat from this Process. (p. 936)

Most CM companies fail to adequately address the topic of farm animal liberation, as well as the fate of farming. While doing his research for his book *Meat Planet*, Wurgaft writes: "While I encountered little or no anti-agrarian sentiment in the cultured meat movement, it is true that most cultured meat research imagines food production from the standpoint of urban consumers rather than rural producers" (Wurgaft, 2019, p. 110). This remains a blind spot of CM today.

Most tend to agree that there will still be a need for free-range meat and that cellular agriculture will never replace plant and animal agriculture. They view cellular agriculture as another means to feed a growing world population. "People want choices. Part of the choice is they want animal products without animals," said Paul Mozdiak, professor of poultry science at North Carolina State University whose research is funded by New Harvest (Hart, 2016). According to Uma Valeti of Memphis Meats: "There will likely always be a role for family farms to play in feeding the planet, and we have no problem with that…But family farms can only produce a tiny fraction of the world's demand for meat" (Fleming, 2017). Valeti seems to imply that the demise of industrial animal farm is inevitable over the next few decades and that cultured meat will supply the meat necessary to feed the world. In addition, recent declarations by Believer Meat founder Prof. Nahmias, when discussing the development of continuous manufacturing as a means to drastically cut the cost of scaling up CM, suggest that CM is viewed as a "viable and sustainable alternative to traditional animal farming" as well as potentially meeting the global demand for animal protein (Hebrew University of Jerusalem, 2024). The notion of phasing out traditional farming persists within the discourse surrounding cellular meat.

Eating hamburgers or steaks made from cell-based agriculture therefore does not invite us to confront "the technologically mediated question of the human, animal, and ecological stresses exacerbated by industrialized systems and capitalist logics" (Galusky, 2014). Instead, it focuses our attention on meat as a technological object, representing the kind of nature we anticipate, one where animal farming may no longer be desirable.

While animal replacement products constitute an important part of this paradigm, cellular agriculture as a whole provides the industry with a set of converging technologies and scientific infrastructures that play into the promissory economy as well as narratives about the salvatory powers of technology. Some might say, cell-based meat narratives play into post-human scenarios that consider the end of nature and our decreased reliance upon it as a form of human-directed evolutionary enlightenment. By contributing to discourse about cellular agriculture, the developers and supporters of cell-based meat have insinuated themselves into narratives about the post-animal, anticipating a future characterized as "after" animality. One is tempted to think that the post-animal concerns factory-farmed animals only, however, the animal replacement economy involves culturing tissues from endangered, extinct, and rare species in an effort to produce not only food but also leather and ivory. In fact, some alt-protein companies and investors aim to stop animal agriculture by 2025.

Reconstituting Nature and Corporeality in the Bioreactor

The idea of human-directed evolution or ability to create novel forms of life plays into what has been called the plasticity of life, that is, the fact that bodies and genomes can be constantly rewritten depending on hereditary, environmental, socioeconomic, and other significant risk factors. Plasticity holds several meanings both in the social sciences and in the natural sciences, applies to inanimate objects as well as living organisms, and it is by nature a dualistic term, its meaning caught between "an ability to change and capacity to retain a shape" (Meloni, 2019, p. 6). Nevertheless, what Meloni calls the new plasticity of modern biology includes the possibility of manipulating our brain, genome, or microbiota in order to improve ourselves and optimize our bodily functions under the guise of individual control via activities such as meditation, healthy diet, or exercise (p. 9). Hence, the biology of plasticity entails the ability to regenerate or repair:

> It builds on and expands an ideology of individual consumption and personal freedom deployed in the service of neoliberal and marketized models of health. Its popular versions emphasize choice, control and reversibility. (Meloni, 2019, p. 9)

The permeable nature of the body to its surroundings and to its history and the biological embedding of social experience and the environment express how "the environment gets inside the body" and describe epigenetics and what has been called the postgenomic body (Meloni, 2019). The era of postgenomics, named after the completion of the Human Genome Project (HGP), signifies that the environment is increasingly viewed as a productive and bioactive force and, according to Meloni, it is no longer relegated to a passive container of phenotypes:

> Even in terms of biological capitalism, postgenomics introduces a different logic that makes not just DNA sequences alone but the "whole spatial and temporal contexts and circumstances surrounding DNA" a new potential source of biovalue. (Meloni, 2019, p. 2)

Postgenomics entail thinking about heredity differently, not simply as the transmission of genetic material but expanding it to cellular architecture, involving complex processes that go beyond DNA expression and genetic causality. It can be described as "more than DNA" whereby processes that go beyond linear models of genetic expression and transmission define heredity. While the environment has become crucial in this new thinking about interactions with living organisms and their development in this postgenomic era, the so-called environmental turn risks being relegated to a mere series of "code-compatible molecular representations" such as "maps, codes or blueprints" for the purposes of postgenomic research (Ibid). Meloni contends that biologists have digitalized or molecularized visions of the social and of the environment. For him and others, "bringing the environment and experience into biological matter" transforms these into data, into informational molecularized forms of environmental and social experience useable for research but completely detached from their innate complexities. Hence, "the social" becomes flattened and "entified and concretized by its experimental capture within a signaling framework" (Landecker quoted in Meloni, 2019, p. 120).

Cellular plasticity then is how organisms respond to environmental changes and how they evolve. This means that the environment plays an active role in biological matter and its regulatory apparatus. Epigenetics means that phenotypes can be expressed irrespective of an organism's experience of trauma or exposure, meaning that the environment is imprinted into biological materials, inducing reactivity of the cell in a non-linear, complex, and dynamic fashion over an organism's spatial and temporal contexts, including its entanglement within a web of relationships with other organisms (Ibid). This discussion about postgenomics and epigenetics raises two important questions for the purposes of our analysis: (1) Without going into the details of cellular architecture and epigenetic changes via chromatin structure and DNA methylation events, which environment is being molecularized or datafied when tissues are grown outside the body? (2) As the animal's body and its environment get molecularized and flattened in order to create "meat" in a bioreactor, what assumptions are we making about the ontological and ontogenic aspects of the living tissues produced and their fate as "food"? The cellular medium used to grow the tissues performs the role of nature, housing the function of the body within the model. This is where an individual animal's cells achieve the process of becoming a de-contextualized, de-historicized and de-localized life system within an artificial human-produced environment. This life system is precisely what this quotation describes:

> The living world is one of complexity, the result of innumerable interactions among organisms, cells, molecules. In analyzing a problem, the biologist is constrained to focus on a fragment of reality, on a piece of the universe which he arbitrarily isolates to define certain of its parameters. In biology, any study thus begins with the choice of a "system." On this choice depend the experimenter's freedom to maneuver, the nature of the questions he is free to ask, and even, often, the type of answer he can obtain. (François Jacob quoted in Rheinberger, 1993, p. 443)

So, the system in question is determined by the possibilities of a constellation of scientific disciplines, types of expertise, and skills coming together. Experimental systems become the scientific hardware within which in vitro tissues become knowable living objects, the 'software of epistemic things' (Rheinberger, 1993), but it is also the means through which they are connected, disconnected, placed, and displaced from their origin story. Experimental systems tend to be unpredictable, they become spaces of representation because they generate "strategies of material signification" and the laboratory is where epistemic things become transformed into technical things (Ibid). The laboratory becomes its own 'culture,' that is, where meanings about things are created and where their future is assigned. The bioreactors where tissues are grown or the bioprinter where they are printed is where cultivated fragments of meat materialize, it is where they become "thing-able" and therefore thinkable, they become machines for fabricating the future (Rheinberger, 2015).

This perspective resonates deeply with Kant's notion of the "thing-in-itself" (das Ding an sich), suggesting that our understanding of nature is inherently limited and mediated by our perceptual and cognitive frameworks. Kant posited that we can never truly know things as they are in themselves, only as they appear to us through the filters of our senses and mental constructs. This philosophical stance is echoed in Heisenberg's principle, which reminds us that what we observe is not nature in its pure, unaltered form, but nature as it is revealed through our specific modes of inquiry and technological interventions. When applied to the study of animals and their bodies within nature, this insight underscores a critical limitation in our scientific and technological approaches. We often interpret and manipulate biological systems through a narrow lens, shaped by our methods, tools, and conceptual frameworks. This leads to a reductive understanding of the intricate and dynamic relationships that constitute natural ecosystems and the lives of the animals within them. For instance, in the realm of cellular agriculture and synthetic biology, the creation of lab-grown meat involves isolating and cultivating animal cells in controlled environments. While this technological feat is remarkable, it also abstracts these cells from the complex web of interactions that define their existence in a natural setting. The laboratory achieves the status of nature as animal tissues materialize.

Philosopher of science, Daniel J. Nicholson criticizes what he calls the machine conception of the cell, which is put into question by recent advances in real-time, single-molecule tracking techniques and manipulations within cells (Nicholson, 2019). He writes:

> Cells are not engineered systems of discrete, interacting computational components, naturally yielding to compositional analysis…which is why they cannot be fully explained reductionistically; and neither do they operate deterministically, which is why their behaviour cannot be perfectly predicted. (Nicholson, 2019, p. 123)

Disrupting the Cow

Among some of cellular agriculture's future claims is the obsolescence of farm animals for food production. In their reporting about the future of cultivated meat, the RethinkX report emphasizes the disruption of the cow as a major form of technological progress. In their description of what that disruption might look like, the authors are referring to replacing all the proteins produced by cows, mostly milk and meat, using micro-organisms rather than extracting it from the macro-organism, that is, the cow. So, they go on to explain the many ways of making cows obsolete. They write:

> The next proteins to be disrupted are those produced by cows, namely those in milk and meat. They will instead be created directly from micro-organisms rather than extracted from the cow (the macro-organism). These individual proteins will then be built up to make the end product, whether it be ground meat, a burger, or a steak. This is a complete reversal of conventional production methods, where the cow is broken down into constituent components and then processed according to which end product is desired. In the conventional system, single molecules such as whey are the hardest and most expensive to produce. In the new system, they are the easiest and cheapest to produce. Crucially, the single protein molecules made using modern production techniques will be superior, purer, and more consistent than those extracted from the cow. (Tubb & Seba, 2020)

They explain that making the cow obsolete does not mean replacing cow products with synthetic alternatives. Instead, it involves disrupting key ingredients—rather than entire products like milk—which can then be used to create the final product. This involves a series of substitutions of ingredients and end products into specific productive entities. As a result, the cow is no longer necessary; in fact the authors repeat that while the animal is limited by its biology and economics, it cannot be improved. On the other hand, the design and biomanufacturing of single molecules offer infinite varieties of possibilities.

While cell meat is an example of one-to-one substitution, conventional meat is replaced by its cellular counterpart. The animal in this scenario is seen mostly as being replaced by muscle and fat. They predict that by 2030, 70% of beef consumed will come from modern production methods (fermentation, plant-based, and cell meat). In addition, cows will no longer be needed for materials such as leather. However, they also predict this complete disruption of the cow, even without the input from cultured meat, because other beef products will be produced in sufficient quantities for humans and for the domestic pet market. They estimate that removing animals from food production will reduce U.S. greenhouse gas emissions by 65% by 2035. Finally, the general idea is best expressed by the CEO of Impossible Foods, the plant-based meat company: "unlike the cow, we get better at making meat every single day" (Ibid, p. 18).

In this vision, the bioreactor replaces the animal body and becomes a closed-looped system for the continuous production of protein. Guthman's analysis of the RethinkX report is that:

> Finally, the report avers that precision technologies, along with these "virtually limitless inputs" will allow for a "move from a system of scarcity to one of abundance," "from a system of extraction to one of creation." (Guthman, 2022, p. 76)

She sees this as extending the promise of efficiency to one of limitlessness which she relates to Melinda Cooper's idea of harnessing the regenerative powers of life, through the bioeconomy, to limitless economic growth. She then makes the connection with pronouncements by the Dutch team of scientists who went on to found Mosa Meat, who said "In theory, one such cell line would be sufficient to literally feed the world" (Bartholet, 2015, p. 5). Mark Post has also stated in several media interviews that the cells of one cow could prevent the killing of 440,000 cows.

Animals have long been engineered to be living factories. The erasure of their body in meat and milk production, the disruption of the cow, essentially turns this metaphor around. It is the living factories that now stand for the animals. The bioreactor has now replaced their corporeality as well as their ecology to produce flesh. Meat could be produced in bioreactors instead of farms. The view that cows are "obsolete technologies" often came up during CM conferences.

Bibliography

Adams, C. J. (2015). *The Sexual Politics of Meat: A Feminist-Vegetarian Critical Theory* (Bloomsbury revelations ed.). Bloomsbury Academic, An imprint of Bloomsbury Publishing Inc.

Aleph Farms. (2024). *Space Meat—The Future of Astronaut Food*. Aleph Farms. Retrieved from https://aleph-farms.com/aleph-zero/

Barrett, H., & Rose, D. C. (2022). Perceptions of the Fourth Agricultural Revolution: What's In, What's Out, and What Consequences are Anticipated? *Sociologia Ruralis, 62*(2), 162–189. https://doi.org/10.1111/soru.12324

Bartholet, J. (2015, June 1). *The Quest for Lab-Grown Meat*. Scientific American. Retrieved from https://www.scientificamerican.com/article/the-quest-for-lab-grown-meat/

Beckett, E. L. (2017, June 27). *Memphis Meats Scientist: The World is Ready for Clean Meat*. Food Dive. Retrieved from https://www.fooddive.com/news/memphis-meats-scientist-the-world-is-ready-for-clean-meat/445892/

Castle, N. (2022). In Vitro Meat and Science Fiction. *Extrapolation, 63*(2), 149–179. https://doi.org/10.3828/extr.2022.11

Evans, A. B., & Miele, M. (2012). Between Food and Flesh: How Animals are Made to Matter (and Not Matter) within Food Consumption Practices. *Environment and Planning D: Society and Space, 30*(2), 298–314. https://doi.org/10.1068/d12810

Fish, K. (2013). *Living Factories: Biotechnology and the Unique Nature of Capitalism*. McGill-Queen's University Press.

Fleming, A. (2017, September 20). Could Lab-grown Fish and Meat Feed the World—Without Killing a Single Animal? *The Guardian*. Retrieved from https://www.theguardian.com/lifeandstyle/2017/sep/20/lab-grown-meat-fish-feed-the-world-frankenmeat-startups

Galusky, W. (2014). Technology as Responsibility: Failure, Food Animals, and Lab-grown Meat. *Journal of Agricultural & Environmental Ethics, 27*(6), 931–948. https://doi.org/10.1007/s10806-014-9508-9

Galusky, W. (2022). *Protein Machines, Technology, and the Nature of the Future*. Springer International Publishing. https://doi.org/10.1007/978-3-031-08717-2

Good Food Institute. (2024). *Alternative Protein Company Database* [Dataset]. Retrieved from https://gfi.org/resource/alternative-protein-company-database/

Guthman, J. (2009). Teaching the Politics of Obesity: Insights into Neoliberal Embodiment and Contemporary Biopolitics. *Antipode, 41*(5), 1110–1133. https://doi.org/10.1111/j.1467-8330.2009.00707.x

Guthman, J. (2022). The CAFO in the Bioreactor. *Environmental Humanities, 14*(1), 71–88. https://doi.org/10.1215/22011919-9481440

Haraway, D. (2015). Anthropocene, Capitalocene, Plantationocene, Chthulucene: Making Kin. *Environmental Humanities, 6*(1), 159–165. https://doi.org/10.1215/22011919-3615934

Hart, J. (2016, October 24). *Cellular Agriculture—Can it Help Feed the World?* Farm Progress. Retrieved from https://www.farmprogress.com/soybeans/cellular-agriculture-can-it-help-feed-world

Hebrew University of Jerusalem. (2024, August 23). Unprecedented Efficiency: New Manufacturing Method Could Make Cultivated Meat Cheaper Than Organic Chicken. *SciTechDaily*. Retrieved from https://scitechdaily.com/unprecedented-efficiency-new-manufacturing-method-could-make-cultivated-meat-cheaper-than-organic-chicken/

Hribal, J. (2003). "Animals are Part of the Working Class": A Challenge to Labor History. *Labor History, 44*(4), 435–453.

Just Egg (Producer). (2018, May 21). *Cultured Meat: A Vision of the Future* [Video recording]. Retrieved from https://www.youtube.com/watch?v=f8Ii3DB6ejE

Landecker, H. (2007). *Culturing Life: How Cells Became Technologies* (First paperback ed). Harvard Univ. Press.

Lee, L. Y. (2005). *Dialectics of the Body: Corporeality in the Philosophy of T.W. Adorno*. Routledge.

Meloni, M. (2019). *Impressionable Biologies: From the Archaeology of Plasticity to the Sociology of Epigenetics*. Routledge. https://doi.org/10.4324/9781315169583

Miller, J. (2012). In Vitro Meat: Power, Authenticity and Vegetarianism. *Journal of Critical Animal Studies, 10*(4), 41–63.

Mol, A. (2007). *The Body Multiple: Ontology in Medical Practice* (3rd print). Duke University Press.

Nicholson, D. J. (2019). Is the Cell Really a Machine? *Journal of Theoretical Biology, 477*, 108–126. https://doi.org/10.1016/j.jtbi.2019.06.002

Pell, R., & Allen, L. (2015). Preface to the Genealogy of the Postnatural. In A.-S. Springer & E. Turpin (Eds.), *Intercalations 2: Land & Animal & Non-Animal* (pp. 29–55). K. Verlag/Haus der Kulturen der Welt.

Porcher, J. (2017). *The Ethics of Animal Labor*. Springer International Publishing. https://doi.org/10.1007/978-3-319-49070-0

Porcher, J. (2020). Incarnation ou incubation?: Les animal studies et la "clean meat". *Zilsel, N° 7*(2), 292–306. https://doi.org/10.3917/zil.007.0292

Rheinberger, H.-J. (1993). Experiment and Orientation: Early Systems of In Vitro Protein Synthesis. *Journal of the History of Biology, 26*(3), 443–471. https://doi.org/10.1007/BF01062057

Rheinberger, H.-J. (2015). Difference Machines: Time in Experimental Systems. *Configurations, 23*(2), 165–176. https://doi.org/10.1353/con.2015.0013

Rodríguez Fernández, C. (2019, February 19). *From Saving Endangered Rhinos to Enabling Lab-Grown Meat*. Labiotech.Eu. https://www.labiotech.eu/food/lab-grown-meat-biobank/

Shukin, N. (2009). *Animal Capital: Rendering Life in Biopolitical Times*. University of Minnesota Press.

Tønnesen, M., Armstong Oma, K., & Rattasepp, S. (Eds.). (2016). *Thinking about Animals in the Age of the Anthropocene*. Lexington Books.

Tubb, C., & Seba, T. (2020). *RethinkX: Disruption, Implications and Choices. Rethinking Food and Agriculture 2020–2050*. RethinkX.

Ung-Lanki, S. (2014). Constructing the Biotech Nonhuman Animal: Instrumentalism and Ambivalence. *Society & Animals, 22*(5), 439–458. https://doi.org/10.1163/15685306-12341349

Weis, T. (2018). Ghosts and Things: Agriculture and Animal Life. *Global Environmental Politics, 18*(2), 134–142. https://doi.org/10.1162/glep_a_00457

Wurgaft, B. A. (2019). *Meat Planet: Artificial Flesh and the Future of Food*. University of California Press.

9

Techno-utopian Imaginaries and Anthropocene Narratives

The narratives contained within cellular agriculture consist of the alignment of economic interests concerned with land scarcity and demographic predictions, the exhaustion of the global agri-food productivist system and its unsustainability, yield losses and the diminishing value of animal and plant production, food scares and zoonotic transmissions of disease, as well as the food crisis of 2007–2008, which significantly highlighted our dependence on land-based, highly financialized, and cartel-controlled industrial food systems. These events have led to crop failures, food insecurity, and economic losses, exacerbating challenges in the global food supply chain. More recently, catastrophic events such as the COVID-19 pandemic, various major zoonoses such as swine and avian flu during and since 2020, the war in Ukraine and droughts in the Horn of Africa (2020–present), India and Pakistan (2022), the Yellow River Basin in China (2022), as well as Thailand, Cambodia, and Myanmar have had a major impact on agricultural supply chains, including meat exports. At the same time and in some of the same regions, massive flooding has affected agricultural lands in Pakistan (2022), China (2020), Brazil (2024), and several parts of Europe (2021, 2022, and 2023). The increasing frequency and intensity of droughts and floods, in grain-growing countries, have severely interrupted global agricultural supply

E. Abergel, *Dead Meat*, https://doi.org/10.1007/978-981-97-9049-4_9

chains and created havoc in terms of market availabilities and pricing. These combined events have seriously impacted the livestock industry not only in terms of diminished numbers of animals for market—the swine flu meant the destruction of millions of pigs—but grain and cereal supply chains needed for animal feed were also affected. As a result, meat prices and scarcity as well as food safety and environmental concerns created opportunities in the meat analogue and alternative protein market starting in the year 2020. These scenarios of food insecurity and climate catastrophes constituting the stories occurring in the past and present serve to shape the meaning of what the future might look like. They constitute narratives of destruction and climate uncertainty and remind us of the fragility of our world. They also serve as a reminder of the utopian potential of technoscience to address food and other forms of insecurity, offering a counternarrative of possible futures.

Acknowledging the role played by ideologies and utopian thinking and their materialization through foods, food practices, their production and consumption provides an opportunity to interrogate how the introduction of new foods not previously encountered in the human diet such as synthetic meat is socially, politically, ethically, and financially negotiated. In that sense, the cell-based meat industry narratives are typically techno-utopian in nature and are typically locked in the grand narrative of capital, displaying all the hype and promise of a better tomorrow in the midst of multiple crises. The concept of sociotechnical imaginaries captures the essence of the messages conveyed through these narratives, expressing particular social and ethical values, ideological tendencies, institutional interests, and strategic commitments. As stories about the future get told and retold in which positivist narratives take hold, what other stories are not being told? What gets translated and what doesn't? What other visions of the future are we not imagining? How do these narratives materialize as cellular agriculture reaches new technoscientific, financial, and regulatory milestones? What are the dynamics of their capitalization? What human and non-human actors get left out? How does nature get framed? Animals? Farming and rurality? Food and eating? Eaters and food cultures? These are some of the questions we will explore in the context of techno-utopias and Anthropocene narratives.

Paul Ricoeur's nuanced understanding of ideology and utopia, inspired by the work of Karl Mannheim's *Ideologie und Utopie*, focuses on their interrelated nature and role in shaping human thought and action. I think it is possible to examine lab-grown meat as ideologies and as utopias and what I have called "meat imaginaries" in the context of techno-utopias. Before discussing techno-utopias, it's important to delve into Ricoeur's definition of ideology as a system of ideas, beliefs, and values that reflect and promote a particular social, political, or cultural worldview (Ricœur, 1984). He claims that ideologies operate as frameworks or narratives that shape our interpretation of reality, influencing perceptions, judgments, and actions. However, he also asserts that ideologies can be both repressive and productive, capable of both enabling or constraining human agency and imagination. On the other hand, utopia goes beyond the familiar notion of a perfect or ideal society. Ricoeur views utopia as a dynamic and imaginative force that challenges the constraints of the present reality and envisions alternative possibilities. Mannheim placed a great deal of importance on the idea of utopia. For him, utopia provides a point of view which makes the present and the past comprehensible. It also opens a possibility for the future. Utopia's transformative vision serves as a powerful critique of existing social structures, norms, and practices. In Mannheim's sociology:

> ideology always reflects the desire of identifiable groups to block change to protect their own status and utopia reflects the desire of identifiable groups to bring about change to enhance their status. Because the concrete determination of what is utopian proceeds always from a certain stage of existence, it is possible that the utopias of today may become the realities of tomorrow. And once that happens, the former utopia can become an ideology. (Sargent, 2017, p. 26)

Ricoeur emphasizes the dialectical relationship between ideology and utopia, conceiving them as two expressions of the social imagination. While ideologies tend to be constraining, reinforcing dominant power relations, utopias challenge the *status quo* and create spaces for social and political transformation, envisioning alternative futures. However, ideology and utopia represent a double social and cultural imaginary because they must be understood relative to one another, as complementary rather than as opposites. He presents various levels of understanding of the concepts of ideology and utopia as a means to identify what he calls "deep correlations" between them. His analysis is nuanced and seeks to uncover how these concepts interplay within societies, contributing to both stability and change. Ideologies often justify the existing power structures and social norms, reinforcing the legitimacy of the current social order. According to Ricoeur, ideology operates by distorting reality to present a biased view that supports the ruling class, providing a sense of identity and cohesion within a society, and legitimizing power by rationalizing existing hierarchies. In contrast, Ricoeur sees utopia as the imaginative reconfiguration of social realities. Utopian thinking disrupts the taken-for-granted assumptions and encourages innovation and social transformation, critiquing the present, inspiring movements for change, and expanding the horizons of social and political thought. Ricoeur proposes that ideology and utopia are dialectically intertwined within the social imagination, existing in a state of conflict and tension that drives societal change. Despite their opposition, they influence each other; utopian ideas can be co-opted by ideologies once their transformative potential is neutralized, while utopias often arise from critical reflections on existing ideologies. Thus, a cycle emerges in which utopian ideas can be undermined by ideologies, while ideological visions can, in turn, lead to the creation of new utopias. This dialectic ensures a balance between stability and change within a society, with ideology providing continuity and cohesion, and utopia injecting creativity and progress so that society encompasses both conservative and transformative elements. Ricoeur places these concepts within the broader context of the social imagination which is the collective capacity to envision and interpret social realities through narrative construction, symbolic mediation, and cultural frameworks (Ricœur, 1984). In today's political climate, it seems that advances in technology have become one of the most powerful conduits for channeling utopian as well as ideological visions, shaping both the aspirations for a better future and the reinforcement of existing power structures and societal norms.

Applying Ricoeur's dialectic of ideology and utopia to the cultivated meat industry provides a critical lens through which to understand the complex interplay of resistance and innovation, in this emerging field. It highlights how cultivated meat both challenges and is shaped by existing societal norms and power structures, driving a dynamic process of transformation within the food system. Hence, from the advocates of cultured meat, ideology can be seen in the traditional meat production and consumption practices that are deeply embedded in societal norms and economic systems. The conventional meat industry is supported by powerful stakeholders, including large agribusinesses, governmental policies, and cultural traditions that emphasize meat consumption as a marker of status and normalcy. These ideological elements perpetuate the *status quo* by maintaining the economic dominance of traditional meat producers and by reinforcing cultural narratives that associate meat with prosperity and health. As a result, the cultivated meat industry or cellular agriculture faces resistance by consumers and others because it challenges established norms. Critics might see this meat as unnatural or unappealing, being anti-nature and anti-farming. These serve to defend the current meat production system by casting doubt on the desirability and viability of the industry. However, as we discussed previously, cultivated meat is an extension and reinforcement of existing society values and norms, particularly those related to carnism, but more importantly, those related to technological progress. The pursuit of cellular meat reflects a belief in the power of human innovation and scientific advancement to solve pressing global issues. This aligns with broader ideologies of techno-optimism or techno-utopianism. Techno-utopianism can be broadly defined as an idealized vision of the future where technology creates a better world, including beliefs about scientific progress, often envisioned through advances in automation, artificial intelligence, sustainable energy, and cellular agriculture. It is projected into the future, while ideology tends to find its roots into the past. Techno-utopias are not simply the product of the imagination, but according to Ernst Bloch's philosophy of utopia, they are rooted in objective reality, they are concrete and emanate from rational thought. Hence, Bloch's concrete utopias are what drive historical and revolutionary transformations. Therefore, what I am suggesting is that techno-utopias enter into contemporary neoliberal ideologies because these ideologies promote market mechanisms for solving the environmental crisis. The basic tenets of neoliberalism, what can be described as the Washington consensus, are free entrepreneurship, private property rights, free markets, and free trade (Harvey, 2005). Markets tend to be the preferred policy instrument and technological innovation plays a major role in maintaining competitiveness and economic growth. Thus, a central idea of neoliberal ideology is that it is also a "marketplace of ideas," becoming an epistemic phenomenon in addition to an economic one (Pellizzoni & Ylönen, 2012). Neoliberalism also entails privatization of public goods and deregulation. Innovation in the field of biotechnology has been a major part of what is known as the bioeconomy, which can be summed up as the tight "coupling of technoscientific knowledge and biological resources within a market framework" (Ibid, p. 6). As was seen in the chapter about cellular economies, the life sciences have been involved in value creation in financial markets, and small-scale biotechnology firms as well as large pharmaceutical firms have been key actors in biological accumulation alongside already established sectors such as factory farming and agriculture (Cooper, 2015). According to Melinda Cooper:

> Neoliberalism and the biotech industry share a common ambition to overcome the ecological and economic limits to growth associated with the end of industrial production, through a speculative reinvention of the future. At the height of the high-tech euphoria of the 1990s, the biotech industry promised to overcome hunger, pollution, the loss of bio diversity, and waste in general, while the ecological and biopolitical problems associated with industrial capitalism only continued to worsen. (Cooper, 2008, p. 19)

It is the speculative reinvention of the future that constitutes the utopian element of biotechnologies linked with neoliberal policies. Cooper is critical of biotechnological solutions to the problems of industrialism, as in the case of herbicide-tolerant crops or oil-eating bacteria:

> The biotech revolution, I argue, is the result of a whole series of legislative and regulatory measures designed to relocate economic production at the genetic, microbial, and cellular level, so that life becomes, literally, annexed within capitalist processes of accumulation. (Cooper, 2008, p. 27)

One could argue that the same logic applies to cultivated meat, since as Cooper claims, cellular agriculture does not overcome the harms caused by industrial agriculture but displaces them elsewhere. It is evident that the commercialization and promotion of cultivated meat are driven by capitalist values and market dynamics that emphasize efficiency, scalability, and profitability. It expresses the values of neoliberal ideology. We discussed the fact that this industry is not the antithesis of the dominant conventional animal agricultural paradigm, but rather, it is the continuation or culmination of its logic of eliminating animals from meat production, a process started with the intensification and industrialization of agriculture. Cultivated meat's ideological form can be thought of as an ideal to which reality will have to conform, while its utopian form consists in proposing an alternative to the *status quo* and putting in place the conditions for the replacement of animals through cell cultivation and producing new definitions of meat while expanding the meaning of agriculture by referring to its cellular components.

Cellular Meat Utopias

Viewing cellular meats as both ideology and utopia enables us to think through synthetic meat as a social project, meaning that it simultaneously embodies a set of technocultural critiques about industrial animal agriculture and a series of technological fictions regarding the material possibilities of growing meat outside animal bodies. While the narratives surrounding cellular agriculture can at times seem transgressive and clearly in opposition to factory farming, the field of cellular agriculture is clearly an extension of the logic of global technocapitalism, as mentioned before, situated within the global flows of resources essential to the accumulation of capital (Hornborg, 2014).

Drawing on various types of discourse (visual and textual) and narratives about the past, present, and future, my analysis will turn to how synthetic meat becomes a vehicle among others for a wider social project of transforming agriculture into its most industrialized form, potentially calling for the displacement of small- and medium-scale farmers and imposing changes on the countryside. Even if this presently occurs in a discursive context, cellular meat has not been commercialized nor has it been normalized. The existing sociomaterial and technological infrastructure of cellular agriculture represents a kind of symbolic capital and a way to position the industry in the market. The 30,000 square ft complex in Singapore built for GOOD Meat,[1] the cultivated division of Eat Just, who had planned to produce thousands of pounds of cultivated chicken meat or Upside Foods' EPIC pilot plant in Emeryville, California, capable of producing 50,000 lbs of cultivated meat per year in its shiny stainless steel facility[2] all represent speculative investments that can attract the interest of investors and the public. The construction of these facilities is motivated by political and economic considerations, as a way of legitimizing CA and creating hype about the industry and its products. These meat factories embody social imaginaries about how meat may be produced in the future, also how we might understand and eat meat differently.

The Anthropocene provides the narrative framework for shaping values, beliefs, and creating a temporal understanding of the world. By closely examining several types of discourse including visual representations of cellular agriculture, I will interpret the stories told and retold about synthetic meat's transformative visions and its power to shape alternative futures in terms of current and projected technoscientific achievements, involving animal-free or animal replacement products and how the urbanization of meat production might alter nature and rural landscapes, potentially leading to the rewilding of the countryside and of farm animals. Included in those narratives are stories of the de-extinction of past animal species that could be eaten as exotic meats.

[1] Although Good Meat's cultivated chicken facility broke ground in June 2022, plans were put on hold and the company has refocused its Singapore operation on producing hybrid meat, containing 3% cultivated chicken meat and 97% plant-based ingredients (Ong Guang-Xi, 2024).

[2] Visit https://upsidefoods.com/innovation for more information on EPIC which stands for Engineering, Production and Innovation Center. The company is projecting an expansion from the current 50,000 lbs to 400,000 lbs of cultivated meat per year. They are also working on an even larger facility projected to produce 10 million pounds (UPSIDE Foods, 2021).

Technofixes Imaginaries: Techno-utopian Foodways

The notion of imaginaries finds its origin not simply in the promise of a better economic future but in the construction of a better future society dependent on the development of total technical systems driven by *utopian* or *ideological* values (Flichy, 2007). It gives a political dimension as was discussed in the previous section on neoliberalism. The idea of imaginaries has been used in science and technology studies by authors such as George E. Marcus (1995), where he first introduced scientific workplaces as the sites of technoscientific imaginaries in the context of how scientists well positioned in their field conceptualize their research. Subsequently, Jasanoff and Kim's work on sociotechnical imaginaries expanded the notion to include other social actors in the process of creating imaginaries (Jasanoff & Kim, 2015). They define sociotechnical imaginaries as "collectively held, institutionally stabilized, and publicly performed visions of *desirable futures*, animated by shared understandings of forms of social life and social order attainable through, and supportive of, advances in science and technology" (p. 19). They therefore link the normativity of the imagination and the materiality of networks or assemblages that come together in order to perform visions of desirable futures (p. 19). In their formulation, the idea of *desirable futures* implies positive social progress as coterminous with technological progress. It follows that meat imaginaries express the collective beliefs and aspirations of a particular social order in which meat-eating persists but also engages us to think about unnatural ways of producing food ostensibly requiring a leap of faith. This means that we must break with our mode of knowing what meat actually is and understanding where it comes from. At the core of social imaginaries—or Flichy's *imaginaires*—lie utopia and/or ideology. This is slightly different from Jasanoff and Kim's use of imaginaries, as their definition hinges on an understanding of a shared sense of collective action based on a common vision. Flichy is inspired by Paul Ricoeur, who views utopia and ideology as a connected process within a continuum, as discussed earlier. He believes that one should not define these according to their correspondence or relation to reality. This is important because unlike the

Marxist tradition which places ideology in the realm of the symbolic or the immaterial, Ricoeur's conception of utopia and ideology operates on both a dialectic symbolic and material level as two poles of social imaginaries. While one tries to maintain the social order (ideology), the other tries to disrupt it (utopia), creating constant tension between stability and change (Flichy, 2007). Whereas utopia is concerned with the construction of alternatives, ideology is where power gets legitimized and group identity is formed. Promissory economies rest on the ability to convince a large public that a better future—even if it is premised on major transformations or important shifts—is possible. In essence, this is what the history of technology is built on. However, the interest of this particular approach is that it allows a biopolitics of post-animal meat to emerge, enabling us to explore it as "both a response to and responsive to the growing importance of the life sciences within the political and economic context of late capitalism" (Cooper, 2015). This means that "these experimental and [tissue] economies have evolved in response to the internal restructuring of the late liberal, post-colonial state, engendering new circuits of embodied work at the lower echelons of the service economy" (Cooper, 2015:257). Utopia and Ideology constitute important aspects of the legitimizing process of new technical systems, articulating a better future linked to technological progress. In order to achieve this, technologies must present themselves as radically novel, yet they must be credible and based on a subversion of what is known, like the concept of agriculture when it is taken up in the idea of cellular agriculture or the metaphor based on farming like the idea of cultivation or using cell feed. Cultured meat promises to disrupt the food system, feed the world, save animals, and solve the climate crisis. As Joly suggests, the more urgent and serious problems make these promises more attractive which in turn will legitimize political decision-making (Joly, 2015).

Calls by cell-based meat producers to 'disrupt' the food system convey the idea of a new sustainable agricultural paradigm based on biotechnological innovations as well as technoscientific convergence, inspired in part by hacking culture, and seems to be signaling new ways of defining, producing, and eating foods far from traditional and customary practice. In fact, it falls under what the World Economic Forum has identified as the *fourth industrial revolution*, as we saw earlier. The WEF claims that we are in a unique situation:

> because of the growing harmonization and integration of so many different disciplines and discoveries. Tangible innovations that result from interdependencies among different technologies are no longer science fiction. Today, for example, digital fabrication technologies can interact with the biological world. Some designers and architects are already mixing computational design, additive manufacturing, materials engineering and synthetic biology to pioneer systems that involve the interaction among micro-organisms, our bodies, the products we consume, and even the buildings we inhabit. In doing so, they are making (and even "growing") objects that are continuously mutable and adaptable (hallmarks of the plant and animal kingdoms). (Schwab, 2016, p. 15)

The *fourth industrial revolution for the Earth* discusses a possible means for managing food systems. The World Economic Forum's White paper entitled "Meat of the Future" asserts that:

> [It] enables the world's leading environmental experts and technology innovators – drawn from across governments, international organizations, civil society, business, research, science and start-up communities – to form unique partnerships that identify, co-design and test new policy innovations and technology applications for improving the world's environment and natural resource security. The project will help shape a new agenda for how companies, governments and international institutions better understand – and manage – nature and our natural resources, making use of the latest science and technology for the benefit of all in society. (WEF, 2018)

Couched in neoliberal rhetoric about management and an understanding of nature and natural resources, the World Economic Forum invokes the Paris Climate Agreements as well as the UN Sustainable Development Goals (SDG) in order to promote this techno-utopian future. This brief description of the fourth industrial revolution reveals how technoscientific imaginaries function when planetary and systemic challenges are used to justify them. According to the White Paper, the current 'future of meat equation' requires that 'protein possibilities' be explored in order to keep with the increasing demand for red meat, a demand driven by "millennial trends for better health and provenance dimensions related to food." We will relate the story of the fourth industrial revolution to ecomodernist thinking in the section dealing with Anthropocene narratives.

According to Flichy, a common vision is articulated and a technical utopia is negotiated via the idea of a new era of convergence, one that challenges the boundaries of disciplines and focuses on the creation of technological platforms as a model for business and scientific innovation. Cultured meat is a prime example of our increasing dependence on technology for the production of the most basic food commodities—and the economic opportunities and prospects that existential threats represent. The concern is that this comes with an ever-greater estrangement from nature (Welin & van der Weele, 2012). The emancipatory potential of technology is consistently reinforcing end of nature narratives, with the promoters of cultured meat focusing on creating new and ever-converging technological platforms that improve old and inefficient technologies, as a way to evade scientific responsibility in ecological collapse.

One of the most important ways to justify investments and developments in CM is the continuous discursive oscillation between techno-dystopian/techno-utopian narratives. It also consists of a constant play between different temporalities, as utopian pasts and futures become enmeshed with dystopian catastrophic scenarios in their discussion of conventional agriculture, rurality, food, animals, meat, and nature. These temporalities operate on different principles, the time frames of environmental degradation on the one hand that take place over a long period of time and are based on the disconnection between time and space resulting in non-linear, disproportionate, synergistic multiple time lags which vary from the nanosecond to the millennial scale (Adam, 2010), from deep time to the future present. On the other hand, the timescapes of industrial society and technoscience function on principles of money and control. They operate on divisions between space and time, everyday knowledge and science, nature and culture, and, according to Adam (Adam, 1998, p. 40), like Newtonian time or what she calls atemporal time, "a time unaffected by the transformations it describes." The logic of a de-temporalizing world based on universal laws and predictability highlights the fact that knowledge, or lack thereof, is always temporary, hence subject to change. The ultimate expression of atemporal time is quantification which further abstracts reality. According to Adam:

> technological products are premised on the Newtonian principles of decontextualisation, isolation, fragmentation, reversible motion, abstract time and space, predictability, and objectivity, on maxims that stand opposed to organic principles such as embedded contextuality, networked interconnectedness, irreversible change and contingency. (Adam, 1998, p. 41)

As agricultural processes become industrialized and increasingly technologically determined, they defy time and space. In the case of industrial agriculture, authors like Barbara Adam have discussed the disembeddedness of complex temporalities of "ex/live foods" and products and control over the processes of life. She questions the way efficiency and productivity are associated with maximum yield in the shortest possible time. She uses the example of milking a cow to show how productivity is measured in terms of liters per day, month, and year instead of over the lifetime of the animal. Maximum output of milk and continuous pregnancy and lactation shrink the time span of the cow. Disembedding beings and temporalities works by focusing on "the economic valorization of intensity of use value in the immediate present, measured in decontextualized units of performance, imposed on an exploitable resource that happens to be a living creature" (p. 142). This means that the productive period of a cow reared in an intensive agricultural system is consistently shrinking, being approximately two years. As a result, the life span of an industrial dairy cow is approximately 5 years as opposed to 15–20 years. By way of comparison, in 2007 the average dairy cow produced more than 20,000 lbs. of milk in one year, more than double the production of 40 years ago. It also means that the quality of the milk and the meat are not questioned in terms of their potential to pose a health threat to humans. As we are learning through pandemics and zoonotics, these issues are important to consider as successive zoonoses require the systematic destruction of millions of sick or disease-exposed animals. The important point here is that the living processes of animals have been consistently abstracted in industrial farming in terms of yields and other functional production issues that guide the technological transformation of farm animals in what Adam calls atemporal time, divorced from "life." For this reason, cellular agriculture and cell-based meat appear to be the culmination of the technologically driven process emanating from the Newtonian logic discussed by Adam. Goodman et al.'s discussion of the industrialization of agriculture suggests that industrial time works by controlling what was once thought to be uncontrollable, as exemplified by the scientific and technological control of biological time and geophysical space in food production.

Anthropocene Narratives

The concept of the Anthropocene is used as a backdrop to these narrative discourses as it sets up the need and the urgency of developing alternatives for resource-intensive animal agriculture. The term and concept of the Anthropocene were first proposed by Crutzen (Crutzen, 2002), essentially establishing that humans have entered a new geological epoch amidst a realization by scientists that human activities were fundamentally altering planetary processes such as the atmosphere, the biosphere, the cryosphere, the geosphere, the hydrosphere, the pedosphere, the technosphere, and the climate (Zalasiewicz et al., 2021).

Climate change and the Anthropocene represent large problems, requiring multiscalar understandings. Although the term is very broad and at times unclear, it helps us to convey the many stories underlying the need for cellular meat, in particular because the Anthropocene articulates visions of the technosphere in ambiguous ways. The term technosphere is used in the field of Earth System Science and can be defined as the Earth's human-made technological infrastructure, which includes technologies and products associated with industrialization, such as chemicals but also vast technological networks that enable rapid resource extraction that create free energy flows (Haff's definition quoted in Picot & Guillaume, 2024). It exists in opposition to natural objects and processes. In this sense, the technosphere as it relates to Anthropocene thinking plays both a dual and contradictory role linked to an ecomodernist worldview that sees technology as the solution to environmental problems as well as humanity's hold on the future. This vision is best expressed by Picot and Guillaume (2024) who, questioning the ability of humans to control the technosphere in its totality, write:

> the technosphere is originally constructed on the assumption of a dualism between humans and the environment (or the other geological spheres). This dualism is then turned into a dualism between the technosphere and the environment, in which both terms are conceived in a relation of exteriority to each other…this opposition serves to describe environmental problems. Pollution consists of flows of (hazardous) substances from the technosphere into the environment. Moreover, our analysis shows that in addition to this dualism, the technosphere is defined by the possibility that humans have to control it. In consequence, flows and stocks that are controllable are part of the technosphere and are not considered pollution anymore. (Picot & Guillaume, 2024, p. 98)

Therefore, technologies such as CM are viewed within this perspective as a kind of pharmakon, where the problems created by technology (within the technosphere) are believed to contain both cure and poison. Human-produced animal meat can therefore become part of an external process, that is, as a capacity to objectively exert influence on the environment at a technological level without the need to consider the technosphere and the environment, as part of the same whole. But more broadly, it can be viewed from its possibilities to exert some form of control over the "uncontrollable" aspects of the Anthropocene. Ecomodernists believe that human-induced or anthropogenic geological and climatic changes can somehow be overcome through technological breakthroughs, considering the environment and the technosphere almost as autonomous entities operating in parallel. These processes can best be encapsulated within the term technonature, a new form of naturalism, which describes the intricate entanglement of technology and nature, often challenging their distinctions. This perspective emphasizes that our understanding and experience of nature are mediated by technology, from agricultural practice and genetic modification to climate engineering or synthetic biology, and, as such, implies that nature itself has become a hybrid construct, where the natural and the technological co-evolve. Moreover, the idea of technonature, a specific understanding of the hybrid qualities of the environment, is emblematic of the Anthropocene Epoch because it naturalizes the range of human impacts on the planet, as an integral part of the Earth's ecological and technological systems. Lab-grown meat illustrates how technological processes reconfigure biological entities, creating new forms of life that defy conventional categories using biotechnological methods to produce meat from animal cells without traditional animal farming. It also speaks to the increasingly technologized materiality of food that culminates in cellular agriculture. As a result, technonature and cellular meat intersect, they reflect broader trends about how human innovations transform the environment and our biosphere, but, more specifically, they exemplify the intricate intertwining between technological and natural processes, creating new forms of ecological and technological hybrids. This is reminiscent of the plastic variants or novel plastic forms, such as plastiglomerate and pyroplastic, found in various parts of the world which combine plastic pollutants with sediments and rocks that interact with the environment, making them part of the Anthropocene geological record (Shruti et al., 2023). These become models of environmental hybridity, the material and tangible impacts of anthropogenic effects; they are part of technonature as they merge plastics, the product of industrialization and petrochemicals, with nature becoming a kind of stratigraphic indicator of the Anthropocene. The important point with this analogy and the short digression into plastiglomerate is that it invites us to think about cellular meat in terms of its material and physical properties, its emergence as a technoscientific object, relationships between human and non-human animals, while at the same time exploring the "full spectrum of problem definitions and suggested responses reflective of human disagreements about the right way to live on Earth" (Liboiron, 2016). In that sense, the Anthropocene enables a sociomaterial exploration of planetary change using cultivated meat as illustrations of new vital forms created intentionally, as well as the by-products of technological progress in biomedical sciences. In his article, *Technology and/or Nature: Denatured/Renatured/Engineered/Artifacted Life?* philosopher Holmes Rolston III asks how Pleistocene inherited diets might be managed in the Anthropocene (2017). He claims that the technosphere might supercede the biosphere, as the world of technological artifacts might no longer be contained within the natural world. Nature, he contends, will become artificialized and less constitutional as it becomes modified as illustrated by things like research into artificial blood, artificial photosynthesis, and artificial foods (Holmes, 2017).

Delving deeper into the notion of the Anthropocene, it can be defined as a scientific concept steeped in technoscientific discourse about its origins and its geological significance, but it also stands for a certain vision of the world, one in which humans have come to dominate the entire planet, leaving their geological imprint (Chakrabarty, 2010; Federau, 2017). As such, it is both catastrophic and represents an existential risk, but as we just said, it can also be interpreted by ecomodernists as an opportunity to make the world "better" through technology.

However, there is a deep contradiction embodied within Anthropocene discourse. Impregnated with a message of interspecies relations and engagement, Anthropocene discourse aims to bring about new subjectivities to construct new political possibilities for envisioning collective futures. But, because human activity is deeply implicated in geospheric and biospheric changes, Anthropocene discourse also assumes a perspective,

> pointing to the organic interconnection of humanity and the natural world, as evidenced by our impact upon the Earth and the coming consequences of this for human society; while demanding that we rise above nature by taking control of our collective species activity and consciously modifying it, in what would be the ultimate demonstration of humanist transcendence and self-mastery. (Nimmo, 2015, p. 195)

While Nimmo views this paradox as a contest between humanist and post-humanist ontologies, where the humanist position posits that human reason, scientific rationality, autonomy, and individualism are paramount. Such a perspective views humans as distinct from and superior to other living beings as well as to technology which is considered as a tool at the service of humanity. Nature is instrumentalized through technology with little regard for environmental harms. A post-humanist ontology, on the other hand, tends to challenge the centrality of human agency, rejecting anthropocentrism and human exceptionalism, focusing instead on the interconnectedness and interdependence of humans, non-humans, and technology. Instead, this approach recognizes the agency of non-human entities and their constant involvement in processes of co-evolution and co-construction with humans. Technology is seen as an integral and embedded part of the human condition, it is no longer simply a tool but a social agent, where the boundaries between humans, non-human animals, and machines tend to be blurred leading to the formation of hybrid entities, with the cyborg as its best-known fig. A humanist perspective tends to view human agency as "unbound by material constraints" (Nimmo, 2015) and capable of achieving autonomy and separation from nature while the post-human is often associated with a new materialist framework for understanding human/non-human relations in relation to the vitality of objects and materials as well as their ability to possess agency (Coole & Frost, 2013). However, critics point out that maintaining the nature/society distinction may in fact be necessary, not to be considered as discontinuous entities but as realms possessing different analytical dimensions. Retaining this distinction between nature and society enables a powerful critique of the role of technoscience as an instrument of domination over humans, non-human others, and nature. Rather than collapsing them, as in Donna Haraway's nature culture concept (Haraway, 2003), which argues for the continuous dynamic co-construction of the world through matter, discourse, and semiotics, thus rejecting the claim of a passive nature over which human ideological forces dominate, Hornborg argues that the distinction must be retained in order to avoid the inclination to naturalize technologies and depoliticize their effect on the world, because they have been and continue to be "instruments of social control and intraspecies exploitation" (Hornborg, 2014). In opposition, Haraway's view is that there is no pure nature underneath culture, they are deeply intertwined.

Because the category of non-human includes any material entity either living or non-living, there is a tendency to flatten the distinction between "things and beings," which in turn implies that materiality automatically confers vitality (Ibid). This material turn, according to post-humanists, by connecting the vital and affective elements of matter recognizes the political potential of direct engagements with the entangled nature of humans to non-humans and living and non-living (Coole & Frost, 2013). However, drawing attention to artifacts and their agency has disregarded in large part that technological systems viewed as naturalized and autonomous agents, as we saw in the discussion of the technosphere, are directly responsible for the materiality of global trade and, therefore, are not independent from their design as instruments of wealth accumulation, global inequality, and ecological unsustainability. This echoes Escobar's phrase (quoted in Hornborg) that "we design our world, and our world designs us back" (Hornborg, 2021, p. 760). As Hornborg points out, artifacts or objects may be charged with meaning, but they fail to possess purposive agency, their meaning is derived from human action, he writes:

> For an artifact to signify something to somebody is not to communicate but to serve as a vehicle for communication between agents that are equipped to do so. (Hornborg, 2021, p. 760)

Herein lies the distinction in his words between "purposive beings and significant objects" as he observes. I often refer to the concept of material semiotics, in this book, to highlight this difference and to analyze the perceptions that objects, such as synthetic meat, might elicit in human and other subjects. This is a way of retaining the distinctions Hornborg and others have pointed to while recognizing the social life of objects and their symbolic and cultural significance drawing from the anthropological material culture approach. Similarly, I believe that affect is a powerful engine for political action and that it evokes a deep sense of belonging on this Earth that is both embodied and subjective. The claim that non-living entities have agency is to Hornborg a way to blur the boundaries between the categories of living and non-living and human and non-human. This carries important ramifications for our understanding of the Anthropocene. As he and others point out, focusing on artifacts or non-living entities to extract political agency suggests that responsibility for planetary degradation gets displaced away from humans. Naturalizing and fetishizing technological objects prevent us from critical engagements with the ideological aspects of technological systems, as key social forces that encourage global inequalities and drive unsustainable ecological practices. Before engaging in the specifics of cultivated meat, the next section details some Anthropocene narratives and how they are enlisted by supporters of the cellular agriculture industry.

There are many ways of engaging with the idea of Anthropocene narratives. Firstly, it should be noted that the Anthropocene has become a cultural concept, starting with its narrative form as a story, a collective story—not shared equally—with sociological significance. Authors such as Dürbeck claim that these narratives "provide orientation in light of an enormous complexity of causal relationships, feedback loops, and unintended side-effects of human activities" (Dürbeck, 2020, p. 275). She identifies five narratives of the Anthropocene:

> (i) the disaster narrative, which considers the Anthropocene as the sum of environmental depletion; (ii) the court narrative, which blames the developed countries in the Global North and the neoliberal socioeconomic system as generators of global environmental destruction, proposing terms like "Technocene," "Capitalocene," or "Plutocene"; (iii) the narrative of the Great Transformation, which advocates for efforts to mitigate the causes of environmental destruction along with reasonable adaptation to Earth system changes, using novel technologies and higher environmental efficiency; (iv) the (bio-)technological narrative, which propagates technocratic interventions such as geoengineering and ecomodernist ideas; and finally, (v) the reflexively oriented interdependency narrative of nature-culture, which presents the Anthropocene as an opportunity to rethink mankind from a posthumanist perspective. (Dürbeck, 2020, p. 275)

While these narratives tend to overlap, they generally propose different political, economic, ethical, and anthropological perspectives. From a structural standpoint, she further suggests that the Anthropocene operates as a "meta-narrative"—a comprehensive framework that encompasses various perspectives and discourses. In this meta-narrative, *anthropos* is presented as a geophysical force within a deep-time dimension, adopting a planetary perspective on global environmental change. This perspective assumes the inseparability of nature and culture and emphasizes humanity's responsibility in addressing these changes. Each story about the Anthropocene reflects different historical temporalities and geographical scales. While geologists debate the start date of the

Anthropocene, often referred to as the golden spike, social scientists have developed various interpretations of this epoch. These interpretations influence how responsibility for ecological destruction is assigned and shape the perceived appropriateness of different interventions and actions (Pellizzoni, 2022). So, it appears the Anthropocene can be narrated in different ways for different purposes. As previously stated, several alternatives to the Anthropocene have been proposed, situating its origins in various historical periods, ranging from the era of colonialism to the advent of the steam engine and the Industrial Revolution. In addition, the story has been told from different perspectives by scholars from the Global South as well as Indigenous and Black thinkers. The Anthropocene concept can be understood in two ways:

> On the one hand, the Anthropocene is a given historical period during which certain events have taken place. On the other, the Anthropocene is a result, the consequence of those socionatural events and processes. In other words, it is both a chronology (marked by the anthropogenic influence on natural systems during time) and a given state of relations between society and nature. One that, in fact, blurs their separation. (Arias-Maldonado, 2015, p. 2)

Dürbeck's five detailed narratives can be grouped into three broader sets of competing discourses which relate to different brands of planetary politics. These are eco-catastrophism, ecomodernism, and planetary realism (Rothe, 2020). Rothe's analysis focuses on the eschatological politics of each Anthropocene narrative, drawing on the idea that the Anthropocene represents an irreversible threat that is spatially and temporally absolute. In particular, he delves into the narratives that depict the end of times and the relative role of human agency in each case. His focus is how each approach negotiates finitude and the time left, within which technological solutions like CM or geoengineering are negotiated. CM is seen as one approach to decelerating climate change and possibly preventing further environmental erosion. It is also envisioned as part of a post-Anthropocenic world, imagined as a time when the dominance of human influence either diminishes or transforms into something different, often through ecological collapse, a radical shift in human behavior, or the emergence of new forms of life and technology.

Eco-catastrophism

Rothe describes eco-catastrophist imaginaries as projections of global collapse and depletion of natural resources that relate to Malthusian thinking in the eighteenth century. Publications such as Erlich's *The Population Bomb* and the Club of Rome's 1972 *Limits to Growth* contributed to the popularization of a kind of neo-Malthusian environmentalism. In this account of environmental change, predictions are dire; humanity's hunger for natural resource will overshoot the Earth's capacity and will inevitably lead to a global collapse in the twenty-first century. The Stockholm Resilience Centre developed concept of planetary boundaries, linking eco-catastrophism with the Anthropocene in the context of Earth system science, a complex system with feedback loops and subsystems which culminates in defining critical boundaries for a 'safe operating space for humanity.' Each of these boundaries has thresholds depending on the level of uncertainty humans reach as well as tipping points which define the moments when critical elements of the Earth system (land, oceans, etc.) tip each other into a 'tipping cascade' or a kind of domino effect:

> Thus, according to the authors, humanity is left with less than two decades to mitigate 'conditions that resemble planetary states that were last seen several millions of years ago'. Eco-catastrophism in relation to the Anthropocene concept – as portrayed by the planetary boundaries model – thus focuses on the risk of acceleration and the threat of 'abrupt, nonlinear change'. (Rothe, 2020, p. 150)

This view requires what the authors call Earth system stewardship, that is, international action, mostly through the U.N. system which breaks down the environment into different governance objects, subject to regulations, conventions, and mechanisms.

> Eco-catastrophism draws on this universal subject position of 'anthropos' and rearticulates it through the figure of the planetary manager: a 'rational', scientific subject that manages the different parts of the Earth system on the basis of a comprehensive knowledge of the Earth system. (Rothe, 2020, p. 151)

The EAT-Lancet Commission report, Food in the Anthropocene, fits into the eco-catastrophic narrative. The term Anthropocene is mentioned eight times in the report and is broadly defined in its glossary as: "A geological epoch in which humans have become the dominating drivers of change on Earth" (Willett et al., 2019, p. 4). It evokes neo-Malthusian scenarios of population growth in light of exceeding the current food system pushing the Earth beyond safe boundaries, hence, the need to operate a global revolution in the global food system starting with shifting diets. Among key messages in the report state that:

> Unhealthy and unsustainably produced food poses a global risk to people and the planet. More than 820 million people have insufficient food and many more consume an unhealthy diet that contributes to premature death and morbidity. Moreover, global food production is the largest pressure caused by humans on Earth, threatening local ecosystems and the stability of the Earth system.
>
> Current dietary trends, combined with projected population growth to about 10 billion by 2050, will exacerbate risks to people and planet. The global burden of non-communicable diseases is predicted to worsen and the effects of food production on greenhouse-gas emissions, nitrogen and phosphorus pollution, biodiversity loss, and water and land use will reduce the stability of the Earth system. (Willett et al., 2019, p. 2)

In this scenario, human action through the changing of global food systems and dietary modifications can reverse or at least manage resources so as not to bring humanity to the brink of extinction. One cannot emphasize the influence this report has had on global institutions and in the public imagination. Armed with the conviction that food systems are major contributors to poor health and environmental degradation, changes in the type and quality of foods produced globally are required to meet the challenges we are facing:

> This Commission concludes that global food systems can provide win-win diets to everyone by 2050 and beyond. However, achieving this goal will require rapid adoption of numerous changes and unprecedented global collaboration and commitment: nothing less than a Great Food Transformation. (Willett et al., 2019, p. 2)

This transformation is to be achieved through scientific targets to "define the safe operating space for food systems" (p. 32). The language of planetary boundaries is used here to signify the urgency of the situation, stating that the dietary changes they are proposing can bring about a fast transition. In fact, scientific research and evidence-based institutions are central to keeping food systems within the Earth's limits. They are calling for the Great Food Transformation to be guided by the principles of SDGs, in particular SDG 17, which calls for partnerships between governments, private actors, and civil society at all levels of governance.

Cellular agriculture, as was mentioned before, is being touted as a sustainable solution to global agricultural problems as well as to the nutritional challenges brought about by social and health inequalities. International organizations such as the FAO and WHO have published reports that engage with stakeholders in order to estimate the viability of cell-based foods as well as to report on the regulatory needs for countries. The launch of their technical report about the safety of cell-based foods together with a webinar (2023) attracted huge interest with over 1000 participants from 75 countries. Similarly, the non-profit organizations that support the cell ag industry refer to international commitments like the Sustainable Development Goals (SDGs) set out by the UN, as well as their ongoing commitments to just food systems transitions. *From Fauna*, the lobby organization for the advancement of CM, released a textbook entitled *Modern Meat*. The introductory chapter of the book refers to the UN's SDG framework in order to, first, discuss the impacts of new technologies on global systems and, second, to introduce CM as the technological solution as it relates to six UN SDGs.

> SDG2: Zero hunger identifies opportunities for cell-cultured meat to mitigate world hunger. SDG3: good health and well-being, examines the ways the meat industry adversely affects our health, and this chapter discusses how cell-cultured meat could address these concerns.
>
> SDG8: decent work and economic growth, outlines the effects of conventional meat on workers' lives and economies.

SDG9: Industry, innovation and infrastructure, delineates opportunities for how cell-cultured meat could spark innovation in a variety of different sectors and transform the industrial landscape of nations.

SDG10: Reduced inequalities details ways in which conventional meat contributes to inequality on racial and socioeconomic lines and this chapter will discuss how cell cultured meat could alleviate this burden

SDG17: Partnerships for the goals surveys corporate and national partnerships that could facilitate the growth of the cell-cultured meat sector. (Anthis et al., 2023, p. 28)

Ecomodernism

Ecomodernism came out of the Californian Whole Earth movement of the late 1960s, a blend of cybernetic culture with hippie counterculture, "an evolving environmental consciousness and enthusiasm for emerging computer technologies" inspired by images of Earth from space (Rothe, 2020). The publication by the California Breakthrough Institute of the *Ecomodernist Manifesto* in 2016 signaled a new environmentalism turn, one that remains highly contested in terms of the environmental movement landscape.

Ecomodernist thinkers claim that human well-being can be decoupled from environmental degradation and that climate mitigation using technological innovations doesn't have to alter the economy. They present a more optimistic picture of the Anthropocene narratives. The Manifesto provides an alternative framework against the apocalyptic scenarios and anti-consumerism of mainstream environmentalism. In fact, the authors support the idea that economic growth can be decoupled from ecological harm using technologies to achieve efficiency gains (higher crop yields, etc.) and the intensification of human activities (less land use), so that a good Anthropocene can be achieved. In their words:

> Natural systems will not, as a general rule, be protectedor enhanced by the expansion of humankind's dependence upon them for sustenance and well-being. […] Intensifying many human activities — particularly farming, energy extraction, forestry, and settlement — so that they use less land and interfere less with the natural world is the key to decoupling human

> development from environmental impacts. These socioeconomic and technological processes are central to economic modernization and environmental protection. Together they allow people to mitigate climate change, to spare nature, and to alleviate global poverty. (Asafu-Adjaye et al., 2015)

The authors of the Ecomodernist Manifesto claim that human technologies will make us less reliant on ecosystems through the development of technological substitutes, thus liberating the environment from the economy. This viewpoint relies on the capacity of humans to reshape the planet (Rothe, 2020). This reshaping may also be understood as the post-Anthropocene discussed earlier.

Critics view ecomodernist visions of technologically driven societies as reductive of the multiscalar complexities of the ecological crisis. Invoking a "good" versus "bad" Anthropocene not only ignores the power inequalities that have led to planetary collapse but, on a philosophical level, it revives the modernist anthropocentric intellectual tradition of nature denialism, at a time when the survival of non-human others is of critical importance (Arias-Maldonado, 2019). Humans and their technological mediations are at the center of their logic. By asserting that nature can effectively be subordinated to human control, ecomodernists discount the intrinsic agency of nature, focusing attention on the fallacy of hyper-control and technological mastery. The Ecomodernist idea that you can protect nature while still meeting all of human needs is highly paradoxical. This, according to Fremaux, conveys blind faith in technology, a domination of nature narrative and a dualism, in which nature is either seen as totally pristine or as not even being real, an "ecology without nature" (Fremaux, 2019). Moreover, this view fails to convey that humans must rethink their place on the planet and that they must "accept the fragility and vulnerability of the human species in the face of complex and unpredictable natural phenomena" (Fremaux, 2019, p. 23).

To illustrate the ecomodernist idea, Vow, an Australian cellular agriculture start-up, recently announced that it produced a mammoth[3] meatball

[3] Woolly mammoths lived during the Pleistocene epoch which extends from 4.8 million to 4000 years ago. They inhabited the icy environments of northern regions in North America, Europe, and Asia and scientists believe the species became extinct due to overhunting by humans and due to the warming of the planet following the ice age.

made from DNA fragments obtained from mammoth myoglobin or muscle protein using African elephant DNA—its closest existing relative—in order to complete the gaps in the genetic sequence and using myoblast cells from a sheep, in order to produce the giant meatball (Carrington & Environment editor, 2023). On their website, they explain the rationale for the mammoth project:

> We made the meatball not to eat, but to show the potential of cultured meat from many different perspectives. For example, we know the large-scale production of agricultural meat needs to reduce in order to help end the climate crisis. But this means we need to greatly cut the amount of meat we consume, which is not likely to happen as it requires a huge behavioural change by the human race. With cultured meat, we don't need to wait for a behavioural shift. We can make meat without the environmental impact. (Vow Food, 2023)

The co-founder of Vow, Tim Noakesmith at the giant meatball's unveiling in Amsterdam's NEMO Science Museum, claimed that:

> You can use the mammoth meatball as a symbol for an exciting future that is better for the planet and avoid the perils of what we had originally with the woolly mammoth and maybe eat our way out of extinction.[4]

The idea of eating our way out of extinction speaks as much for the extinct animal cells his company is trying to use in synthetic meat as it does for the human species having to fend off its own extinction. The company believes that climate change is reversible—including extinction—if we change our dietary habits. Faced with the Anthropocene as an existential threat, cellular agriculture companies aim to provide a "viable alternative" to the destructive practices currently used to produce meat. Noakesmith explains further: "We chose woolly mammoth meat because it is a symbol of loss, wiped out by climate change…we face a similar fate if we don't do things differently" (AFP, 2023). He claims that

[4] The original source for this quotation is no longer accessible. It came from an Australian newspaper found at https://www.news.com.au/world/woolly-mammoth-meatball-will-an-australian-innovation-help-create-the-future-of food/video/49a60d5254dc69faeca06aa5a5eac1f1

a mammoth meatball was a good way to bring attention to the future of food and how it might be more sustainable. According to Neil Stephens, this meatball does more than just show how cultured meat can become normalized, for him, it shows that the technology enables new ways of envisioning eating meat, producing completely novel meats "made from species we've never eaten before" (Ibid), aiming to redefine what meat is. Noakesmith expresses it clearly when he states that:

> Because with new technology it means that the food that we can have doesn't have to replicate what we've had before,…It can be more exciting, it can have better flavor profiles, better nutrition profiles. And so we wanted to create something that was totally different from anything you can get now. (Chasan, 2023)

The scientists who worked on the mammoth project revealed that they wouldn't eat the meatball because humans haven't been ingesting this protein for 4000 years (Ibid), adding that they had no idea how safe it would be to consume. The mammoth goes from being a "symbol of loss" to a "beacon of hope" according to Vow's promotional video (Vow Food, 2023).

The focus of the mammoth meatball project was not to bring back extinct animals, clearly their value lies in their exoticism and showing the potential of cellular meat technology. De-extinction projects, as they are known, are not designed for food but for conservation purposes, and they aim to bring back extinct traits by editing them into living species in order to adapt them to different habitats. The idea that extinct species might be resurrected is controversial as some scientists believe that extinction is permanent and identical copies of the originals cannot be recreated in the laboratory (Hughes-Castleberry, 2023). Furthermore, the mammoth meatball project is ironic since it is used as a symbol of climate change; it is an animal whose carcasses are found embedded in permafrost, yet it is being "revived" in an era of accelerating global warming, the very phenomenon that threatens the preservation of these ancient remains. For the purposes of starting a conversation about cell meat, the use of mammoth DNA for human consumption might divert from more pressing issues such as species conservation and biodiversity loss.

By creating a space for the future of extinct meats, the Vow Foods' project juxtaposes deep time with biotechnological advancements and futuristic possibilities. This transgression of temporal boundaries reflects a spectral interplay where the past is never entirely gone but is reanimated in new forms. This also represents instrumental reason, the use of scientific knowledge not to understand the natural world but to exert mastery over it, treating extinct species as commodity forms for human purposes. Clearly produced as a kind of spectacle, this project is aimed at captivating public attention while generating some media hype around this particular cultured meat start-up. By transforming a prehistoric creature into a culinary novelty, the project commodifies and trivializes both the mammoth's historical significance and the material scientific resources involved, reducing them to mere entertainment. The meatball project not only commodifies nature but time itself. By presenting mammoth protein as meat, as a potentially new food, the project collapses historical time into a commodified present and potential future. Time becomes a thing to be manipulated and controlled through technologies. Bringing the prehistoric into the present reifies temporal processes, treating them as objects that can be manipulated and that disrupt historical continuity. The extinction of the mammoth and its preservation in permafrost are part of a historical narrative that speaks to the Earth's climatic and ecological changes over millennia. By intervening in this narrative, the mammoth meatball project undermines the social and historical consciousness of the Anthropocene by equating technological novelty and culinary experimentation with the systemic causes and profound ethical implications of agricultural unsustainable practices and their impact on global ecosystems and future generations. By presenting technology uncritically as a panacea, this form of appropriation of the past constitutes an ideological commitment to technological progress and human domination over nature. This potentially fosters a shallow environmental engagement with the current environmental crisis through an ecomodernist belief in technology as savior.

Planetary Realism

Planetary realism suggests that the End of time is already taking place, acknowledging that:

> human history is embedded into broader temporal rhythms beyond the species' control. In this understanding, humanity should give up its attempts to accelerate or decelerate the course of history and instead embrace the uncertainty of a world marked by the temporalities of complex socioecological systems. (Rothe, 2020, p. 153)

This vision of the Anthropocene includes Latour's Gaia concept, for whom the idea of Earthboundness signifies that humans cannot exert control on the dynamics of the Earth system because of our complete entanglement and dependence within it. Bruno Latour's Gaia politics are conceived as a theoretical framework that reconfigures our understanding of the Earth and its political implications in the context of the Anthropocene. Central to Latour's argument is the concept of Gaia, derived from James Lovelock's Gaia hypothesis, which posits that the Earth functions as a self-regulating system. Latour extends this idea to emphasize the interconnectedness and agency of both human and non-human actors in shaping planetary conditions. In Gaia politics, Latour challenges traditional dichotomies between nature and society, advocating for a recognition of the Earth as an active participant in political processes, a kind of democracy. This perspective is crucial in the Anthropocene, a geological epoch characterized by significant human impact on the Earth's systems, as it calls for a reevaluation of human responsibilities and the development of new political institutions that can address the complex, intertwined challenges of environmental and social change (Latour, 2017). By framing the Earth as a political actor, Latour's Gaia politics seek to foster a more integrated and responsive approach to governance that acknowledge the planetary limits and the urgent need for sustainable practices.

Another prominent author of planetary realism is the philosopher, Timothy Morton. Central to Morton's argument is the concept of "hyperobjects," entities so massively distributed in time and space that they transcend human comprehension and reshape our understanding of reality (Morton, 2013). Examples include climate change, nuclear waste, and global warming, which are not merely external threats but deeply interwoven with human existence. Morton's narrative challenges anthropocentric perspectives, emphasizing that the "end of the world" does not signify a literal apocalyptic event but rather the collapse of the human-centric worldview that has dominated for centuries. Rothe quotes Morton when he says: "Clearly Planet Earth has not exploded," writes Morton, "but the concept world is no longer operational" (Rothe, 2020, p. 154).

In this narrative vision of the Anthropocene, the key concept is one of resilience whereby there is a common understanding that modernity's top-down control and prediction are challenged. The subject is now faced with adapting to an uncertain world instead of controlling it. In the world of the adaptive subject, as Rothe describes it, resilience is seen as:

> a policy model to cope with uncertainty and complexity thus works through the (Earth-)bounded rationality of resilient subjects, and mobilises their established forms of local knowledge and self-organising capacities for their own survival. (Rothe, 2020, p. 155)

The Intergovermental Panel on Climate Change (IPCC) with its series of alarming reports essentially calls for massive adaptation efforts from vulnerable populations worldwide in the face of inevitable climate change, as a way of preventing predictable catastrophic impacts. It is assumed that vulnerable communities possess appropriate forms of local knowledge required for successful adaptation. Rothe describes how indigenous populations have been singled out as "promising agents of adaptation policies and resilience programming due to their (perceived) embeddedness into their surrounding ecosystems" (Rothe, 2020, p. 155). Traditional knowledge becomes linked with an imaginary of indigenous resilience which relies on the ability of indigenous communities to detect changes in their environment. In this context, because of the failure of scientific methods to anticipate and predict environmental changes at the local level, planetary realism discourse:

> turns to indigenous forms of knowledge and related cosmologies. As indigenous cosmologies are perceived as being established through interaction with nature's own rhythms, they are considered more appropriate for local practices of adaptation. (Rothe, 2020, p. 155)

This view is held with the intent to appropriate this knowledge in order to guarantee the survival of "all." So, this view works by essentializing indigeneity and romanticizing suffering, and it has been criticized by Indigenous scholars like Kyle Whyte (Whyte, 2018), Heather Davis and Zoe Todd (Davis & Todd, 2017), and non-Indigenous authors like Katherine Yussof (Yusoff, 2018), who claim that in fact for many "local peoples" the world has already ended or the Anthropocene has already happened. Rothe contends that by universalizing the experience of Black and Indigenous people, planetary realists "dramatize white people living under the conditions they have forced upon others" (Rothe, 2020, p. 156).

So, for most Indigenous people, the struggle means fighting to keep what's left of wild nature and their territories. Within this narrative of the Anthropocene, it's difficult to see how Indigenous foodways can be related to cellular agriculture, in a planetary realism framework. It is clear that the bioprospecting of culturally important and endangered animal and plant species falls into this category. While this is not specific to the CM industry, the use of patenting and intellectual property regimes have been used to organize the global appropriation of local resources across the world. In particular, in the global south these regimes of ownership over biological materials reproduce colonial relations. Vandana Shiva has written extensively about this process and the biotechnology industry's role in perpetuating these inequalities. But, more than this, it is clear that as far as food systems are concerned, cellular agriculture reenacts forms of food colonialism. I came across an initiative by New Harvest to engage Indigenous communities in a dialogue about cell meat, in the Fraser Valley of British Columbia, in an effort to develop responsible innovation. I also found an article entitled *Could lab-grown meat ever be Indigenized?* This article referenced a three-day conference held with native communities. In the piece, we learn that the start-up Wildtype Foods, a company that produces sushi-grade cultivated salmon, seeks out

Indigenous collaborations "to consider the social implications of its lab-grown salmon" (Eschner, 2022). Because they specialize in the production of cell-based wild species like salmon and even though they do not depend on wild fish specimens to grow their product, they feel a responsibility toward preserving wild spaces. They also seem to be actively involved in conservation efforts, in Alaska for instance, where the return of the sockeye salmon holds great cultural and economic importance to the Indigenous communities in the region. The fact that sacred animals in Indigenous culture have been the subject of experimentation in cellular agriculture (we learn that elk, bison, moose, salmon, etc. were studied as potential CM), and the idea that the use of the cells or genetic materials of these species might happen outside of Indigenous participation seems to run counter to the idea that one day cell meat could be indigenized. It's not clear how cell-based meats could contribute to Indigenous food sovereignty, their ecosystems, or to the resilience of their communities. There are no specifics about what decolonizing CM might look like. So, according to the planetary narrative, the plant and animal species that are considered as kin to Indigenous people and their colonial appropriation via patent regimes, at a cellular and technoscientific level, can be seen as perpetuating and enacting a kind of colonial violence. Trying to engage Indigenous peoples with CM also disregards Indigenous people's role in preserving biodiversity, focusing instead on the ways cell meat will save the planet by replacing animals used for food. This implies that CM would help conserve species in the wild or at least in the spaces inhabited by Indigenous communities. Indigeneity and Indigenous people are instrumentalized as witnesses of meat futures. They are enrolled to perform a legitimating role for CM as Earthkeepers and spiritual guides. A critical approach to indigenizing cultured meat must begin with the decolonization of science and technology—a complex and essential undertaking, if not an impossible one.

Transitioning from an exploration of Anthropocene narratives, as stories that implicate CM as a potential solution to industrial meat in the context of climate change, I want to examine how these themes are visually represented in the speculative futures of CM. In this context, imagery plays a key role in shaping our understanding and expectations of technological solutions for both food production and environmental transformation.

The Visual Language of CA: Visions of Meat Futures

Cultured meat has, to date, existed predominantly in promissory narratives rather than in tangible, material forms (Jönsson, 2016; Stephens et al., 2018, 2019; Stephens & Ruivenkamp, 2016). This is why, in this chapter, the social and political framing of Anthropocene, slaughter-free, rural versus urban, and health narratives contained within the promise of laboratory meat will be explored, as these narratives are seen as performative of specific modernist utopian and ideological visions of the future. It is therefore imperative to interrogate the social, political, and ethical contours of cell-based meat and ask how they might fit into particular visual environmental narratives. From what we saw in the previous sections, cell meat adheres to the ecomodernist narrative, with some aspects of the eco-catastrophism to rationalize the urgency of investing and commercializing products. Planetary realist approaches offer a less direct association but this Anthropocene narrative still provides valuable insight into how CM narratives can be understood and conveyed. It's important to clarify that Anthropocene narratives, in their dialectical complexity, offer a critical framework within which the promises and perils of cellular meat (CM) are both articulated and obscured; simultaneously, CM narratives, in their engagement with the discourses of ecological crisis and technological salvation, are themselves enmeshed in and perpetuate the multifaceted dimensions of Anthropocene thought, thus complicating any simplistic resolution between innovation and environmental redemption.

Narratives of new types of urban food factories are also significant as they tell us something about a determined future. As Bhat et al. (2015) write:

> Even with difficulties of progress so far, many scientists have great hopes and determination in regards to in vitro meat production and hypothesize that in vitro meat may soon have the ability to become a universal medium and envision a world where, instead of mega farms across the country, there is a local meat laboratory on each block (Kaplan 2012). Thus, *in vitro* meat production is the inescapable future of humanity. (p. 53)

The indeterminacy of cultured meat is resolved in the particular purpose of the technology. Making sense of in vitro meat via narratives is therefore an important part of the bioeconomy process; it is also part of "doing responsible innovation" in order to create the spaces for its realization. Narratives of responsibility for saving the world, for instance, feature prominently, as do several others, as we have already seen. For example, Mark Post, the founder and chief scientist of Mosa Meats, claims that "one cell line, once it has been stabilized could produce as much as 20,000 tons of culture meat," hinting that it may one day feed the world (Szondy, 2013).

As the prospects of sustainable and slaughter-free meat become invested with social expectations about the future, we are confronted with questions about scientific research as a market and industrial-scale tissue biofabrication in terms of "the interpretation of laboratory practices in the life sciences as ethical and political practices, and about the material techniques of caring for, exchanging and using biological matter that characterize contemporary attitudes to the living" (Landecker, 2002). Consequently, the anthropocenic narratives driving various conceptions of sustainability, visions of food security, and agricultural practices reveal much about how food orders societies, environments, and cultures. These visual narratives also uncover how sociotechnical trajectories linked to certain forms of techno-utopianism drive future dietary habits and culinary tastes. For instance, they may combine images of what the fourth industrial revolution or the second domestication might look like, as a way to bring into view and give meaning to these extremely vague concepts.

Meat Imaginaries

Meat imaginaries work not by just promising to feed people's bodies but by changing their understanding and their perspective of what food is and can become in the future. This is achieved by proposing new possibilities for changing foodways but also through shared understandings and imaginaries of the promises presented by the technosciences. As we navigate, the intricate interplay between Anthropocene narratives and cellular meat (CM), envisioning the future of eating, reveals a landscape both familiar and alien. Stories told by CM actors about what eating might look like in the future are conveyed in this scenario:

> Post—along with numerous others, such as Isha Datar, Paul Mozdziak, and Marie Gibbons—envisions a future in which meat is homebrewed similarly to beer, or cell lines are shared like sourdough starters: a "time when people like him will be 'selling teabags of stem cells from tuna, tiger, cows, pigs, or whatever meat you want, and from the comfort of your own kitchen, you could grow your own meat'". (Castle, 2022, p. 154)

CA imaginaries work by envisioning the sociotechnical and ever-shifting possibilities of life, living tissues, and their edibility. Furthermore, while, as we explained, the redefinition of life processes within a context of planetary decline underpins and guides certain narratives, these emerge from and build on older narratives about the utopian nature of technology and human ingenuity. The progress narrative, most notably as Federau explains, is central to an understanding of the Western values embodied within our visions of the future (Federau, 2017) but, equally relevant, is the relationship between biology and culture as a fundamental to our understanding of what the latest developments in life technologies and industries reveal about how Nature and life are conceptualized.

These narratives also work by focusing our attention away from "things," the objects of science, the manipulation of life, and the creation of edible tissues, by creating a context for these to find a place in the world. Studying these narratives provides us with an opportunity to uncover the complex relationships and connections that make these edible tissues appear as a rational project. Moreover, it is also a way of studying the complexity of ways in which life and the living as categories of being are being redefined, renegotiated, and repurposed. As cell-based biotechnologies gain in prominence and in vitro tissue culture and stem cell techniques advance at a fast pace, in particular in the field of food production, the narratives and the narrative construction surrounding the promise of edible tissues and post-animal agriculture can provide some insights into social change. If we could examine the advent of edible tissues as a way of understanding social change and social processes, what could we understand about the world we are living in? If as Beckert states, images of the future shape present decisions, it then follows that narratives of the promise of cultured meat are deeply anchored in networks and institutions (Beckert, 2013). They create collective expectations but "hold insights that are also relevant for understanding contingent representations of future states" (p. 224) that are important to analyze because so much is placed on the phenomenon at stake and because the aim is to materialize them. Narratives will help us unravel the technological future-making at work in the stories that shape and establish edible tissues as rational solutions to various contemporary problems while also anchoring their realization in the realm of concrete actions and investments into the future, premised on making these narratives cr/edible. Hence the production of futures (Adam, 2010), which lies at the core of the narratives we will be studying, involves, as Adam explains, an interplay between future-making and future-taking. Accordingly, the continuity of humanity has been threatened due to rapid technological developments, such as nuclear technologies to amplified climate emergencies. This is particularly important because as Barbara Adam states:

> In a context where we are deprived of the taken-for-granted fundamental assumption that successor generations will carry on where we leave off, faith is required to maintain belief in this particular fiction. (Adam, 2010, p. 368)

The following section draws from statements by several cultured meat start-ups as well as the images used to tell their story. These will reveal the underlying narrativity of cultivated meat technologies as well as their telling of the story of the future.

What the field of cellular agriculture, as a legitimate industrial sector, tells us about the fate of food remains uncertain; however, it does create opportunities for an organizing structure to take shape, for a network of like-minded actors to connect via the Cell Agriculture Society (CAS)—renamed *From Fauna*—and as a site for building the future of food. With its very slick and well-crafted imagery, the From Fauna website provides a snapshot of a utopian future. Through the use of two images as one enters the website, CAS sets out the scene (both images can be viewed at https://fromfauna.org). While there is much meaning to unpack from these images and others surrounding the visual culture of cultured meat, in my view, the CAS website imagery offers a very condensed idea of a utopian future, one where food comes from urban carneries or meat factories and where urbanscapes and rewilded natures and cows having reverted to some 'wild state' join seamlessly. Rather than utopian we might consider the idea of heterotopia as a way to understand the discursive mechanisms at work in these images. The idea of heterotopia comes from philosopher Michel Foucault, seen as a complex assemblage of diverse symbols that define it. When introduced by Foucault, heterotopia was used to describe a space that disrupts the continuity and normality of common everyday places. These spaces work by representing and at the same time inverting or distorting other spaces. In these two scenes, fictional and "real" ordinary spaces are juxtaposed.

The Imaginary Landscapes of Cellular Agriculture

In the first image that looks like a pastoral painting or a mythical bucolic scene, two foregrounded cows look down on a cityscape from an idealized green valley. This image is not static; the wind blows leaves from the nearby forest, the trees and tall grasses sway, clouds are moving, water ripples, wind turbines are turning, and a monorail train leaves and enters the city briefly stopping at what appears to be a train station. The calm breath of the two cows watching this scene offers a strange viewpoint clearly reconciling both rural and urban landscapes, as they melt into one. From this romantic vision of the countryside as a pristine environment of a past, it is suggested that we must return to Arcadia. The glistening cityscape with its white high-rise buildings and clean energy wind turbines seen from a distance appears as an ideal of the urban industrial future we must attain. In this image, the key aspirations of the CAS members—a beautiful clean environment and cows saved from slaughter—are reconciled with the ultra-modernity and clean technologies of urban life into one "visualized system" (Mirzoeff, 2016). This is enhanced with the following words: "For the people. For the animals. For the world" (CAS, 2020). In such a system, categories such as urban/rural, past/future, wild/domesticated nature, culture/nature, etc. are no longer antithetical but meld into one in a kind of dialectical image. Walter Benjamin's idea of the dialectical image refers to a way of understanding history and reality, where a single image or moment captures a complex, contradictory relationship between the past and the present. This image brings together opposing ideas or forces—such as progress and decline, hope and despair—in a way that sparks critical reflection and reveals deeper truths about society. It's like a snapshot that shows both what something was and what it has become, urging us to think about the tensions and conflicts that shape history. In this case, the idea is to juxtapose a kind of nostalgia about nature and animals with the promise of technology.

Another interesting juxtaposition is the wild landscape versus the hypermodern architecture of the cityscape where the meat factory is situated. This brings up the notion of wilderness or rewilding as another trope. The concept of rewilding has come to be a popular solution to the problems of the Anthropocene. Its origin can be traced to ecological restoration and conservation biology and it involves the (re-)introduction of species in order to compensate for biodiversity loss and ecosystemic services, that is, the functions and services provided to humans by biodiversity (Jørgensen, 2015; Nogués-Bravo et al., 2016). It also includes habitat restoration, potentially bringing back lost species but also at its base it calls for limited human presence and intervention (Martin et al., 2021). The distinction between ecological restoration and rewilding seems to be a matter of debate. Jørgensen notes that the controversy centers around expanding the concept of "wild" beyond a fixed historical baseline, thereby complicating the question of which time period we should be aiming to restore. Instead, rewilding among European advocates includes what Jørgensen calls "moving forward to move back" by allowing the introduction of ancient species that thrived before human habitation, including "wild" animals obtained through genetic manipulation or back-breeding (2015). Species like the Heck cattle, a human creation, was cattle back-bred by two German zoologists, to recreate the aurochs, and have been promoted and introduced throughout Europe as a hardy species for conservation grazing with minimal management and as part of rewilding projects (Lorimer & Driessen, 2013). Hence, the debate between conservation and rewilding is also one which pits conservationist ideals of managing extinction with a focus on equilibrium and fixity, with the desire to preserve or recreate existing biodiversity, versus a vision that includes the Heck cattle, where rewilding offers an experimental space for biodiversity without a clear plan, targets, or monitoring (Ibid).

The World Economic Forum defines rewilding as "letting nature do its own thing" equating it to nature repairing and restoring itself (Masterson, 2022). This constitutes what is known as passive rewilding, and it often involves abandoned farmland. There is still some confusion surrounding the science and politics of rewilding. Today rewilding projects include returning grazing lands to forests or letting trees and other forms of vegetation grow on pastures, this is called agricultural wilding or rewilding (Corson et al., 2022; Vogt, 2021), and it is often recommended to counteract the erosion of biodiversity from livestock systems. Livestock production requires 77% of global agricultural land in order to produce animal feed and this is having a major impact on agricultural land diversity more generally (Corson et al., 2022). An example of agricultural rewilding cited in the literature is the Knepp Wildland project in the U.K.,[5] where a couple transformed their troubled dairy farm:

> The Knepp Wildland project was not driven by specific goals or target species. Instead, it introduced rustic breeds of domestic herbivores (i.e. longhorn cattle, Exmoor ponies) and pigs (i.e. Tamworth) and wild herbivores (i.e. red, fallow and roe deer) to establish an ecosystem in which nature was given as much freedom as possible. Herbivores and pigs began to be introduced gradually in 2002, and their density by 2010 was such that, due to the absence of large predators and the existence of regulations prohibiting leaving dead animals to decompose, some of them had to be relocated or harvested, which yielded premium organic meat. (Corson et al., 2022, p. 5)

Some rewilding projects involving the reintroduction of large predators or carnivores, such as wolves or bears, often create conflicts between

[5] The project is described in a bestselling book entitled *Wilding: The Return of Nature to a British Farm* by Isabelle Tree (2018). It is quoted in Corson et al. (2022).

farmers, ecologists, and between local cultures and traditions (Ibid). There is no doubt rewilding projects are beneficial to the environment and to climate change. The return of biodiversity contributes to new ways of enriching the agricultural landscape and managing livestock systems. Rewilding projects may come at a cost to farmers, because of lower livestock productivity and because they feel like their traditional role as custodians of the land may be threatened (Wynne-Jones, 2022). The politics of rewilding are complex, and their social and material realities are poorly understood. The return of ecosystem services—although difficult to quantify—is also seen as a positive impact of rewilding the countryside.

Rewilding can also be interpreted as a form of biopolitics that aims to impose a human ordering of nature. It is a form of negative domestication in a sense because it provides the spaces in which 'wild' nature is able to exist and regain its agency. Spaces that are rewilded tend to be spaces that were affected by industrialization. If we look at the role of technology in rewilding efforts, there are of course the biobanks we've discussed before that serve as repositories for the genetic materials of endangered, extinct, or threatened species. The same technologies that are used for saving these species are used for making cellular meat. Whether it is the black rhinoceros, the passenger pigeon, or the woolly mammoth, there are projects underway to de-extinct lost species. De-extinction restores 'new versions' of historic species to their former habitats (Novak, 2018). According to the International Union for the Conservation of Nature (IUCN):

> de-extinction is the ecological replacement of an extinct species by means of purposefully adapting a living organism to serve the ecological function of the extinct species by altering phenotypes through means of various breeding techniques, including artificial selection, back-breeding and precise hybridization facilitated by genome editing. The goal of de-extinction is to restore vital ecological functions that sustain dynamic processes producing resilient ecosystems and increasing biodiversity and bioabundance. (Novak, 2018, p. 5)

This highly technologized vision of rewilding, the making of anthropogenically produced animal species, may not be compatible with the goals of a more traditional view of ecological restoration and an understanding of what a wild species might be. This is reminiscent of William Cronon's article *The Trouble with Wilderness or Getting Back to the Wrong Nature* which analyzes the rift between humans and wilderness and the fact that our view of wilderness is a fiction, a myth built around ideologies of humans existing outside of a pristine, far away nature (Cronon, 1996). This myth precludes us from considering the environments that we live in, our own backyard, and that sustain us, including the relationships between human and non-human subjects within those ecosystems. De-extinction falls into ecomodernist ideas about restoring the Earth, as a form of Earth stewardship.

In their book *Mutant Ecologies,* Borg and Policante provide an analysis of the Mammoth de-extinction project by the Colossal start-up. Besides showing how bioengineering might present itself as a way to recover lost or degraded global ecologies, they write:

> In the salvation drama that infuses environmental technoscience in the Anthropocene, the mammoth is offered as a katechonic icon. The pressing, future-oriented temporality imposed by climate change – melting arctic ice, leading to increased methane emissions, leading to increased concentration of greenhouse gases, leading to further climatic change – is short-circuited by the salvific image of a mammoth resurrected to save humanity. It is a dazzling promise that is as much religious as it is scientific: an ecological theology that presents technoscience as a creative spirit – a force that will not only bring back the dead, but also offer new means of Earthly salvation – and capital as the medium that enables the spirit to become embodied. Without financial backing, the mammoth chimaeras remain science fiction, trapped in the imaginations of scientists, venture capitalists, and amateurs. (Borg et al., 2022, p. 163)

Meat Factories: The Urban Carnery

In the second image posted on the *From Fauna* website, with the caption "Our Vision for Meat Production in 2040," we see a modern space-age brewery as the meat plant of the future. There, the scene is of a circular factory reminiscent of a spaceship (or a giant hamburger) with a series of bioreactors on either side of the image and in the foreground, what appears to be packages of meat coming down large conveyor belts being produced out of a large white cylindrical structure. This image expresses the idea of an urban carnery, a type of meat brewery, with its clinical and sterile environment and pure aesthetics. Everything is clean and tidy, aseptic and clinical. There is even a viewing gallery where visitors can observe the factory from above and see its few workers, some humans in lab coats and a few robots, engaged in the various activities of meat production. In the factory, machines predominate—bioreactors, robots, computers, steel, glass, conveyor belts, etc.—there is not one hint of animals depicted here nor are there any hints as to what scientists' work involves besides operating machines. This is in sharp contrast with the chaos, noise, blood, and guts one might see in a slaughterhouse. This depicted new reality engages us in an alternative space where food emerges out of machines, out of generic clinically sterile places. This vision might appeal to anxious consumers in pandemic times or as we discussed earlier in a food system viewed as 'dangerous' due to disease outbreaks, zoonoses, and talks of transmissions of infectious agents from animals to humans. But it also seems so removed from the current realities of animal meat production. Project CMF describes a vision for a Cell-based Meat Facility, and on their website, *From Fauna* explains the design process for the image described here. They write:

> Our Founder, Kris Spiros, had been developing a vision for many years on what a cultivated meat-driven meat production facility would look like. He felt the design had to strike a crucial balance between the old and the new. It couldn't feel too futuristic, and hence, out of reach, but also had to confidently stand within its innovativeness. After all, a revolutionary architecture effort like CMF was indeed the first of its kind. (From Fauna, 2024)

The CMF design described as a "slaughterhouse of the future" conjures up science fiction visions of futurism. In science fiction narratives, lab-grown meat and human-generated foods tend to center around futurism, environmental toxicity, and general food scarcity.

Project CMF is reminiscent of an article that appeared in the magazine *Anthropocene* where the idea of the carnery, the meat factory, was explained. The authors detail what carneries might look like, they would resemble breweries but instead of beer, in vitro meat would be produced. They explain the logic of meat production from animals as living factories to the making of in vitro meat:

> If we're comfortable treating meat animals like bioreactors, and engineering them strictly for the purpose of maximal protein production, then perhaps we can go a step further. Meat is simply a collection of muscle, fat, and connective tissues. Rather than raise an entire complex organism only to harvest these tissues, why not start at the basic unit of life, the cell, to produce meat? *In vitro* meat is meat, created in a bioreactor rather than in an animal. (Datar & Bolton, 2017)

They add:

> At the carnery of the future, large stainless-steel tanks house the biological processes that are transforming organic ingredients into food products. Conditions such as temperature and pressure are controlled and manipulated. Inputs and outputs are carefully measured. The work environment is clean and safe. But it doesn't feel like sterile science. It feels crafted, artisanal—because it is. (Ibid)

While the artisanal aspect of the meat factory is not present in the visual rendition of the future meat factory, the authors propose that there will be mid-range in vitro meat factory in local carneries located in urban areas. This could mean locating a factory in a skyscraper because it does not require land. They also ask us to imagine a future where "carniculturists" could create all kinds of meats, using rare or extinct species or co-culturing multiple cell types or even, creating totally new cell lines never experienced before. The article envisions a community of carniculturists federated around the carnery, where people might even produce their own cellular meat in their own homes. Moreover, the claim is that by making our own meats, new sets of values will emerge, and we will "grow" more "authentically connected to the origins and creation stories of what we eat" (Ibid).

Both images illustrate heterotopia, the city of the future and the carnery as both ordinary places as well as "other" spaces. By invoking the imagery of a brewery like meat factory and integrating it into these futuristic visual narratives, the CM plant of the future becomes more than a physical space; it is a manifestation of an idealized future. The description of a new CM manufacturing plant resonates with the idea of the carnery. When the cultured meat company Believer Meats (formerly Future Meats) announced its plans for building a large cell meat factory in the U.S., the unprecedented scale of the new facility and its potential to manufacture large quantities of quality product was emphasized:

> The bioreactors planned for the US facility will be over 40 feet tall and will hold 250,000 liters (that's 66,043 gallons) of meat. This is a massive scale-up from existing technology; the same manufacturer that's making the US equipment, ABEC, is also making a 6,000-liter bioreactor for a facility in Singapore, and when it goes online in 2023 it will be the biggest of its kind installed to date. Multiplying that by more than a factor of 40, then—and making sure the quality of the final product is still the same—will be no small feat. (Ramirez, 2022)

Building on the visual imaginaries of CM that emphasize sleek, futuristic carneries, as sterile spaces where safe, high-quality, uncontaminated meat is produced, it is crucial to explore how these representations intersect with narratives of disease outbreaks and food safety, particularly as CM is invoked as a solution to preventing zoonoses and safeguarding public health.

Cell-Based Meat and Disease Outbreak Narratives

Another aspect of industrial agriculture that has recently come under intense scrutiny has been around issues of biosecurity and threats from emerging infectious diseases, zoonotic and food-borne illnesses. As the Covid-19 virus brings to light the connections between animal factory farms and the consumption of wild animals, it is important to highlight some key points that will be relevant for our discussion around cell-based meat. Disease narratives and cellular meat (CM) are deeply intertwined with Anthropocene narratives, highlighting the interconnectedness of environmental and health crises. In the Anthropocene, ecosystem disruptions driven by human activities increase the risk of zoonotic diseases, emphasizing the need for integrated solutions. CM is framed as a response to these crises by offering a sustainable alternative to traditional meat production, which reduces environmental impact and minimizes disease transmission risks. The image of the sterile meat factory of the future underscores the technological optimism contained within an ecomodernist in relation to global health risks; the scene depicting the merging urban/rural landscapes highlights how CM can address environmental challenges, by organizing the space where meat is produced and separating it from the natural spaces where bioinsecurities occur. Of course, these images ignore the problems of contamination that might occur during the manufacturing of CM, as discussed earlier.

Firstly, as Hinchliffe et al. remark, biosecurity has been represented as the separation of diseased from healthy life, the containment of disease agents and outbreaks, as well as the policing of movements of potentially dangerous agents (Hinchliffe et al., 2013). By conceptualizing disease as relational, that is "both integral to, and always part of, an entangled interplay of environments, hosts, pathogens and humans" (p. 2), the authors suggest that we should not be paying attention to securitizing the *borderlines* separating healthy from diseased life so much as to engage with infected life constituting the *borderlands* within a particular mutable disease environment. This is characterized by contrasting perspectives on how to conceptualize and act upon the circulation of pathogens through territories and bodies. The idea that health is equated with being disease-free leads to an exclusively Cartesian spatial mapping of outbreaks as opposed to a networked spatial understanding of disease marked by varying intensities and densities of relationships and intra-actions within a topological landscape of embeddings and disembeddings. Thus, borderlands, topological spaces where "contrasting elements working to different rhythms and logics… intra-act" (p. 8), become the tipping points of disease outbreaks rather than breach points across borderlines. This is where the intra-actions between human and non-human worlds take place and where emergencies begin, already potentially present. The idea of tipping points provokes us to think about the complex interplays and structural causes of disease outbreaks. Reframing biosecurity discourse around a different kind of spatiality can help us trace the interlaced trajectories of disease within multiple spatio-temporal scales and their dynamic intra/interactions across biocultural domains. This geo-ecological perspective renders the question of the structural origins of disease outbreaks possible. The multidimensional relational disease model enables us to gain new insight into the production and geographies of pathogens in a way that doesn't view biosecurity as an apolitical project but as one which actively seeks to exert technocratic control.

According to Robert Wallace, an evolutionary epidemiologist and author of *Big Farms Make Big Flu: Dispatches on Influenza, Agribusiness, and the Nature of Science*, political power shapes and is shaped by infectious diseases and what we know about them. This is what Neel Ahuja would describe as the use of neoliberal immunological knowledges, in order to legitimate the interventions and interests of powerful state and capital forces (Ahuja, 2016). In fact, the biology of influenza is closely enmeshed with the political economy of food (Wallace, 2016, p. 103). Disease outbreak narratives miss how "some pathogens emerge right out of centers of production. Food-borne bacteria such as Salmonella and Campylobacter come to mind. But many like COVID-19 originate on the frontiers of capital production. Indeed, at least 60 percent of novel human pathogens emerge by spilling over from wild animals to local human communities (before the more successful ones spread to the rest of the world)" (Wallace et al., 2020). What this shows is that the industrial or intensive meat production system may have contributed and will continue to contribute to global pandemics, such as the one experienced in 2020. This has led to many questions about the implication of the consumption of animals as food, opening up the possibility of plant-based or cell-based "alternative meats" to gain momentum. While these food trends were already well underway before the pandemic, the situation has accelerated the development and demand for these products.

Agribusiness has been spreading its entire meat production line across the world according to Wallace engaging in corporate strategies which expand their geographic reach. When an outbreak of the African Swine Fever (ASF) decimated Chinese hog production in 2019, ironically the year of the pig according to the Chinese calendar, Hong Kong-based WH Group Ltd, the largest pork producer in the world and owner of Smithfield Foods Inc., culled its hog herd by half which caused the price of pork to double driving China's top producers to post their highest-ever profits leading up to September 2019, in spite of massive animal losses and the ruin of small hog farmers throughout Asia (The World Staff, 2019). The company's global platform allowed for 410,000 metric tons of U.S. pork versus 125,000 the previous year to be imported into China. The African Swine Fever exacerbated existing shortages within the hog industry. The Covid-19 pandemic of 2020 severely impeded hog slaughtering, processing, and packaging operations in the U.S., causing larger shortages, as Smithfield Foods closed three of its plants in April 2020 due to Covid-19 contamination among workers (Einhorn, 2020). But, the logic of agribusiness means that WH Group Ltd. will likely continue profiting from an outbreak they failed to control, as they engage in particular forms of globalized material exchanges supported by corporate, labor, and regulatory regimes (Allen & Lavau, 2015; Wallace, 2016). As a result of the double epidemic, ASF and Covid-19, meat shortages including pork are causing the sale of plant-based alternative meats to soar. The Financial Times reports that in the U.S. sales of plant-based meat substitutes have increased by 200% in April 2020, compared to the same period last year (Meyer & Terazono, 2020). While this represents a small portion of the U.S. protein market, it is nevertheless significant for three main reasons: Firstly, big meat producers such as Smithfield Foods and Tyson Foods control the supply chain for key ingredients such as pea proteins and genetically engineered soy which they use for animal feed production and food processing. Secondly, they also control the global distribution channels for fast-food restaurants, institutional buyers, and retailers. Thirdly, meat shortages in China mean that plant-based meat substitutes are becoming popular in that country and elsewhere. Beyond Meat started selling its products through Starbucks across China and KFC added Cargill's plant-based fried chicken nuggets to its Asian menu as of 2020

(Jack, 2020). Smithfield sells soybean-based burgers, meatballs, and sausages and also market-blended beef burgers and chicken nuggets (part meat and plant). However, since pork remains the most consumed meat globally, China being the largest pork consumer, these companies are hoping to benefit from a growing demand for meat substitutes and in particular plant-based pork. Whether Asian consumers substitute pork with plant-based alternatives or go to another type of meat such as poultry or fish is difficult to predict; however, large agribusiness companies are still profiting from biosecurity failures.

In fact, the purity of cultured meat is one of its key selling features, early on discussions focused on the term "clean meat" to describe cell-based meat which implied that flesh from live animals might be unclean, dirty, and inherently unsafe. The claim by proponents of cellular meat regarding purity or cleanliness is embedded within a binary classification system which posits material purity of food, animals, nature, science, consumers as desirable while implying the impurity and dangers of animal agriculture, conventional meat production, and meat. However, this form of dualism limits the space of possibilities between industrial animal and cellular agriculture; it excludes alternative forms of meat production and eating and essentially precludes critical engagement over the future of food. The use of the adjective clean is also reminiscent of the Cleantech bubble a decade ago, which, as it turns out, was an unfortunate association since Clean Energy technology companies failed to meet expectations despite large initial investments in green energy. This bubble emerged in the mid-2000s and marks the time when governmental authorities across the world implemented policies and incentives for the development of clean energy projects and technologies. The rising price of oil at that time made clean energy economically competitive, attracting investors to pour money into cleantech start-ups, thus creating a speculative bubble. The subsequent collapse of the clean energy sector highlights the risks associated with speculative excesses and unrealistic expectations linked with untested technologies and emerging industries.

Culture Wars

The following example serves to illustrate the ideological contours of CM, revealing past, present, and future narratives, in order to defend particular ideological positions among those who reject it. It also speaks of CM as being counternatural, invokes a sense of loss for an idealized past and a vision of the world where CM has no place. Objections started arising about the use of language on product labels for "fake meats," mostly plant-based at first since lab-grown meats are not commercially available, but also products such as plant-based milk, cheeses, and alt-dairy products.

In 2023, Italy's nationalist party introduced a bill that would essentially ban the production and use of food and feed from cell cultures or tissues derived from vertebrate animals, claiming the protection of culinary tradition (Speciale, 2023). According to Bloomberg reporter Alessandro Speciale, one minister claimed that Italy introduced the first law of its kind in order to "stop the decadence." Invoking the precautionary principle as its rationale, Italy's move to suppress lab-grown meat might be seen as ideologically motivated; nevertheless, it stands in total contrast to Vow's claims about the future of meat with its mammoth meatball concept, defending what it sees as a threat to its culinary past and cultural heritage. Here we have divergent temporalities: on the one hand, the use of an extinct species as part of a "proof of concept" approach that places the problem of extinction within a series of cutting-edge technologies as a way of securing the future; on the other hand, Italy's views on lab-grown foods question the loss and potential extinction of its traditions and culinary heritage. While Vow aims to transcend time completely, Italy's proposed bill frames cellular foods as a rupture within the sociocultural fabric, as a type of decadence and failure to preserve the past and present, signaling that Italy's culture would in fact be "eating its way towards extinction," thus opening up different questions about the future and temporality. So, Italy's proposed law stands to preserve a mythical past through its culinary traditions while holding on to specific ideas about what meat is and what it signifies in terms of defending its culture in nationalistic terms.

However, as the U.S. Department of Agriculture (USDA) and the Food and Drug Administration (FDA) discussed federal regulations for this new technology, several states across the U.S. began mandating special labels for cell-cultivated meat. In 2018, Missouri became the first state to pass such a law. The following year, Alabama, Arkansas, Kentucky, Maine, Mississippi, Montana, North Dakota, Oklahoma, South Carolina, South Dakota, and Wyoming followed suit. Kansas, Louisiana, Georgia, and Texas have since joined them (Nowell, 2024). Lab-grown meat, which received federal approval for sale in the U.S. last year, is currently facing opposition from Florida Republicans due to safety concerns. Legislative bills aimed at prohibiting the sale or distribution of this product are progressing in both the House and Senate, despite assertions from biotech industry experts that commercial retail availability is not imminent (Ibid) and the product is not available for consumers anywhere in the U.S.

Specifically, Florida legislators have introduced bills House Bill 1071 (HB1071) and Senate Bill 1084 (SB1084), which seek to ban the cultivation and sale of cell-cultivated meat within the state. A significant development occurred on March 6, 2024, when the state's House of Representatives passed SB1084 which sets "a $500 U.S. fine, corporate penalties and potentially even 60 days in jail for making, selling or distributing lab meat in his state" (Panetta, 2024). It appears that three other states—Alabama, Arizona, and Tennessee—are contemplating similar legislation banning the production and sale of cell-based meat. In Alabama, SB23 introduced by Republican Senator Jack Williams was signed into law in May 2024. The bill essentially "prohibits the manufacture, sale, distribution, or holding for sale of cultivated food products—defined as any food product produced from cultured animal cells—in the state" (Monger, 2024). In fact, according to the Guardian Newspaper:

> Sixteen states plus the federal government have already instituted regulations on labeling cell-cultivated meat, such as prohibiting companies from using the word "meat" in their marketing, or requiring them to print a disclosure explaining that the product contains cell-cultured products. (Nowell, 2024)

It's interesting to note that these laws do not go as far as prohibiting scientific research on cell-based meat, in fact universities and federal institutions, can continue their research. The prohibition of cell-based meat in certain regions is ostensibly aimed at safeguarding the traditional meat industry, particularly cattle ranching and animal farming. It is therefore unsurprising that these prohibitions are predominantly enacted in states where these industries constitute significant economic activities and where Republicans are in power. According to Sparsha Saha, a lecturer in meat politics at Harvard, quoted in the Guardian, what transpires is that politicizing meat turns out to be an interesting strategy for attracting male conservative voters (Nowell, 2024). Clearly, the issue of fake meats and the push toward banning cultivated meat goes beyond partisan politics. Much of the debate involves contradictory statements and positions by politicians, as it relates to free markets and individual freedoms, namely, the values embodied in free enterprise, technological innovation. and people's right to choose what they purchase. In addition, some arguments for banning cultured meat revolving around health risks tend to fall flat as the sale and manufacture of many unhealthy foods, with clearly demonstrated adverse health effects are allowed. The bans might very well cast doubt about the safety of the cell meat as well as its quality (Rao, 2024). Even the North American Meat Association, the country's largest meat packing association, defends the idea that choices regarding consumption and purchasing should be determined by the market and consumers, rather than being constrained by legislation that impedes progress and competition (Ogles, 2024). Pepin Tuma of the non-profit Good Food Institute, a pro-cell meat industry thinktank, questions whether governments should dictate consumer food choices, arguing for market-driven decisions (Rao, 2024). However, this stance overlooks the historical and ongoing involvement of the U.S. government in shaping consumer food choices. Despite free market rhetoric, the government has always played a pivotal role in determining food policies, influencing prices, and subsidizing agriculture. From setting nutritional guidelines and implementing food safety regulations, to providing substantial subsidies to certain crops and industries, the government's interventions have significantly influenced what is available and affordable for consumers. Such policies are designed to stabilize food markets, support domestic agriculture, and address public health concerns, underscoring the paradox in Tuma's argument. The intricate relationship between government actions and consumer choices reveals that the notion of a purely free market in food consumption is more ideological than practical and both sides of the debate are reinforcing the inherently political nature of food and agriculture.

Furthermore, opponents of the ban believe that:

> the proposed bans are part of a longtime strategy by the politically powerful agribusiness lobby and its allies in Congress and statehouses to further entrench factory farming as America's dominant source of protein. (Rao, 2024)

Saha remarks that science and agriculture have always had strong ties. However, dismissing big agriculture's involvement in the cellular meat industry would be misleading, given the established relationship between scientific advancements and agricultural practices. Furthermore, opponents of the prohibition of cell meat argue that such measures contradict policies aimed at promoting innovation in emerging technologies, particularly biotechnology. This tends to ignore the deliberate effort by agribusiness to align with and support the development of cellular agriculture. Given the real or assumed potential of cellular agriculture to affect traditional farming methods and the growing interest in alternative protein sources, it is reasonable to conclude that big ag and agribusiness entities are actively engaged in the cellular meat industry, whether through direct investment or strategic partnerships with relevant stakeholders. Detractors of the bans view the economic fallout on the country's biotechnology and biomanufacturing sectors that leverage "biological systems to produce goods at a commercial scale, offering innovative solutions across various sectors including plastics, fuels, foods and pharmaceuticals" as disastrous according to a letter addressed to Florida's legislators and signed by 38 venture fund leaders (Ogles, 2024). For opponents, research and commercial activities related to cellular meat cultivation would likely be relocated to other states, analogous to how bans on animal confinement in California and Massachusetts compelled certain agricultural enterprises to relocate (Johnson, 2024). This would also discourage export-oriented companies from choosing Florida as a reliable market, and it would create uncertainty about the future of this industry and leave that state out of "the myriad of benefits this burgeoning industry offers both the public and private sectors" (Ogles, 2024). Faraz Harsini, an employee with a non-profit organization advocating for alternatives to conventional meat, derided the notion that a state exemplified by Florida's Kennedy Center would jeopardize its position in long-term space exploration. He questioned the feasibility of transporting conventional livestock to sustain astronauts during prolonged space missions, highlighting how the banning of cell meat impedes progress (Panetta, 2024). Of course, Harsini's point also takes for granted that progress entails space exploration while overlooking the broader implications of adopting cell meat on Earth. Once again, space exploration and colonization act as powerful themes in mainstream culture, indicative of speculative and promissory technoscience, and entrepreneurial ventures (Tutton, 2018). Moreover, it takes for granted particular ways that will make up the future, including questions about what a desirable future might be and how it might be achieved and for whom (Jasanoff & Kim, 2015).

Florida's and Alabama's bans and their criminalization of cell meat products are viewed as anti-progress and anti-science. These are seen as an obstacle to economic growth and innovation but also as going against climate-friendly solutions. However, Governor DeSantis of Florida, who identifies as "not a global warming person," enacted legislation effective in July 2024 that prohibits references to climate change in state law. This legislative move redirects the state's energy policy away from the goal of transitioning to clean energy by 2050. Instead, the focus is now on "reducing reliance on foreign energy sources and strengthening the energy infrastructure here against…natural and manmade threats" (Inskeep & Green, 2024).

Beyond the realms of electoral politics and economic interests, cultivated meat appears to serve as a symbol within the broader context of what might be characterized as a cultural war. Headlines often refer to CM as "woke meat." Those advocating against cellular meat at the state level in the U.S. are clearly not supporting organic agriculture or veganism, as they tend to deny climate change. Unlike, the Italian government, they are not defending nationalism anchored in localist discourse but are clearly focused on maintaining the supremacy of the U.S. cattle industry worldwide.

Bibliography

Adam, B. (1998). *Timescapes of Modernity: The Environment and Invisible Hazards* (1. publ). Routledge.

Adam, B. (2010). History of the Future: Paradoxes and Challenges. *Rethinking History, 14*(3), 361–378. https://doi.org/10.1080/13642529.2010.482790

AFP. (2023, March 29). Meatball from Extinct Mammoth Unveiled by Food Tech Firm. *News.Com.Au*. Retrieved from https://www.news.com.au/breaking-news/meatball-from-extinct-mammoth-unveiled-by-food-tech-firm/news-story/81ddb6d5aca637050375d36fb5df085f

Ahuja, N. (2016). *Bioinsecurities: Disease Interventions, Empire, and the Government of Species*. Duke University Press.

Allen, J., & Lavau, S. (2015). 'Just-in-Time' Disease: Biosecurity, Poultry and Power. *Journal of Cultural Economy, 8*(3), 342–360. https://doi.org/10.1080/17530350.2014.904243

Anthis, J. R., Berenstein, P., Beurrier, Biehl, C., Birdie, P., Burley, J., Chantre, C., Davis, J., Dohadwala, S., d'Origny, G., Espirito Santo, V., Espitia, F., Gonzalez, G., Grandic, I., Green, C., Groenewegen, L., Guvenilir, A., Hanyu, Y., Kane, M. H., & Zimmerman, J. (2023). In K. Spiro, S. Hall, J. Darling, & M. Benami (Eds.), *Modern Meat: The Next Generation of Meat from Cells* (1st ed.). Cellular Agriculture Society.

Arias-Maldonado, M. (2015). The Anthropocenic Turn: Theorizing Sustainability in a Postnatural Age. *Sustainability, 8*(1), 10. https://doi.org/10.3390/su8010010

Arias-Maldonado, M. (2019). Towards a Good Anthropocene? In *Rethinking the Environment for the Anthropocene: Political Theory and Socionatural Relations in the New Geological Epoch* (1st ed.). Routledge.

Asafu-Adjaye, J., Blomquist, L., Brand, S. B., Brook, B. W., DeFries, R., Ellis, E. C., Foreman, C. A., Keith, D. A., Lewis, M. J., Lynas, M., Nordhaus, T., Pielke, R. A., Priztker, R., Roy, J., Sagoff, M., Shellenberger, M., Stone, R. W., & Teague, P. (2015). *An Ecomodernist Manifesto*. The Breakthrough Institute. Retrieved from https://static1.squarespace.com/static/5515d9f9e4b04d5c3198b7bb/t/552d37bbe4b07a7dd69fcdbb/1429026747046/An+Ecomodernist+Manifesto.pdf

Beckert, J. (2013). Imagined Futures: Fictional Expectations in the Economy. *Theory and Society, 42*(3), 219–240. https://doi.org/10.1007/s11186-013-9191-2

Bhat, Z. F., Kumar, S., & Fayaz, H. (2015). In Vitro Meat Production: Challenges and Benefits over Conventional Meat Production. *Journal of Integrative Agriculture, 14*(2), 241–248. https://doi.org/10.1016/S2095-3119(14)60887-X

Borg, E., Policante, A., & Skou, A. (2022). *Mutant Ecologies: Manufacturing Life in the Age of Genomic Capital.* Pluto Press.

Carrington, D., & Environment editor (2023, March 28). Meatball from Long-Extinct Mammoth Created by Food Firm. *The Guardian*. Retrieved from https://www.theguardian.com/environment/2023/mar/28/meatball-mammoth-created-cultivated-meat-firm

CAS (Cellular Agriculture Society). (2020). Introduction to cellular agriculture. [Now known as From Fauna]. https://fromfauna.org/

Castle, N. (2022). In Vitro Meat and Science Fiction. *Extrapolation, 63*(2), 149–179. https://doi.org/10.3828/extr.2022.11

Chakrabarty, D. (2010). Le climat de l'histoire: Quatre thèses (trad. De Charlotte Nordmann) (C. Nordmann, Trans.). *Revue Internationale Des Livres et Des Idées, 15*, 15.

Chasan, A. (2023, March 28). *Startup makes Giant Meatball Out of Lab-grown Mammoth Meat—CBS News*. Retrieved from https://www.cbsnews.com/news/giant-meatball-lab-grown-mammoth-meat-netherlands/

Coole, D. H., & Frost, S. (2013). *New Materialisms: Ontology, Agency, and Politics*. Duke University Press.

Cooper, M. (2008). *Life as Surplus: Biotechnology and Capitalism in the Neoliberal Era*. University of Washington Press.

Cooper, M. (2015). In Vivo Economies: Temporalities of Life and Value. *Journal of Cultural Economy, 8*(3), 257–259. https://doi.org/10.1080/17530350.2015.1039461

Corson, M. S., Mondière, A., Morel, L., & van der Werf, H. M. G. (2022). Beyond Agroecology: Agricultural Rewilding, a Prospect for Livestock Systems. *Agricultural Systems, 199*, 103410. https://doi.org/10.1016/j.agsy.2022.103410

Cronon, W. (1996). The Trouble with Wilderness: A Response. *Environmental History, 1*(1), 47–55. https://doi.org/10.2307/3985063

Crutzen, P. J. (2002). Geology of Mankind. *Nature, 415*(6867), 23–23. https://doi.org/10.1038/415023a

Datar, I., & Bolton, R. (2017, July 21). *The Carnery*. Retrieved from https://www.anthropocenemagazine.org/2017/07/the-carnery/, https://www.anthropocenemagazine.org/2017/07/the-carnery/

Davis, H., & Todd, Z. (2017). On the Importance of a Date, or Decolonizing the Anthropocene. *ACME: An International Journal for Critical Geographies, 16*(4), 761–780.
Dürbeck, G. (2020). Narratives of the Anthropocene: From the Perspective of Postcolonial Ecocriticism and Environmental Humanities. In *Post-colonialism Cross-examined: Multidirectional Perspectives on Imperial and Colonial Pasts and the Neocolonial Present, Monika Albrecht Ed* (pp. 271–288). Routledge, Taylor & Francis Group.
Einhorn, B. (2020, April 22). The World's Top Pork Processor Is Battling Two Epidemics at Once. *Bloomberg.Com.* Retrieved from https://www.bloomberg.com/news/articles/2020-04-22/coronavirus-swine-fever-threaten-world-s-biggest-pork-producer
Eschner, K. (2022, February 7). Could Lab-grown Meat ever be Indigenized? *Canada's National Observer.* Retrieved from https://www.nationalobserver.com/2022/08/02/news/could-lab-grown-meat-ever-be-indigenized
Federau, A. (2017). *Pour une philosophie de l'Anthropocène* (1re éd.). Presses universitaires de France.
Flichy, P. (2007). *The Internet Imaginaire.* The MIT Press.
Fremaux, A. (2019). The Return of Nature in the Capitalocene: A Critique of the Ecomodernist Version of the "Good Anthropocene". In *Rethinking the Environment for the Anthropocene: Political Theory and Socionatural Relations in the New Geological Epoch.* Routledge.
From Fauna. (2024). *What We Do.* Retrieved from https://fromfauna.org/work/
Haraway, D. (2003). *The Companion Species Manifesto: Dogs, People, and Significant Otherness.* Prickly Paradigm Press.
Harvey, D. (2005). *A Brief History of Neoliberalism.* Oxford University Press.
Hinchliffe, S., Allen, J., Lavau, S., Bingham, N., & Carter, S. (2013). Biosecurity and the Topologies of Infected Life: From Borderlines to Borderlands. *Transactions of the Institute of British Geographers, 38*(4), 531–543. https://doi.org/10.1111/j.1475-5661.2012.00538.x
Holmes, R. I. (2017). Technology and/or Nature: Denatured/Renatured/Engineered/Artifacted Life? *Ethics and the Environment, 22*(1), 41. https://doi.org/10.2979/ethicsenviro.22.1.03
Hornborg, A. (2014). Technology as Fetish: Marx, Latour, and the Cultural Foundations of Capitalism. *Theory, Culture & Society, 31*(4), 119–140. https://doi.org/10.1177/0263276413488960
Hornborg, A. (2021). Objects Don't Have Desires: Toward an Anthropology of Technology beyond Anthropomorphism. *American Anthropologist, 123*(4), 753–766. https://doi.org/10.1111/aman.13628

Hughes-Castleberry, K. (2023, April 21). The Woolly Mammoth Meatball Could Kick Off a Trend of Eating Extinct Meats. *Discover Magazine*. Retrieved from https://www.discovermagazine.com/planet-earth/the-woolly-mammoth-meatball-could-kick-off-a-trend-of-eating-extinct-meats

Inskeep, S., & Green, A. (2024, May 17). *Florida Gov. DeSantis Signs Bill that Deletes Climate Change from State Law*. NPR. Retrieved from https://www.npr.org/2024/05/17/1252012825/florida-gov-desantis-signs-bill-that-deletes-climate-change-from-state-law

Jack, J. H., Simon. (2020, April 23). "Fake" Meat on the Menu as China Reopens Restaurants. *BBC News*. Retrieved from https://www.bbc.com/news/business-52391517

Jasanoff, S., & Kim, S. H. (2015). *Dreamscapes of Modernity: Sociotechnical Imaginaries and the Fabrication of Power*. University of Chicago Press. Retrieved from https://www.press.uchicago.edu/ucp/books/book/chicago/D/bo20836025.html

Johnson, A. (2024, February 23). *What is Lab-grown Meat, and Why does Ron DeSantis want to Ban it in Florida?* Miami Herald. Retrieved from https://www.miamiherald.com/news/politics-government/state-politics/article285829481.html

Joly, P.-B. (2015). Le régime des promesses technoscientifiques. In M. Audétat (Ed.), *Pourquoi tant de promesses?* (pp. 31–48). Harmann Éditeurs.

Jönsson, E. (2016). Benevolent Technotopias and Hitherto Unimaginable Meats: Tracing the Promises of In Vitro Meat. *Social Studies of Science, 46*(5), 725–748. https://doi.org/10.1177/0306312716658561

Jørgensen, D. (2015). Rethinking Rewilding. *Geoforum, 65*, 482–488. https://doi.org/10.1016/j.geoforum.2014.11.016

Landecker, H. (2002). New Times for Biology: Nerve Cultures and the Advent of Cellular Life In Vitro. *Studies in History and Philosophy of Science Part C: Studies in History and Philosophy of Biological and Biomedical Sciences, 33*(4), 667–694. https://doi.org/10.1016/S1369-8486(02)00026-2

Latour, B. (2017). *Facing Gaia: Eight Lectures on the New Climactic Regime*. Polity Press.

Liboiron, M. (2016). Redefining Pollution and Action: The Matter of Plastics. *Journal of Material Culture, 21*(1), 87–110. https://doi.org/10.1177/1359183515622966

Lorimer, J., & Driessen, C. (2013). Bovine Biopolitics and the Promise of Monsters in the Rewilding of Heck Cattle. *Geoforum, 48*, 249–259. https://doi.org/10.1016/j.geoforum.2011.09.002

Marcus, G. E. (Ed.). (1995). *Technoscientific Imaginaries: Conversations, Profiles, and Memoirs* (Vol. 2). University of Chicago Press.

Martin, A., Fischer, A., McMorran, R., & Smith, M. (2021). Taming Rewilding—From the Ecological to the Social: How Rewilding Discourse in Scotland has Come to Include People. *Land Use Policy, 111*, 105677. https://doi.org/10.1016/j.landusepol.2021.105677

Masterson, V. (2022, October 10). *What is Rewilding?* World Economic Forum. Retrieved from https://www.weforum.org/agenda/2022/10/what-is-rewilding-nature/

Meyer, G., & Terazono, E. (2020, April 26). *Pandemic Accelerates Shift to Meat Substitutes*. Retrieved from https://www.ft.com/content/0127984a-6def-4040-9bca-002b6ffd4e0a

Mirzoeff, N. (2016). *How to See the World: An Introduction to Images, from Self-portraits to Selfies, Maps to Movies, and More*. Basic Books, a member of the Perseus Books Group.

Monger, C. (2024, April 27). *House Approves Senate Bill to Ban the Sale of Lab-grown Meat* |.... 1819 News. Retrieved from https://1819news.com/news/item/house-approves-senate-bill-to-ban-the-sale-of-lab-grown-meat

Morton, T. (2013). *Hyperobjects: Philosophy and Ecology after the End of the World*. University of Minnesota Press.

Nimmo, R. (2015). Apiculture in the Anthropocene: Between Posthumanism and Critical Animal Studies. In *Animals in the Anthropocene: Critical Perspectives on Non-human Futures*. Sydney University Press. Retrieved from https://www.academia.edu/8976378/Apiculture_in_the_Anthropocene_Between_Posthumanism_and_Critical_Animal_Studies

Nogués-Bravo, D., Simberloff, D., Rahbek, C., & Sanders, N. J. (2016). Rewilding is the New Pandora's Box in Conservation. *Current Biology, 26*(3), R87–R91. https://doi.org/10.1016/j.cub.2015.12.044

Novak, B. J. (2018). De-Extinction. *Genes, 9*(11), 548. https://doi.org/10.3390/genes9110548

Nowell, C. (2024, April 9). 'Political Efforts': The Republican States Trying to Ban Lab-grown Meat. *The Guardian*. Retrieved from https://www.theguardian.com/environment/2024/apr/09/us-states-republicans-banning-lab-grown-meat

Ogles, J. (2024, March 4). Meat Institute Argues Cellular Agriculture Ban will Hurt Florida's Food Economy. *Florida Politics—Campaigns & Elections. Lobbying & Government*. Retrieved from https://floridapolitics.com/archives/662944-meat-institute-argues-cellular-agriculture-ban-will-hurt-floridas-food-economy/

Ong Guang-Xi, J. (2024, May 17). *After a Promising Beginning, is Singapore's Cultivated Meat Industry on the Way to the Slaughterhouse?* CNA. Retrieved from https://www.channelnewsasia.com/singapore/labgrown-cultivated-meat-technology-eat-just-acceptance-4332656

Panetta, A. (2024, May 8). In Budding Battle over Lab-Grown Meat, Florida takes Opening Stab. *CBC News*. Retrieved from https://www.cbc.ca/news/world/florida-ban-analysis-canada-1.7197136

Pellizzoni, L. (2022). Anthropocene. In P. Rebughini & E. Colombo (Eds.), *Framing Social Theory* (pp. 40–54). Routledge.

Pellizzoni, L., & Ylönen, M. (2012). *Neoliberalism and Technoscience: Critical Assessment*. Ashgate.

Picot, P., & Guillaume, B. (2024). The Controllability of the Technosphere, an Impossible Question. *The Anthropocene Review, 11*(1), 91–109. https://doi.org/10.1177/20530196231187242

Ramirez, V. B. (2022, June 1). The World's Biggest Cultured Meat Factory Will Soon Be Built in the US. *Singularity Hub*. Retrieved from https://singularityhub.com/2022/06/01/the-worlds-biggest-cultured-meat-factory-will-soon-be-built-in-the-us/

Rao, D. (2024, March 25). *Lab-grown Meat Might be About to Meet its Maker*. Theweek. Retrieved from https://theweek.com/science/lab-grown-meat-controversy

Ricœur, P. (1984). L'idéologie et l'utopie: Deux expressions de l'imaginaire social. *Autres Temps, 2*(1), 53–64. https://doi.org/10.3406/chris.1984.940

Rothe, D. (2020). Governing the End Times? Planet Politics and the Secular Eschatology of the Anthropocene. *Millennium, 48*(2), 143–164. https://doi.org/10.1177/0305829819889138

Sargent, L. T. (2017). Ideology and Utopia: Karl Mannheim and Paul Ricoeur. In *Utopian Horizons: Ideology, Politics, Literature in Z. Czigànyik (Ed.)* (pp. 19–40). Central European University Press.

Schwab, K. (2016). *The Fourth Industrial Revolution*. World Economic Forum.

Shruti, V. C., Kutralam-Muniasamy, G., & Pérez-Guevara, F. (2023). New Forms of Particulate Plastics in the Anthropocene. *Earth-Science Reviews, 246*, 104601. https://doi.org/10.1016/j.earscirev.2023.104601

Speciale, A. (2023, March 28). Italy Wants to Ban Cultivated Food to Protect Culinary Heritage. *Bloomberg.Com*. Retrieved from https://www.bloomberg.com/news/articles/2023-03-28/italy-wants-to-ban-cultivated-food-to-protect-culinary-heritage

Stephens, N., & Ruivenkamp, M. (2016). Promise and Ontological Ambiguity in the *In vitro* Meat Imagescape: From Laboratory Myotubes to the Cultured Burger. *Science as Culture, 25*(3), 327–355. https://doi.org/10.1080/09505431.2016.1171836

Stephens, N., King, E., & Lyall, C. (2018). Blood, Meat, and Upscaling Tissue Engineering: Promises, Anticipated Markets, and Performativity in the Biomedical and Agri-food Sectors. *BioSocieties, 13*(2), 368–388. https://doi.org/10.1057/s41292-017-0072-1

Stephens, N., Sexton, A. E., & Driessen, C. (2019). Making Sense of Making Meat: Key Moments in the First 20 Years of Tissue Engineering Muscle to Make Food. *Frontiers in Sustainable Food Systems, 3*. https://doi.org/10.3389/fsufs.2019.00045

Szondy, D. (2013, August 6). *First Public Tasting of US$330,000 Lab-grown Burger*. New Atlas. Retrieved from https://newatlas.com/cultured-beef/28584/

The World Staff. (2019, October 31). *Pork Prices Soar as Millions of Pigs Die due to African Swine Fever*. Public Radio International. Retrieved from https://www.pri.org/stories/2019-10-31/pork-prices-soar-millions-pigs-die-due-african-swine-fever

Tree, I. (2018, August 15). If you want to save the world, veganism isn't the answer. *The Guardian*. https://theguardian.com/commentisfree/2018/aug/25/veganism-intensively-farmed-meat-dairy-soya-maize

Tutton, R. (2018). Multiplanetary Imaginaries and Utopia: The Case of Mars One. *Science, Technology, & Human Values, 43*(3), 518–539. https://doi.org/10.1177/0162243917737366

Vogt, M. A. B. (2021). Agricultural Wilding: Rewilding for Agricultural Landscapes through an Increase in Wild Productive Systems. *Journal of Environmental Management, 284*, 112050. https://doi.org/10.1016/j.jenvman.2021.112050

Vow Food (Director). (2023). *Forged by Vow—Introducing the Mammoth Meatball* [Video recording]. Retrieved from https://www.forgedbyvow.com/the-feed/the-mammoth-meatball/

Wallace, R. G. (2016). *Big Farms make Big Flu: Dispatches on Infectious Disease, Agribusiness, and the Nature of Science*. Monthly Review Press.

Wallace, R., Liebman, A., Chaves, L. F., & Wallace, R. (2020, April 1). Monthly Review | COVID-19 and Circuits of Capital. *Monthly Review*. Retrieved from https://monthlyreview.org/2020/04/01/covid-19-and-circuits-of-capital/

WEF, W. E. F. (2018). *Meat the Future: Time for a Protein Portfolio to Meet Tomorrow's Demand*. World Economic Forum. Retrieved from https://www.weforum.org/publications/meat-the-future-time-for-a-protein-portfolio-to-meet-tomorrow-s-demand/

Welin, S., & van der Weele, C. (2012). Cultured Meat; Will it Separate us from Nature? In *Climate Change and Sustainable Development. Ethical Perspectives on Land Use and Food Production* (pp. 348–351). Wageningen Academic Publishers. Retrieved from http://library.wur.nl/WebQuery/wurpubs/479212

Whyte, K. P. (2018). Indigenous Science (Fiction) for the Anthropocene: Ancestral Dystopias and Fantasies of Climate Change Crises. *Environment and Planning E: Nature and Space, 1*(1–2), 224–242. https://doi.org/10.1177/2514848618777621

Willett, W., Rockström, J., Loken, B., Springmann, M., Lang, T., Vermeulen, S., Garnett, T., Tilman, D., DeClerck, F., Wood, A., Jonell, M., Clark, M., Gordon, L. J., Fanzo, J., Hawkes, C., Zurayk, R., Rivera, J. A., De Vries, W., Majele Sibanda, L., et al. (2019). Food in the Anthropocene: The EAT–Lancet Commission on Healthy Diets from Sustainable Food Systems. *The Lancet, 393*(10170), 447–492. https://doi.org/10.1016/S0140-6736(18)31788-4

Wynne-Jones, S. (2022). Rewilding: An Emotional Nature. *Area, n/a*(n/a). https://doi.org/10.1111/area.12810

Yusoff, K. (2018). *A Billion Black Anthropocenes or None*. University of Minnesota Press.

Zalasiewicz, J., Waters, C. N., Ellis, E. C., Head, M. J., Vidas, D., Steffen, W., Thomas, J. A., Horn, E., Summerhayes, C. P., Leinfelder, R., McNeill, J. R., Gałuszka, A., Williams, M., Barnosky, A. D., de B. Richter, D., Gibbard, P. L., Syvitski, J., Jeandel, C., Cearreta, A., et al. (2021). The Anthropocene: Comparing Its Meaning in Geology (Chronostratigraphy) with Conceptual Approaches Arising in Other Disciplines. *Earth's Future, 9*(3), e2020EF001896. https://doi.org/10.1029/2020EF001896

10

Phenomenology of Repair: Negotiating the Social Order in the Anthropocene

Technologies of Repair as a Form of Planetary Redemption

The integration of the idea of repair into specific sets of scientific practices and social discourse pertaining to the powers of technosciences is what links together the main themes of this book. Cell meat is a form of repair work. On a direct level it regenerates tissues from a simulated wounded tissue to stimulate cell growth and proliferation and, on a larger scale; it aims to resolve some of our most pressing and urgent planetary problems, that is, industrial meat production and its deleterious effects on the environment. In addition, much of biotechnology is concerned with the tools used in cell repair (i.e. genome editing), regeneration, and DNA fixes to restore gene function, for instance. As such, repair is viewed as a form of (re)vitalization in the cellular meat project as well as in molecular biology in general. Seen as a paradigm, repair can help us unravel the types of material interventions and metaphors that become the imperative for reversing past mistakes and damages and that stand in for greater moral and financial investments in the technosphere.

What we call the technoscientific project can be conceived as a 'repair project' on a grand scale. It embodies modernist ideas of control and

E. Abergel, *Dead Meat*, https://doi.org/10.1007/978-981-97-9049-4_10

domination as discussed earlier. The idea of repair, in the sense of fixing or making right functionally or in the sense of restoration meaning to recreate what was lost or ultimately in the sense of reparations, entails a totally different set of issues. However, repair can also signify recuperating the losses imposed by specific conditions or destruction. It can restore bodily integrity and a sense of wholeness after an accident, some form of trauma, or to something that has been ruptured. This also applies to reparations linked to different forms of dispossession. Catastrophic events such as wars or Earthquakes or extreme climate events also provide fertile ground for repair work. Seen from the standpoint of repair as a material process rather than its psychic or psychological aspects which go beyond our analysis, we can analyze how science and technology can perform 'repair acts' in order to respond to a series of different crises, namely, geological, sociocultural, and existential. The concept of repair serves as an entry point to understanding how ideas about reconstructing life, animals, and nature have become such a dominant feature of biomedical and, more broadly, life technologies. Repair also leads to the idea of reconstruction, regeneration, and even recovery either from very much localized trauma, illness, or from wider historical, sociopolitical, pandemic, and geological disasters. Reconstruction, regeneration, and recovery are ends in themselves but they also underlie a need to restore past forms intact by acting as a mode of healing. However, when something is lost, has become extinct or is on the verge of extinction, is dying or being lost forever, reparative efforts serve to construct a present and a future in order to avoid discontinuity, and in the case of the Anthropocene to provide a sense that something is being done to fix things. The reparative impulse wards off the threat of destruction, death, disease, and aging. It also saves us from thinking we might disappear. It then enables a belief in self-repair, the idea that, through the powers of technology and consumer capitalism, lifestyles can turn individuals into moral beings by setting new collective behaviors about health, self-care, food, eating, and the killing of animals, for instance. These new behaviors are dictated by the Anthropocene as we saw in the narrative chapter, but they are also very much guided by the notion of 'disruption' and the possibilities offered by new technologies. In this case, 'disruption' is identified by its promoters with emancipation from old modes of production and the idea of constant renewal of knowledge, the freedom of civilizational progress in light of what seem like desperate times for the human and other species calling for not desperate measures but for radical technologies. It is a form of chaos that is future-oriented but very much anchored in present-day interests and practices. As Stiegler states, disruption moves fast, as it is filled with both hope and fear (Stiegler, 2019). Repair also entails recovering memories, good and bad, and "the creative destruction of brokenness" (Spelman, 2002). In that sense, it is a call to action filled with hope and possibilities, promises of better futures.

The example of cellular meat is instructive. As we saw, many narratives that inform the ways in which the planet, the treatment of animals, the industrial meat complex and ultimately our bodies, start from a premise of repairing wounds. CM came about because of progress in regenerative medicine. The goal of regenerative medicine is to restore, replace, and eventually regenerate injured or lost tissues and organs to their normal function using the body's own repair mechanisms or using advanced technologies such as stem cells, tissue engineering, and gene therapy. Life and environment are seen as an object of constant improvement, experiment, and in need of constant repair and enhancement through technology. So is the human body. Prosthetic and aesthetic, anti-aging, and performance enhancement all constitute forms of corporeal repair. Similarly, using the cell's repair mechanisms has opened up new spaces for the advent of techs like CrisprCas9 or the use of stem cells in regenerative medicine and biomedical research.

The application of stem cells to construct edible tissues illustrates the harnessing of cellular healing mechanisms to repair damaged tissues, thereby stimulating cellular repair and orchestrating the formation of a living structure to be consumed as dead meat tissue. By simulating a tissue injury or a muscle tear, the process involved in the production of cell-based meat is to induce and direct cellular healing and proliferation. By promoting healing, the process of repair not only is used to heal a torn piece of muscle but it also helps to regenerate and restore tissue architecture as well as its function (Krafts, 2010). Tissue engineering aims to simulate natural bodily processes without bodies, according to what we think we know about them; in a sense this type of repair work is premised on making meat about human design where human desires and intentions are reflected (Galusky, 2014).

I also use the term phenomenology of repair to discuss cell meat as a kind of estrangement from the body of the eaten and of the eater. For Malabou discussing what she calls the ontology of the accident and the post-traumatic plasticity that emerges from such an event, phenomenology is relevant as it signifies:

> Something shows itself when there is damage, a cut, something to which normal, creative plasticity gives neither access nor body: the deserting of subjectivity, the distancing of the individual who becomes a stranger to herself, who no longer recognizes anyone, who no longer recognizes herself, who no longer remembers her self. These types of beings impose a new form on their old form, without mediation or transition or glue or accountability, today versus yesterday, in a state of emergency, without foundation, bareback, sockless. (Malabou, 2012, p. 6)

Malabou claims that traumatic events affect beings in psychic ways but also in physical ways where not only can accidents alter appearance but, more importantly, in terms of the nature and inner architecture of the body. What she calls destructive plasticity is split identities; it is the sudden interruptions that have come because of attacks on the individual, it is the loss of subjectivity after a traumatic event. The reconstruction of beings following trauma necessitates what she calls various forms of destruction, thus creation comes after destruction. Destructive plasticity is repair work of another sort; it does not repair the original form but allows for altered forms of self to emerge as unrecognizable. In her essay Malabou discusses events that ultimately render individuals unrecognizable to themselves, becoming a new 'species' due to an infidelity, the loss of a job, war, or a terrible accident. I use her work as an analogy to explain why cell-based meat renders animals into new species that is non-identifiable as either beef, chicken, or fish. While the tissues may originate from these beings' stem cells, they have nothing in common with the animals presumably saved from slaughter. Their nature is radically different in the sense that they not only fail at becoming animal but, also, I would argue at being meat. Cell-based chicken is a form of dissociated form which has little to do with the true nature of being. It is, as she writes, the total and radical displacement of a body—or parts thereof—in space-time. Far from suggesting that the promoters of in vitro meat are trying to recreate actual chickens or are concerned with the metaphysical processes of 'being chicken,' they are nevertheless actively engaged in bio-fabricating, reproducing, and regenerating some form of a chicken or a cow, that is, one they endeavor to make recognizable as such, either by taste, texture, or other physical or symbolic attributes. From an ethical standpoint, one might view this as an effort to repair our relationships with chickens or cows or fish by replacing them with bioassembled edible tissues. On the surface, CM might also be seen as a way to repair the environmental and health damage done by industrial agriculture.

Narratives work because they convey the essence of being that is absent in the manufacturing of cell meat. As we saw, they help to construct a parallel reality. Because destructive plasticity fails to return to the original life form and being, in this case if we take the analogy further, fabricated edible tissues are unlikely to take us back to familiar ways of eating and knowing 'meat.' It is a form of flight, where flight from our conditions of existence and from ourselves has become impossible. This is the opposite of conservation efforts to preserve natural habitats or free farm animals, acculturating us instead to the idea of loss. CM symbolizes a form of appeasement of cultural anxieties about the future. As Shierry Weber Nicholsen discusses in her book *The Love of Nature and the End of the World: The Unspoken Dimensions of Environmental Concern* (2003), the psychological impact of environmental degradation, what she calls the unified sense of threat to our world, impedes our capacity to think about time and the future, the psychological groundwork necessary for thinking about alternatives, and ultimately results in our inaction, condemning us to continue as we are. The solutions we envision to address the environmental crisis often end up reinforcing and deepening modernist ideals. This means that reality under climate change is at once terrifyingly different and completely ordinary; we live our lives in parallel states (Nicholsen, 2003). This means that technologies like cellular meat give us a sense of normality in light of existential threat; they enable the pursuit of the mundane. They make possible the continuation of daily activities while allowing them to coexist with the utterly strange new reality of the environmental crisis. We can continue eating 'meat' a familiar food which does not compromise our everyday life without surrendering to despair and dread about the future. It is a kind of scapegoating where technological futures become cultural containers that give form to our existential dread about 'the end of world' (Ibid). Nicholsen writes about the proliferation of technology and how the machine ideal leads humans to identify with the forces that destroy the world:

> No individual can "save the Earth." In trying to work against environmental destruction each of us is just one small person working with countless others. The fantasy of omnipotence speaks to our wish to escape such protracted and arduous efforts. Destruction offers itself to our fantasy as a form of omnipotence. We imagine it to be effortless and instantaneous. (p. 137)

She goes on to say that technology represents "the vehicle of our contemporary denial of death" (p. 138) because it is non-human and thus represents immortality. The denial of death through the machine ideal results in the rejection of life (Ibid).

Repairing meat has to happen locally, it has to involve farmers and ecologists as well as consumers. Rather than expanding efforts and investments in ecological land and ocean regeneration and the preservation of species biodiversity, and encouraging conservation within natural settings and changing ecologies, the production of meat and other animal products without animals commits us to a false rupture within an inevitable continuity. This represents a false rupture because, while the technological ability to reproduce animal tissues and products like 'meat' or 'milk' suggests an ideal of animal liberation and the end of suffering associated with industrial farming, it simultaneously perpetuates the inherent logic of carnivory. This continuity reinforces our planetary dependence on meat, rather than challenging or transforming it, thus maintaining and even entrenching the destructive dynamics of our consumption habits and global food system. Cell technology might be able to guide and manage living processes to imitate functional animal tissues, but it is no substitute for Nature. "Fake" meat or milk or other fake animal products provide a sense of hope, technologically diverting the vitality of living systems and reorganizing them as a means of facing environmental and social breakdown (Waldby, 2002).

An important part of the aspect of repair is the relationship developed between the agent and the object being repaired. This is what Bryan Kibbe calls 'mindful mending' or what he elaborates as 'cognitive-agentic repair' which involves not only a basic understanding of the causes of damage in order to gather the "knowledge, tools, and skillsets" of the person undertaking the repair of things, spaces, and places but, more importantly, involves developing the relationship of trust/reliance between individuals and cognitively significant objects. He claims that:

> Cognitive-agentic repair does not only target the broken technology or arrangements of technologies, but instead also aims to address the relationship of reliance or trust between a person and the object. (Kibbe, 2014)

If the planet, food systems, and our relationship to animals are what we are trying to repair via the production of non-animal products, then Kibbe's proposition demands that we establish the idea that humans and non-human animals are entangled in the repair process. Hence, there is a moral dimension to the idea of repair, one that extends the repair of the planet, animals, and ourselves as extensions of one another and as being deeply embedded (Ibid). This demands scrutiny about how we situate and devise how we might repair things materially but cognitively as well. What/who do we repair? For whom? According to which cognitive paradigm? And does the discourse of making better and ethical meat right the wrongs of the past; does it make us escape culpability for perpetuating environmental destruction? How does the reordering of animal bodies into edible tissues, as a product of a whole technological and technoscientific infrastructure that uses energy, money, labor, knowledge, logics, etc., move us away from the capitalist urges that create the predicament we find ourselves in?

As Malabou states, the idea of reprogramming "teleological rigidity" originates from the possibilities of life itself, from the depths of its own memory and not as a result of technological violence (Malabou, 2006). Consequently, the locus of disruption is displaced from 'out of the box' thinking in order to spark social change, firmly establishing technology as the main mechanism of 'rupture' within the continuity of capitalist relations. We saw how tissue and genetic banking efforts become virtual farms that produce the parts and repair materials to repair ecologies. As we also mentioned, these solutions do not constitute a way to avert disaster and collapse from the Anthropocene. In fact, they may do more harm than good. It seems to fit into a kind of redemptive project of reassembly and reanimation which tries to bring order to disorder and reorder that which has been deranged. There is also a sense that preserving animal cells, genomic and genetic material in biobanks for producing meat or repairing extinction may figure as a kind of collecting effort which as Baudrillard would say is a way of organizing the experience of possession (Baudrillard, 1988), in this case, of the possession of nature, ensuring immortality. He writes that the collected object is reparative because it facilitates an integration between objectivity and subjectivity: humans collect reparative objects in an attempt to overcome loss. Accumulating possessions defies death because a person literally survives through their collection. Baudrillard implies there may be something alive in the objects humans collect and something dead in the collector. Collecting has de-animating power on the collector. Walter Benjamin saw the art of collecting or the collection as achieving a "particular category of completeness," accomplished through the integration of parts into a historical system (Lin, 2017). So, it seems the impulse of collection is fundamentally reparative. In the case that interests us, this is achieved through cryopreservation as a form of imbrication of death in life, a suspension of life in liquid nitrogen. Piecing together fragments of life, restoring life, reviving memories of the animal whose tissue is about to be reconstructed/cultured, it would seem serves to appease our anxieties about death and dying. Collections of cryopreserved animals, farm or wild animals, precede and anticipate ecological crises.

Other understandings of repair consider it as a social and collective, participatory process and as the collective practice of mending. In vitro meat is none of those things; it is the product of techno-utopian thinking. It creates forms of unexamined imaginaries, shifting focus away from what is possible here and now toward fabricated futures. This creates a kind of spillage into science fiction, the idea that we can produce animal flesh without relying on nature or that we can harness this technology for space exploration—as another form of flight from ourselves. This approach effectively depoliticizes the critical questions about the food systems we wish to support, develop, and engage with as citizens, communities, and eaters. It also shifts the focus away from nature, bodies, and agriculture—encompassing animals, farmers, and farms—and positions the laboratory as the primary site for planetary "repair." In addition, discourses about the artisanal nature of CM hide its industrial and technological complexity and undermine the potential for the kind of systemic change needed to repair food systems.

At the heart of the repair impulse are survival instincts that are a search for "ways to repair the damage inflicted on human and non-human nature by the unfettered sway of these same instincts" (Cook, 2011).[1]

Bibliography

Baudrillard, J. (1988). The System of Objects. *Art Monthly (Archive 1976–2005), 115*, 5–8.

Cook, D. (2011). *Adorno on Nature*. Acumen.

Galusky, W. (2014). Technology as Responsibility: Failure, Food Animals, and Lab-grown Meat. *Journal of Agricultural & Environmental Ethics, 27*(6), 931–948. https://doi.org/10.1007/s10806-014-9508-9

Kibbe, B. (2014). *Mindful Mending: The Repair of Thought and Action Amidst Technologies* [Doctoral Dissertation, Loyola University Chicago]. Retrieved from https://ecommons.luc.edu/luc_diss/1093/?utm_source=ecommons.luc.edu%2Fluc_diss%2F1093&utm_medium=PDF&utm_campaign=PDFCoverPages

Krafts, K. P. (2010). Tissue Repair. *Organogenesis, 6*(4), 225–233. https://doi.org/10.4161/org.6.4.12555

Lin, L. (2017). *Freud's Jaw and Other Lost Objects: Fractured Subjectivity in the Face of Cancer*. Fordham University Press.

Malabou, C. (2006). Les régénérés: Cellules souches, thérapie génique, clonage. *Critique, 709–710*(6), 529. https://doi.org/10.3917/criti.709.0529

Malabou, C. (2012). *Ontology of the Accident: An Essay on Destructive Plasticity*. Polity Press.

[1] In her book entitled *Adorno and Nature*, Deborah Cook describes how Adorno might engage with various forms of environmental activism (2011).

Nicholsen, S. W. (2003). *The Love of Nature and the End of the World: The Unspoken Dimensions of Environmental Concern*. MIT Press.
Spelman, E. V. (2002). *Repair: The Impulse to Restore in a Fragile World.*
Stiegler, B. (2019). *The Age of Disruption: Technology and Madness in Computational Capitalism* (English ed.). Polity Press.
Waldby, C. (2002). Stem Cells, Tissue Cultures and the Production of Biovalue. *Health: An Interdisciplinary Journal for the Social Study of Health, Illness and Medicine, 6*(3), 305–323. https://doi.org/10.1177/136345930200600304

11

Competing Vitalities: Separating the Real from the Fake in the Search for Authenticity

This chapter will expand the concept of competing vitalities included in the title of the book. The concept of competing vitalities is central to understanding the dynamics of food, particularly in the context of cell-based meat (CM) and traditional farming. Vitality refers to the life force or energy inherent in living organisms, whether it be the plants, animals, or microbes that make up our food systems. In the case of food, this vitality exists before consumption—food is alive in various ways before it is harvested, processed, or prepared for eating. In traditional farming, these vitalities interact within complex ecosystems. Plants and animals, along with the humans who cultivate and care for them, participate in a web of life that is both cooperative and competitive. For instance, in traditional farming, livestock are part of a broader ecosystem, interacting with pastures, predators, and human caretakers. These interconnected vitalities form a dynamic system in which the life force of one organism is often dependent on, or in competition with, others. When it comes to CM, the notion of competing vitalities becomes even more pronounced. CM involves the cultivation of animal cells in a lab, where the vitality of these cells is manipulated, managed, controlled, and appropriated. This process abstracts the vitality of the living animal, reducing it to a form of biological material that can be manipulated outside of its natural ecosystem. In

E. Abergel, *Dead Meat*, https://doi.org/10.1007/978-981-97-9049-4_11

doing so, CM creates a new kind of vitality—one that is disconnected from the traditional life processes of farming and the broader ecosystems in which it is embedded.

Furthermore, the cells used to produce cell-based meat are cultivated in a highly controlled environment, requiring nutrients and energy that are sourced from elsewhere, often through industrial processes and industrial agriculture. These inputs are products of other vitalities—plants and animals that have been harvested and transformed to sustain the lab-grown cells. The potential increased demand for these products means that the vitality of CM is thus in direct competition with the traditional forms of life and ecosystems from which these inputs are derived. Moreover, the people involved in this process—from farmers to scientists—also represent different forms of vitality. In exploring food and CM, the idea of competing vitalities highlights the tensions between different ways of producing and understanding food, including the communities that sustain life and ecosystems.

CM still lives in the realm of promises, expectations, and imaginaries. The competing vitalities in this context are not merely biological or social, but also ideological and utopian. CM competes with traditional farming and the bodies of animals, not only for resources but also for the symbolic and material dominance over the future of food. The lab-grown cells' vitality is dependent on inputs extracted from other living systems, including other animals, yet it is presented as a more 'rational' and 'sustainable' solution within a techno-utopian framework. This framing obscures the social and ecological implications of such extraction and the alienation of both nature and labor from the food production process.

The next two sections will discuss specific ways in which vitalities are seen to compete: one which is a continuation of the legacy of Enlightenment thinking which abstracts bodies and nature and a second one in the forms of postnatural/technologized vitalities constructed within a technoscientific context for specific purposes (bio-objects or hybrids) compete with naturally occurring organisms and living entities. The first competing frame involves an understanding of how CM can be viewed from a critical theory perspective, setting up the idea of the jargon of authenticity and counterfactual hope. In the second scenario, the "boundary work" involved in asserting the authenticity of lab-grown meat sets the stage for discussing not only the limits of biology, but also the ideological implications contained within these claims. Ideological thinking and competing vitalities were already considered in the section dealing with Meat Culture Wars but, here, the focus is on the dichotomy between real and fake meat.

Laboratory-Grown Life Forms and the Emancipation from Nature

Let me start this section with a quotation that encapsulates the mode of thought to which the ideology underlying technologies like CM commits us:

> In Dialectic of Enlightenment, Adorno and Horkheimer had claimed that human emancipation from the terror and mindlessness of our enslavement to nature was achieved only at the cost of self-mutilation, and, eventually, self-negation. We gain our freedom from nature only by dominating our own natural being. Self-determination is therefore simultaneously self-domination; and the metaphysical apotheosis of the self ends in its dissolution. The rationality we use to wrest ourselves free of nature becomes the instrumental reason that organizes all of human society. This is the ironic denouement of the history of enlightenment, a pattern of thought traceable to the very origins of human consciousness and culture. Even our most primitive systems of magic and mythology, for example, were originally attempts to anthropomorphize and thereby gain some measure of control over our natural surroundings. But as it grew in power and sophistication the Enlightenment gradually dismantled these anthropomorphic projections as well, until nature was fully disenchanted. The fully enlightened world is therefore also a fully administered world: nature presents itself to the modern subject as little more than a meaningless and fungible substratum upon which reason can thoughtlessly exercise its authority. But the triumph of unreflective reason locks us into gestures of fatalistic repetition that ironically resemble the myths reason aimed to destroy. The Enlightenment that once promised freedom as its highest ideal has ended by creating a human world in which freedom is little more than a counterfactual hope. (Gordon, 2016, p. 103)

In Peter E. Gordon's quotation, the idea of counterfactual hope brought on by Enlightenment thinking is clearly laid out. The instrumentalization of nature and its deleterious consequences as the products of "unreflective reason" have resulted in what Adorno and Horkheimer refer to as the "ironic denouement of the history of enlightenment" where human freedom and desire for emancipation from nature have led to a situation of the imminent collapse of the Earth's vital systems. Our emancipation from nature has led to a state of self-mutilation and self-negation—the price of wanting to refute the nature within us.

This quotation, emblematic of Frankfurt School thinking, expresses all the difficulties that resorting to "laboratory meat" as a potential solution implies. Let me reiterate that it is not only the technology itself that is problematic but also the mode of thinking it represents. By presenting itself as a solution to pressing issues like climate change, with the promise of synthetically producing "meat" to mitigate the destructive impact of industrial animal agriculture, the production of edible tissues outside of animals or without animals is embedded within visions of liberation from nature. This aligns with the concept of counterfactual hope. Contained within the in vitro meat narrative are animal welfare and human health as important components of its storytelling. While the logic and the reasons for producing synthetic meat might be compelling, signaling a new paradigm of disruption and a cultural shift, they work within a particular kind of logic and trajectory that are inherently part of enlightenment thinking and doing. The beauty of CM, as a panacea and as a solution to alleviating the climate crisis, is the simple almost childlike nature of its enunciation: replace meat from animals with a laboratory-synthesized equivalent and resolve a whole host of issues at the same time. Its simplicity lies in the idea of shifting production away from living animals into the culturing of tissues while interrupting the devastation wrought by large agro-industrial complexes that supply meat across the globe.

Industrial agriculture is portrayed as a kind of death culture, in terms of turning living comestible animals into consumable flesh, meat, and in terms of the killing of other species and ecosystems that make its deployment possible on a global scale. Laboratory-synthesized flesh, on the other hand, is viewed within the realm of the possibilities of life, for cells, unsacrificed animals and the environment and hence humanity as well as all the energy and matter spared to produce meat. This works, as we've discussed previously, through the process of creating hope for the future mostly through the presentation of new technologies and the potentialities associated with animal-free products as important elements of this hopeful future. This in turn functions through specific types of representations of life, through techniques of harnessing various forms of vitality and viability, each with their distinct value (biovalue). Clearly, these types of value are inscribed within a political economy of *life itself* but, more precisely, through what we attempted to describe as the domain of cellular economies. These then become performative and socially embedded within technologies, markets, and eating cultures which enact these narratives as desirable and true. Within the utopian narratives about inventing the best possible futures, animal-free products become an important element in the stories. These stories are anchored in larger imaginaries about overcoming the challenges posed by climate change and the looming prospects of multiplanetary futures. Underlying these narratives about recreating life and vitality in the laboratory through science, the illusion of the possible detachment from nature endures. It is reinforced by images such as the iconic images of Earth from space, the "Blue Marble" photograph taken during the Apollo 17 mission, giving us a god-like view, an illusion of omnipotence and control over the natural world. This perception fosters a sense of separation and dominion, suggesting that humanity is not an integral part of the Earth's ecosystems but rather an external force capable of manipulating and mastering nature. As Latour (2010) argues, the modernist separation of nature and society underpins much of Western scientific thought, leading to a dualistic worldview where nature is seen as an object to be controlled and engineered. This perspective is exemplified in cell meat, synthetic biology, and bioprinting, where life itself is reconstituted in the lab. Is the attempt to recreate a human-dominated world, a process contained in the postnatural, truly a form of liberation from nature? Or does it merely exacerbate the detachment from natural systems? What are the detrimental effects of this increasing separation from nature, especially when framed as a form of "freedom"? Does this desire to transcend nature reflect an escape from natural constraints as a form of denial of the inherent risks associated with technological progress?

It would appear that the drive to break free of nature, as a "meaningless and fungible substratum," reflects a desire to assert our power and Promethean prowess. The simultaneous drive to preserve and destroy nature reveals deep contradictions in our approach to environmental and technological issues. Can technologies like CM genuinely address the root causes of environmental crises by creating the conditions for the synthesis and artificialization of vital processes? Or, as Gordon suggests, are we locked into a "fully enlightened world" where the rationality we use "to wrest ourselves free of nature becomes the instrumental reason that organizes all of human society." Needless to say, this corresponds to counterfactual hope.

Counterfactual hope refers to wishful thinking as a form of simulacrum, like the meat CM is trying to create. This means that technologies like CM rely on a utopian vision of how we might save ourselves from impending disaster or catastrophe, perpetuating the same processes and modes of thinking characteristic of Modernity. False hope is perpetuated through the language and discursive frameworks, the jargon of authenticity, used to describe cellular meat, including its potential to replace animal life.

Cultivated Meat Is Real Meat

Repeatedly, companies emphasize that lab-grown meat is real meat like any other, as it is composed of animal cells. They also insist that it doesn't come from a laboratory but from a production facility or a meat factory. According to CM advocates, cultivated meat, which is as palatable as that currently available on the market, is simply produced differently: "Cultivated meat is normal meat based on animal cells, but the muscle tissue is grown in bioreactors instead of in animals" (Micro Meat). It offers the same mouthfeel, texture, taste, smell, and flavor. As stated by Vital Meat: "Cultivated meat is meat. Therefore, do not expect to be surprised by its taste."

In 2017, Josh Tetrick of the company Eat Just (then called Hampton Creek) explains when asked if their product is real chicken: "We don't like to talk about "synthetic" chicken, because it isn't synthetic. It's real chicken meat. It is just produced in a different way" (Clifton, 2017). While acknowledging that chickens are complete animals, Tetrick affirms that cultured meat will "challenge lexicographers" because it challenges what meat is. He's also not clear whether CM should be considered vegetarian or vegan, due to the "fact" that no animal was involved in its production.

Similarly, these companies argue that the meat they produce is natural to the extent that they are merely replicating outside the animal's body what naturally occurs within it. Some companies even compare lab-grown meat production to greenhouse vegetable cultivation or a baker maintaining sourdough starter. Others claim that the production of cultivated meat is as natural as current mass meat production (which they argue is far from natural), except that it does not contain antibiotics, hormones, etc. This distinction is sometimes highlighted by companies to support the idea that cultivated meat is, in essence, even more natural than conventional meat.

Another aspect of authenticity is linked with the purity of the manufactured product as we saw in the section on disease outbreak narratives. This was also reflected in the name Clean Meat which was eventually abandoned because it was a stark reminder of Clean Tech but also because it implied that conventional agriculture was somehow unclean. Besides, it also conjured up images of meat being washed or cleaned, not an idea that would appeal to consumers.

When defending the naturalness and authenticity of CM, companies are in fact often reacting to the perceived opposition between the authenticity of their products and their technicity (Miller, 2012). The tension derives from the fact that even though cells from 'real' animals are used to produce meat, the technological interventions involved to create the final product tend to compromise its authenticity. As noted by Miller, this involves addressing consumer anxieties about the technological nature of food by saying that everything we eat is cell-based, for instance. This also involves reaffirming the value of CM within the continuum of traditional foods, like the idea of "craft meat" and the association of production facilities or precision fermentation with breweries.

In its educational publication, *From Fauna* recounts that in 2018, the start-ups UPSIDE Foods and the North American Meat Institute

(NAMI) petitioned the White House, arguing that cell-based meat (CM) should be considered real meat. They used this argument to advocate for the USDA and FDA to establish regulations and guidelines for the pre-market evaluation of CM products, similar to those applied to conventional meat.

As the continuity with industrial meat production clearly demonstrates, 'real meat' in the context of CM actually presents us with some difficult questions. The transition from traditional industrial meat production to CM extends beyond conventional categorizations and advocates call for a new lexicon and a new definition of what meat can be, including what it can do. Inquiring into the nature of 'real meat' intersects with epistemological concerns about authenticity and identity. This is where the notion of ontological ambiguity discussed earlier applies. This discussion also reveals cultural and ideological dynamics. The term "real" meat carries cultural weight, often associated with traditional values, sensory experiences, and dietary norms. By labeling CM as real meat, proponents seek to integrate it into existing cultural frameworks and consumer expectations, framing new technologies and new foods in familiar and traditional language. Drawing a parallel with Sexton's remarks about the biopolitics of alternative proteins, not only is the edibility and authenticity of meat being renegotiated at the molecular level, but as we've shown, it actively mobilizes the Anthropocene as a new framework for managing our visceral and sensory understanding of food (Sexton, 2018). In this way, the production of cellular meat not only redefines what is considered edible but also reframes the cultural and sensory experiences of eating within the broader context of environmental crises and techno-scientific interventions.

Bibliography

Clifton, M. (2017, July 8). *Hampton Creek Founder Tetrick Pledges his "Clean Meat" will be Vegan*. Animals 24–7. Retrieved from https://www.animals24-7.org/2017/07/08/hampton-creek-founder-tetrick-pledges-his-clean-meat-will-be-vegan/

Gordon, P. E. (2016). *Adorno and Existence*. Harvard University Press.

Latour, B. (2010). *Nous n'avons jamais été modernes: Essai d'anthropologie symétrique* (Nachdr.). Editions La Découverte [u.a.].
Miller, J. (2012). In Vitro Meat: Power, Authenticity and Vegetarianism. *Journal of Critical Animal Studies, 10*(4), 41–63.
Sexton, A. E. (2018). Eating for the Post-Anthropocene: Alternative Proteins and the Biopolitics of Edibility. *Transactions of the Institute of British Geographers, 43*(4), 586–600. https://doi.org/10.1111/tran.12253

12

Conclusion

This book is an attempt to engage with current and past studies about nature and technology, in a deep and critical way, through the language, images, and power of food in the context of the Anthropocene. Throughout my inquiry into CM, I have explored the structure and formation of various narratives, focusing on their sociological significance (Arnold, 2018). I have examined different uses and interpretations of concepts such as life, living processes, living matter, bio-objects, forms of vitality, and livelihoods. These concepts were discussed in the context of interactions between humans and non-humans, the crisis of capitalism, and the Anthropocene.

My exploration of CM has led me to develop intertwined themes contained in the following questions: (1) What does cultured meat say about the state of human/animal/nature relations? (2) Is "cultured meat" separable from its mode of species life and being? (3) Are animals ontologically present in meat, eggs, milk, etc.? (4) What values and embodied power relations are constituted in the creation, production, and consumption of these edible tissues? (5) How might we understand the present by looking into the sociotechnical imaginaries surrounding cultivated meat? (6) What do narratives of food futures and Anthropocene diets tell us about the role of technology in the climate emergency? (7) How does

E. Abergel, *Dead Meat*, https://doi.org/10.1007/978-981-97-9049-4_12

agency circulate during our bodily engagements with these new forms of life as they become food? (8) What role do Anthropocene narratives play in the determination of what or who becomes edible?

I have used different approaches and ways of tackling these questions by conveying both the stories that CM tells and those that remain untold. By engaging with the complexities of CM, I seek to illuminate not only the narratives of progress and innovation but also the underlying tensions and contradictions inherent in our attempts to reconcile technological advances with pressing ecological and existential challenges. This book is the result of a conviction that contemporary analyses of food and agriculture, on the one hand, and science and technology, on the other, have reached a critical juncture. In a sense, one could say that the Enlightenment project has placed us at an impasse which engages us to critique the logic of Rationality and its instrumentalization of Nature inasmuch as it dictates its social and biological organization. As a result, it is clear that it is not the details of the technologies by themselves but the material objects they produce and their social mode of production which deserve analysis. Since technology and scientific rationality have become universalized, they "project a historical totality" (Marcuse, 1991) which focuses on operationalizing and instrumentalizing biological processes such as the cellular reproduction of meat. Operationalism is at the core of the scientific project according to Marcuse, and it makes use of rationality as objective in order to create the idea of "methodical construction; organization and handling of matter as the mere stuff of control, as instrumentality which lends itself to all purposes and ends—instrumentality per se, 'in itself'" (p. 159). As Nature is increasingly instrumentalized by science, it becomes a scientific enterprise conceptualized and redefined within a purely instrumental horizon (Ibid, p. 161). Marcuse contends that technological rationality redefines human reason. As instrumental reason dominates our understanding of the world, humans become the ultimate object of instrumentalization, perpetuating old forms of domination while creating new ones. It is therefore my position that meat or its analogues are manifestations of long-standing historical processes of the industrialization of nature according to Goodman, Sorj, and Wilkinson's model (1987) (via appropriationism and substitutionism) and its redefinition of life and its rational subversion (via operationalism and instrumentalism) following the logic of late capitalism. Drawing on Simondon's work on the mode of existence of technical objects (Simondon, 2016), one could argue that cell-based meat comes at the end of a long technological evolutionary process that not only mediates human relationships with Nature but is a significant barometer of human reality in the context of the Anthropocene. Rather than opposing accounts of technoscience, one where instrumentality is viewed as an autonomous apparatus of domination and the other which seeks to embed technology within material culture, I see cultured meat as existing within a continuum between these two poles. I tried to situate cultured meat, within a deterministic view of technology which prefigures humans becoming technologies, on the one hand, and within the indeterminacy of emergent technological forms that create new types of sociality, on the other. In the sociology of science and technology, the concept of technological determinism is often challenged by social constructivist perspectives, which emphasize that technoscientific trajectories are shaped by social, cultural, and political contexts, rather than being inherently predetermined. However, despite the ontological indeterminacy of future technological developments, there is a recognition that, at a certain stage, technologies can become "locked-in." This lock-in occurs when technologies become deeply integrated into and are sustained by existing systems of values, institutional norms, historical narratives, and socioeconomic networks (Selin, 2008). Once embedded in these structures, technologies tend to perpetuate the logic of the prevailing social order, reinforcing established practices and limiting the scope for alternative pathways. Thus, while the initial development of technology may be open to various influences and interpretations, its eventual stabilization often aligns with and reinforces the dominant sociotechnical paradigms. This is because the future (as present-future) is not totally indeterminate and technologies such as CM contain predetermined elements that create a bridge between the past, present, and imagined future (Selin, 2008).

The reductionist approach, wherein the technosciences attempt to replicate or alter parts of nature in isolation, fails to capture the full spectrum of ecological and biological interactions that are essential to understanding life in its entirety. By acknowledging the vitality and autonomy of cells, we move away from an anthropocentric view that sees life simply as a resource to be exploited. The very essence of an animal, its body, and its role within an ecosystem are emergent properties that arise from a myriad of interconnected factors—genetic, environmental, behavioral, and more. By focusing on isolated components, we risk missing the holistic and often subtle interdependencies that are crucial for the resilience and sustainability of natural systems. Moreover, this approach reflects a broader ideological commitment to technological mastery over nature, where complex living systems are seen primarily as resources to be engineered and optimized for human benefit. This mindset can lead to unintended consequences, as the manipulation of one aspect of nature often triggers unforeseen effects within the broader system. It is imperative, therefore, to adopt a more integrative and reflective stance that acknowledges the limitations of our knowledge and the inherent complexity of natural systems.

This reductionist approach to cellular meat mirrors our broader reliance on technological solutions to address complex problems such as climate change. Just as geoengineering and carbon capture are seen as quick fixes for the symptoms of climate disruption, cellular agriculture is often presented as a panacea for the ethical and environmental issues associated with traditional livestock farming. However, these technological fixes can overlook the root causes of the problems they aim to solve and may lead to unforeseen consequences. For instance, the energy consumption and ecological impacts of large-scale cellular meat production remain uncertain, paralleling concerns about the unintended effects of climate interventions.

Addressing these challenges requires a shift toward a more integrative and reflective approach, recognizing the limitations of our current knowledge and the interconnectedness of natural systems. In the context of cellular meat, this means considering not only the technological feasibility of lab-grown products but also their broader ecological and social implications. It calls for a holistic perspective that takes into account the sustainability of production methods, the well-being of ecosystems, and the ethical treatment of animals. These ways of improving animal agriculture are all within the realm of the possible, whereas technofixes make us believe that all the problems related to agriculture are technological. By doing so, we are foreclosing solutions that already exist and that can address the structural failures within industrial agriculture and our global food system. These solutions, some traditional some new, are being implemented all over the world on small and large farms, and in various ecosystems. Rather than focusing our attention on systemic changes that promote long-term sustainability and ethical responsibility, climate change or animal agriculture are constructed as engineering problems, preventing us from moving beyond narrow technological solutions. I recently wrote the following statement in a popular environmental magazine:

> What characterizes technofixes is that they are often linked to hubris, the idea that humans can transform life and nature into an extension of human design. Industry leaders claim that they can make meat better, more efficiently and faster than the animal or that they can "trick" cells into behaving as though they are still part of the animal to optimize growth. As animals disappear from agriculture and are progressively dematerialized, nature also disappears as something that human action cannot affect as it gets absorbed into a series of technological projects (cell meat, geoengineering, etc.), alienating us further from the environment we seek to protect. (Abergel, 2022)

Consequently, it is important to not view cellular meat as simply a comestible but as a product of a particularly reduced and narrow view of nature, creating assumptions about its ability to provide necessary nourishment to consuming bodies while feeding the future. This perspective entails questions of response-ability, as discussed by Haraway when she calls for deep awareness of the ways in which our actions—especially those involving biotechnological interventions—impact the world and the beings we share it with. Later, Galusky, uses the concept to question the consequences of our technological choices and challenges us to consider what kind of nature we are engineering when we engineer animals out of their bodies. Who will consume this meat and why? Or "The question is, when we create meat in the lab, what stories about us and our appetites will our creations tell?" (Wurgaft, 2019, p. 120) This is relevant, especially in light of the fact that environment influences health and environmental degradation transits through the bodies of eaters and eaten.

Food and eating are at the core of this book, CM is about re-imagining meat and animal agriculture. I have drawn from a diverse set of food and agricultural writers and thinkers in order to make sense of the *how, what, who, and why* of eating practices. I have, as have food geographers and anthropologists, focused on the idea of edibility, because what makes something like CM edible involves not only material and immaterial qualities but symbolic ones as well (Douglas, 2001; Roe, 2006; Sexton, 2018). Eating as identity politics is a topic that runs through the book, as does the theme of meat as biopolitics because meat catalyzes a series of politics such as environmental, sexual, gender, body, maternal, class, religious and science and technology politics to name a few (Chatterjee & Subramaniam, 2021). The title of a paper I once encountered, *You Are What You (Don't) Eat? Food, Identity and Resistance*, perfectly encapsulates the profound power of food on the body and the land as contested sites of cultural struggle (Cooks, 2009).

The proliferation of 'bio-objects' (Vermeulen et al., 2017), that is, the products of technoscientific and/or medical interventions in research that defy the boundaries between living and non-living, organic and inorganic, as well as natural and artificial, provide a framework for exploring how stem cell research in biomedicine creates the material conditions for living edible muscle tissues to emerge as an agricultural commodity. The idea of bio-objects, inspired by Latour's concept of hybrids: entities straddling the natural and the social world, is also useful for understanding how new forms of in vitro vitality can mediate new relations as they not only create novel and subversive categories of life but they also generate new embodied forms of sociability or social relations (Tamminen & Vermeulen, 2012). The process of bio-objectification, which describes the passage of bio-objects from the research laboratory to the commodity form, gives us a glimpse into the various phases of proliferation, circulation, and commercialization met by cell-based meat as it challenges and blurs institutional, social, and cultural boundaries. In vitro meat and laboratory-based future foods necessitate that we explore tensions between phenomenological questions surrounding the reality and the hype of technologies, as biotechnologies redefine our relationship to life and living systems. Doing so means paying attention to the promises and imagined futures these new ventures take for granted as they insert themselves into particular environmental narratives. As this quote suggests:

> For cultured meat to move the needle on climate, a sequence of as-yet-unforeseen breakthroughs will still be necessary. We'll need to train cells to behave in ways that no cells have behaved before. We'll need to engineer bioreactors that defy widely accepted principles of chemistry and physics. We'll need to build an entirely new nutrient supply chain using sustainable agricultural practices, inventing forms of bulk amino acid production that are cheap, precise, and safe. Investors will need to care less about money. Germs will have to more or less behave. It will be work worthy of many Nobel prizes—certainly for science, possibly for peace. And this expensive, fragile, infinitely complex puzzle will need to come together in the next 10 years.
>
> On the other hand, none of that could happen. (Fassler, 2021)

Whether sacred or secular, the topic of meat confronts us with our own animality and our desire for self-preservation. The term post-animal suggests the possibility of producing animal flesh, as well as milk and eggs, without animals. This raises questions about the technological reproducibility of animals and their parts in the laboratory to be used as food as well as revealing how 'animality' and nature enter into technological systems, and, inversely, how these categories are manifested into technological systems.

As I was writing this book, I asked myself if being critical of cellular agriculture meant necessarily adopting a conservative point of view. Similarly, does calling for 'natural' meat or no meat at all constitute an essentialist, privileged position? Or, does a critique of technological progress, in the name of ecological sustainability and a call to return to more land based, humane farming using existing knowledge and current technologies, echo alt-right ideologies about the purity and virtues of traditional farming? Some have pointed out that the environmental movement's emphasis on the local can project forms of conservatism, provincialism, xenophobia, and even anti-immigrant sentiment because it tends to romanticize autochtonous relationships with the territory (Stanescu, 2011). Clearly the views expressed in this book do not align with the nationalistic, anti-environmental and climate change denialism that some of these movements represent.

Far from arguing in favor of factory farming and animal suffering, I question the underlying significance of potentially achieving the biofabrication of meat tissues in the laboratory. While I do not object to making industrial agriculture better for animals and people and more climate friendly, I would, however, reject schemes that promote large-scale technological solutionism and opt instead for investments in small farm futures, low-tech foods, and strengthening local and rural economies (Smaje, 2020, 2023). Food analyst Holt-Giménez writes that despite the push to industrialize agriculture and "capital's avoidance of actual farming," the "petty producer"—peasants, smallholders, and family farms—persists on the margins of capitalist agriculture and is often more productive than industrial farms (Holt-Giménez, 2017). Considering that peasant agriculture is a major food provider for 70% of the world's population, while the industrial agricultural system feeds less than 30%, and that peasant foodwebs are responsible for breeding and maintaining 8774 rare breeds of livestock species compared to fewer than 100 commercial breeds, it is crucial to contextualize the contributions of these food systems as playing a major role in global food security, biodiversity conservation, and climate resilience (ETC Group, 2017). Because, in my view, the ontological ramifications of eliminating animal bodies out of the meat production process surpass our understanding of our actions.

The concept of ontological collapse as explained by Weisberg, in the context of animal biotechnology is a good way to end this book. The scientific achievements and the benefits promoted by high-tech projects and technoscientific projects like cellular agriculture represent an ontological collapse between animals and the technical-economical apparatus (Weisberg, 2015). Her definition applies to the breakdown and dematerialization of animals as cellular substrate as well as the isolation of cell lines derived from selected individuals. She writes:

> By ontological collapse, I mean the elimination of fundamental ontological tensions between embodied subjects and the principles of scientific, technological, and economic rationalization. Biotechnology imposes this col-

> lapse in various ways: by genetically "reprogramming" animals to serve as uniform commodities, by abstracting them into data and code, and, in some cases, by literally manipulating their movements with computer technologies. These and other forms of ontological violence not only lead to profound physical suffering for the animals involved, but also distort the phenomenological basis of their existence, especially their perceptual experience and expression of subjective time and space. In subordinating nonhuman animals to the logic of "technological rationality" or "technique," to borrow Herbert Marcuse and Jacques Ellul's respective terms, biotechnology perpetuates the productive extermination of animals. (p. 41)

Their elimination is achieved through their integration as materials for the production of cultured meat, further eroding the subjecthood of animals. There is a shift in terms of the ontological register that determines how whole animal categories become CM. By erasing their bodies and reducing them to cells, they become pure materiality. Hence, biological life gets transformed, its temporal and spatial organization continues to be appropriated and substituted in the agricultural industrialization process. Animal identity is usurped through the workings of food tech companies in order to resolve and repair planetary ills. Through the stories we tell ourselves about the Anthropocene, the boundaries between life and death get redrawn reifying the postnatural and post-biological shape the technological contours of possible future existence. However, couched in interplanetary living, space colonization, and futuristic scenarios about future foods and what they may be, it is important to think about how CM may reinforce ideas about our need for nature and alienate us further from the land and the animals that sustain us, neglecting the fact that eating is deeply rooted in the land, the animals, and the farming communities that produce our food. In the end, our pursuit to engineer a new future by altering life, we risk severing the very ties that bind us to the natural world, crafting not only synthetic food but also synthetic lives, alienated from the wisdom of the Earth and ensnared in the fictions of our own making.

Bibliography

Abergel, E. (2022). Cellular Agriculture: The Next Technofix Promises Synthetic, Cruelty-Free "Meat." *The Watershed Sentinel, Summer*, 20–22.
Arnold, A. (2018). *Climate Change and Storytelling: Narratives and Cultural Meaning in Environmental Communication*. Springer.
Chatterjee, S., & Subramaniam, B. (Eds.). (2021). *Meat! A Transnational Analysis*. Duke University Press.
Cooks, L. (2009). You are What You (Don't) Eat? Food, Identity, and Resistance. *Text and Performance Quarterly, 29*(1), 94–110. https://doi.org/10.1080/10462930802514388
Douglas, M. (2001). *Purity and Danger: An Analysis of the Concepts of Pollution and Taboo* (Repr). Routledge.
ETC Group. (2017). *Who Will Feed Us?* Retrieved from https://www.etcgroup.org/sites/www.etcgroup.org/files/files/etc-whowillfeedus-english-webshare.pdf
Fassler, J. (2021, September 22). *Lab-grown Meat is Supposed to be Inevitable. The Science Tells a Different Story.* The Counter. Retrieved from https://the-counter.org/lab-grown-cultivated-meat-cost-at-scale/
Goodman, D., Wilkinson, J., & Sorj, B. (1987). *From Farming to Biotechnology: A Theory of Agro-industrial Development.* Basil Blackwell. Retrieved from https://trove.nla.gov.au/version/45809245
Holt-Giménez, E. (2017). *A Foodie's Guide to Capitalism: Understanding the Political Economy of What We Eat.* Monthly Review Press: Food FirstBooks.
Marcuse, H. (1991). *One-Dimensional Man: Studies in the Ideology of Advanced Industrial Society* (1st ed.). Beacon Press.
Roe, E. J. (2006). Material Connectivity, the Immaterial and the Aesthetic of Eating Practices: An Argument for How Genetically Modified Foodstuff Becomes Inedible. *Environment and Planning A: Economy and Space, 38*(3), 465–481. https://doi.org/10.1068/a3835
Selin, C. (2008). The Sociology of the Future: Tracing Stories of Technology and Time. *Sociology Compass, 2*(6), 1878–1895. https://doi.org/10.1111/j.1751-9020.2008.00147.x
Sexton, A. E. (2018). Eating for the Post-Anthropocene: Alternative Proteins and the Biopolitics of Edibility. *Transactions of the Institute of British Geographers, 43*(4), 586–600. https://doi.org/10.1111/tran.12253
Simondon, G. (2016). *On the Mode of Existence of Technical Objects.* Univocal Pub..

Smaje, C. (2020). *A Small Farm Future: Making the Case for a Society Built Around Local Economies, Self-provisioning, Agricultural Diversity, and a Shared Earth*. Chelsea Green Publishing.
Smaje, C. (2023). *Saying No to a Farm-free Future: The Case for an Ecological Food System and Against Manufactured Foods*. Chelsea Green Publishing.
Stanescu, V. (2011). "Green" Eggs and Ham? The Myth of Sustainable Meat and the Danger of the Local. In J. Sanbonmatsu (Ed.), *Critical Theory and Animal Liberation* (pp. 239–255). Rowman & Littlefield Publishers Inc..
Tamminen, S., & Vermeulen, N. (2012). Bio-objects and Generative Relations. *Croatian Medical Journal, 53*(2), 198–200. https://doi.org/10.3325/cmj.2012.53.198
Vermeulen, N., Tamminen, S., & Webster, A. (2017). *Bio-objects Life in the 21st Century*. Routledge, Taylor & Francis Group.
Weisberg, Z. (2015). Biotechnology as End Game: Ontological and Ethical Collapse in the "Biotech Century". *NanoEthics, 9*(1), 39–54. https://doi.org/10.1007/s11569-014-0219-5
Wurgaft, B. A. (2019). *Meat Planet: Artificial Flesh and the Future of Food*. University of California Press.

Bibliography

Abbots, E. J. (2017). *The Agency of Eating. Mediation, Food and the Body*. Bloomsbury Academic.

Abergel, E. (2011). Climate-ready Crops and Bio-capitalism: Towards a New Food Regime? *The International Journal of Sociology of Agriculture and Food, 18*(3), 260–274. https://doi.org/10.48416/IJSAF.V18I3.248

Abergel, E. (2022). Cellular Agriculture: The Next Technofix Promises Synthetic, Cruelty-Free "Meat." *The Watershed Sentinel, Summer*, 20–22.

Abrell, E. (2023). The Empty Promises of Cultured Meat. In C. J. Adams, A. Crary, & L. Gruen (Eds.), *The Good it Promises, the Harm it Does: Critical Essays on Effective Altruism* (pp. 149–159). Oxford University Press.

Adam, B. (1998). *Timescapes of Modernity: The Environment and Invisible Hazards* (1. publ). Routledge.

Adam, B. (2010). History of the Future: Paradoxes and Challenges. *Rethinking History, 14*(3), 361–378. https://doi.org/10.1080/13642529.2010.482790

Adams, C. J. (2015). *The Sexual Politics of Meat: A Feminist-Vegetarian Critical Theory* (Bloomsbury revelations ed.). Bloomsbury Academic, An imprint of Bloomsbury Publishing Inc.

AFP. (2023, March 29). Meatball from Extinct Mammoth Unveiled by Food Tech Firm. *News.Com.Au*. Retrieved from https://www.news.com.au/breaking-news/meatball-from-extinct-mammoth-unveiled-by-food-tech-firm/news-story/81ddb6d5aca637050375d36fb5df085f

E. Abergel, *Dead Meat*, https://doi.org/10.1007/978-981-97-9049-4

Ahuja, N. (2016). *Bioinsecurities: Disease Interventions, Empire, and the Government of Species*. Duke University Press.

Aleph Farms. (2024). *Space Meat—The Future of Astronaut Food*. Aleph Farms. Retrieved from https://aleph-farms.com/aleph-zero/

Allen, J., & Lavau, S. (2015). 'Just-in-Time' Disease: Biosecurity, Poultry and Power. *Journal of Cultural Economy, 8*(3), 342–360. https://doi.org/10.1080/17530350.2014.904243

Anthis, J. R., Berenstein, P., Beurrier, Biehl, C., Birdie, P., Burley, J., Chantre, C., Davis, J., Dohadwala, S., d'Origny, G., Espirito Santo, V., Espitia, F., Gonzalez, G., Grandic, I., Green, C., Groenewegen, L., Guvenilir, A., Hanyu, Y., Kane, M. H., & Zimmerman, J. (2023). In K. Spiro, S. Hall, J. Darling, & M. Benami (Eds.), *Modern Meat: The Next Generation of Meat from Cells* (1st ed.). Cellular Agriculture Society.

Anusci, V. (2015, February 17). *I'm the Future of 3D Printing, Nice to "Meat" You*. All3DP. Retrieved from https://all3dp.com/meat-3d-printing/

Arias-Maldonado, M. (2015). The Anthropocenic Turn: Theorizing Sustainability in a Postnatural Age. *Sustainability, 8*(1), 10. https://doi.org/10.3390/su8010010

Arias-Maldonado, M. (2019). Towards a Good Anthropocene? In *Rethinking the Environment for the Anthropocene: Political Theory and Socionatural Relations in the New Geological Epoch* (1st ed.). Routledge.

AristarKhova, I. (2021). Eating the Mother. In Meat! A Transnational Analysis. Chatterjee and Subramaniam Eds. (pp. 39–60). Duke University Press.

Arnold, A. (2018). *Climate Change and Storytelling: Narratives and Cultural Meaning in Environmental Communication*. Springer.

Asafu-Adjaye, J., Blomquist, L., Brand, S. B., Brook, B. W., DeFries, R., Ellis, E. C., Foreman, C. A., Keith, D. A., Lewis, M. J., Lynas, M., Nordhaus, T., Pielke, R. A., Priztker, R., Roy, J., Sagoff, M., Shellenberger, M., Stone, R. W., & Teague, P. (2015). *An Ecomodernist Manifesto*. The Breakthrough Institute. Retrieved from https://static1.squarespace.com/static/5515d9f9e4b04d5c3198b7bb/t/552d37bbe4b07a7dd69fcdbb/1429026747046/An+Ecomodernist+Manifesto.pdf

Attard, G. (2022). Robots in Livestock Management. In Q. Zhang (Ed.), *Encyclopedia of Smart Agriculture Technologies*. Springer International Publishing.

Atwood, M. (2013). *Oryx and Crake*. Virago Press.

Audétat, M., et al. (2015). *Sciences et technologies émergentes: Pourquoi tant de promesses?* Hermann.

Awati, R., Bernstein, C., & Wigmore, I. (2022). *What is a Moonshot in Technology?* WhatIs. Retrieved from https://www.techtarget.com/whatis/definition/moonshot

Bach, A. D., Stern-Straeter, J., Beier, J. P., Bannasch, H., & Stark, G. B. (2003). Engineering of Muscle Tissue. *Clinics in Plastic Surgery, 30*(4), 589–599. https://doi.org/10.1016/S0094-1298(03)00077-4

Balasubramanian, B., Liu, W., Pushparaj, K., & Park, S. (2021). The Epic of In Vitro Meat Production—A Fiction into Reality. *Foods, 10*(6), 1395. https://doi.org/10.3390/foods10061395

Barclay, E. (2013, August 5). *Long Awaited Lab-Grown Burger Is Unveiled In London | KQED*. Retrieved from https://www.kqed.org/bayareabites/67631/long-awaited-lab-grown-burger-is-unveiled-in-london

Barrett, H., & Rose, D. C. (2022). Perceptions of the Fourth Agricultural Revolution: What's In, What's Out, and What Consequences are Anticipated? *Sociologia Ruralis, 62*(2), 162–189. https://doi.org/10.1111/soru.12324

Bartholet, J. (2015, June 1). *The Quest for Lab-Grown Meat*. Scientific American. Retrieved from https://www.scientificamerican.com/article/the-quest-for-lab-grown-meat/

Barzee, T. J., El Mashad, H. M., Cao, L., Chio, A., Pan, Z., & Zhang, R. (2022). Cell-Cultivated Food Production and Processing: A Review. *Food Bioengineering, 1*(1), 4–25. https://doi.org/10.1002/fbe2.12009

Baudrillard, J. (1988). The System of Objects. *Art Monthly (Archive 1976–2005), 115*, 5–8.

Beckert, J. (2013). Imagined Futures: Fictional Expectations in the Economy. *Theory and Society, 42*(3), 219–240. https://doi.org/10.1007/s11186-013-9191-2

Beckett, E. L. (2017, June 27). *Memphis Meats Scientist: The World is Ready for Clean Meat*. Food Dive. Retrieved from https://www.fooddive.com/news/memphis-meats-scientist-the-world-is-ready-for-clean-meat/445892/

Benjaminson, M. A., Gilchriest, J. A., & Lorenz, M. (2002). In Vitro Edible Muscle Protein Production System (mpps): Stage 1, Fish. *Acta Astronautica, 51*(12), 879–889. https://doi.org/10.1016/S0094-5765(02)00033-4

Benke, K. (2024, February 14). *By Growing Animal Cells in Rice Grains, Scientists Dish Up Hybrid Food*. EurekAlert! Retrieved from https://www.eurekalert.org/news-releases/1033922

Bennett, C. E., Thomas, R., Williams, M., Zalasiewicz, J., Edgeworth, M., Miller, H., Coles, B., Foster, A., Burton, E. J., & Marume, U. (2018). The Broiler Chicken as a Signal of a Human Reconfigured Biosphere. *Royal Society Open Science, 5*(12), 180325. https://doi.org/10.1098/rsos.180325

Bensaude-Vincent, B. (2015). Promesses et régime d'historicité en technosciences. In M. Audétat (Ed.), *Pourquoi tant de promesses?* (pp. 49–68). Éditions Herman.

Bhat, Z. F., Kumar, S., & Fayaz, H. (2015). In Vitro Meat Production: Challenges and Benefits over Conventional Meat Production. *Journal of Integrative Agriculture, 14*(2), 241–248. https://doi.org/10.1016/S2095-3119(14)60887-X

Bicudo, E., Faulkner, A., & Li, P. (2021). Patents and the Experimental Space: Social, Legal and Geographical Dimensions of 3D Bioprinting. *International Review of Law, Computers & Technology, 35*(1), 2–23. https://doi.org/10.1080/13600869.2020.1785066

BIOMILQ. (2024). *Home | BIOMILQ | Human Milk Ingredients | Research Triangle Park, NC, USA*. BIOMILQ. Retrieved from https://www.biomilq.com

Blanton, J. R., Jr., Grant, A. L., McFarland, D. C., Robinson, J. P., & Bidwell, C. A. (1999). Isolation of Two Populations of Myoblasts from Porcine Skeletal Muscle. *Muscle & Nerve, 22*(1), 43–50. https://doi.org/10.1002/(SICI)1097-4598(199901)22:1<43::AID-MUS8>3.0.CO;2-O

Blaustein-Rejto, D. (2022). *Livestock Don't Contribute 14.5% of Global Greenhouse Gas Emissions*. The Breakthrough Institute. Retrieved from https://thebreakthrough.org/issues/food-agriculture-environment/livestock-dont-contribute-14-5-of-global-greenhouse-gas-emissions

Blum, M., & Ott, T. (2018). Xenopus: An Undervalued Model Organism to Study and Model Human Genetic Disease. *Cells, Tissues, Organs, 205*(5–6), 303–313. https://doi.org/10.1159/000490898

Blythman, J. (2018, February 12). The Quorn Revolution: The Rise of Ultra-processed Fake Meat. *The Guardian*. Retrieved from https://www.theguardian.com/lifeandstyle/2018/feb/12/quorn-revolution-rise-ultra-processed-fake-meat

Bogueva, D., & McClements, D. J. (2023). Safety and Nutritional Risks Associated with Plant-Based Meat Alternatives. *Sustainability, 15*(19), 14336. https://doi.org/10.3390/su151914336

Bonk, L. (2024, February 15). *You've Tried Plant-Based Meat, but here Come Meat-Based Plants*. Engadget. Retrieved from https://www.engadget.com/youve-tried-plant-based-meat-but-here-come-meat-based-plants-163654564.html

Borchers, A., & Pieler, T. (2010). Programming Pluripotent Precursor Cells Derived from Xenopus Embryos to Generate Specific Tissues and Organs. *Genes, 1*(3), 413–426. https://doi.org/10.3390/genes1030413

Borg, E., Policante, A., & Skou, A. (2022). *Mutant Ecologies: Manufacturing Life in the Age of Genomic Capital*. Pluto Press.

Borup, M., Brown, N., Konrad, K., & Lente, H. V. (2006). The Sociology of Expectations in Science and Technology. *Technology Analysis & Strategic Management, 18*(3–4), 285–298. https://doi.org/10.1080/09537320600777002

Boseley, S. (2018, February 14). Ultra-processed Foods may be Linked to Cancer, Says Study. *The Guardian*. Retrieved from https://www.theguardian.com/science/2018/feb/14/ultra-processed-foods-may-be-linked-to-cancer-says-study

Boswell-Penc, M. (2006). *Tainted Milk: Breastmilk, Feminisms, and the Politics of Environmental Degradation*. State University of New York Press.

Boyle, A. (2013, August 6). *Google's Sergey Brin Explains Why He Paid $330,000 for Lab Burger*. NBC News. Retrieved from http://www.nbcnews.com/technolog/googles-sergey-brin-explains-why-he-paid-330-000-lab-6C10853442

Brekke, O. A., & Sirnes, T. (2011). Biosociality, Biocitizenship and the New Regime of Hope and Despair: Interpreting "Portraits of Hope" and the "Mehmet Case". *New Genetics and Society, 30*(4), 347–374. https://doi.org/10.1080/14636778.2011.592012

Brown, N., & Michael, M. (2003). A Sociology of Expectations: Retrospecting Prospects and Prospecting Retrospects. *Technology Analysis & Strategic Management, 15*(1), 3–18. https://doi.org/10.1080/0953732032000046024

Bryant, C., & Dillard, C. (2019). The Impact of Framing on Acceptance of Cultured Meat. *Frontiers in Nutrition, 6*. https://doi.org/10.3389/fnut.2019.00103

Cadloff, E. B. (2022, February 28). Steak's Next Frontier: Customized and Lab-Grown to Order. *Modern Farmer*. Retrieved from https://modernfarmer.com/2022/02/custom-cell-grown-meat/

Carrington, D., & Environment editor (2023, March 28). Meatball from Long-Extinct Mammoth Created by Food Firm. *The Guardian*. Retrieved from https://www.theguardian.com/environment/2023/mar/28/meatball-mammoth-created-cultivated-meat-firm

CAS (Cellular Agriculture Society). (2020). Introduction to cellular agriculture. [Now known as From Fauna]. https://fromfauna.org/

Castle, N. (2022). In Vitro Meat and Science Fiction. *Extrapolation, 63*(2), 149–179. https://doi.org/10.3828/extr.2022.11

Catts, O., Salter, C., & Zurr, I. (2021). Living and Semi-living Artefacts on Display. In I. L. Muller & C. S. Langill (Eds.), *Curating Lively Objects* (1st ed., pp. 115–131). Routledge. https://doi.org/10.4324/9780429053481-10

Chakrabarty, D. (2010). Le climat de l'histoire: Quatre thèses (trad. De Charlotte Nordmann) (C. Nordmann, Trans.). *Revue Internationale Des Livres et Des Idées, 15*, 15.

Chapelle. (2014, November 18). *Et si le but ultime de l'industrie agroalimentaire était de se débarrasser des animaux d'élevage?* Basta! Retrieved from https://basta.media/jocelyne-porcher-industrie-agroalimentaire-disparition-animaux-elevage-viande-de-synthese

Chasan, A. (2023, March 28). *Startup makes Giant Meatball Out of Lab-grown Mammoth Meat—CBS News*. Retrieved from https://www.cbsnews.com/news/giant-meatball-lab-grown-mammoth-meat-netherlands/

Chatterjee, S., & Subramaniam, B. (Eds.). (2021). *Meat! A Transnational Analysis*. Duke University Press.

Ćirković, M. M. (2019). Space Colonization Remains the Only Long-term Option for Humanity: A Reply to Torres. *Futures, 105*, 166–173. https://doi.org/10.1016/j.futures.2018.09.006

Clark, N., & Yusoff, K. (2017). Geosocial Formations and the Anthropocene. *Theory, Culture and Society, 34*(2–3), 3–23. https://doi.org/10.1177/0263276416688946

Cleaver, L. (2024, May 30). *BioCraft's Cultivated Mouse Meat Reaches Price Parity with Premium Pet Food*. PetfoodIndustry. Retrieved from https://www.petfoodindustry.com/ingredient-issues/proteins/news/15676448/biocrafts-cultivated-mouse-meat-reaches-price-parity-with-premium-pet-food

Clifton, M. (2017, July 8). *Hampton Creek Founder Tetrick Pledges his "Clean Meat" will be Vegan*. Animals 24–7. Retrieved from https://www.animals24-7.org/2017/07/08/hampton-creek-founder-tetrick-pledges-his-clean-meat-will-be-vegan/

Cohen, M., & Cassidy, T. (2022). Milk from Mars. The Challenges of Regulating Lab-Produced (Human) Milk. *Food and Drug Law Journal*, 77.

Colebrooke, L., & Miele, M. (2017). Eating Art and the Art of Eating: Unsettling the Practices of Taste. *Performance Research, 22*(7), 102–108. https://doi.org/10.1080/13528165.2017.1353203

Coller, J. (2023). *Council Post: How Business Innovation Can Turn The Food Crisis Into A New Green Revolution*. Forbes. Retrieved from https://www.forbes.com/sites/forbesbusinesscouncil/2023/07/10/how-business-innovation-can-turn-the-food-crisis-into-a-new-green-revolution/

Cook, D. (2011). *Adorno on Nature*. Acumen.

Cooks, L. (2009). You are What You (Don't) Eat? Food, Identity, and Resistance. *Text and Performance Quarterly, 29*(1), 94–110. https://doi.org/10.1080/10462930802514388

Coole, D. H., & Frost, S. (2013). *New Materialisms: Ontology, Agency, and Politics*. Duke University Press.

Cooper, M. (2008). *Life as Surplus: Biotechnology and Capitalism in the Neoliberal Era*. University of Washington Press.

Cooper, M. (2015). In Vivo Economies: Temporalities of Life and Value. *Journal of Cultural Economy, 8*(3), 257–259. https://doi.org/10.1080/17530350.2015.1039461

Corson, M. S., Mondière, A., Morel, L., & van der Werf, H. M. G. (2022). Beyond Agroecology: Agricultural Rewilding, a Prospect for Livestock Systems. *Agricultural Systems, 199*, 103410. https://doi.org/10.1016/j.agsy.2022.103410

Crawford, E. (2024, April 29). Plant-based 3.0: Industry insiders share strategies for pushing alternative proteins forward. *Foodnavigator.com*. https://www.foodnavigator.com/Article/2024/04/29/Plant-based-3.0-Industry-insiders-share-strategies-for-pushing-alternative-proteins-forward

Cronon, W. (1996). The Trouble with Wilderness: A Response. *Environmental History, 1*(1), 47–55. https://doi.org/10.2307/3985063

Crownhart, C. (2023). Here's What We Know About Lab-grown Meat and Climate Change. *MIT Technology Review*. Retrieved from https://www.technologyreview.com/2023/07/03/1075809/lab-grown-meat-climate-change/

Crutzen, P. J. (2002). Geology of Mankind. *Nature, 415*(6867), 23–23. https://doi.org/10.1038/415023a

Datar, I., & Betti, M. (2010). Possibilities for an In Vitro Meat Production System. *Innovative Food Science & Emerging Technologies, 11*(1), 13–22. https://doi.org/10.1016/j.ifset.2009.10.007

Datar, I., & Bolton, R. (2017, July 21). *The Carnery*. Retrieved from https://www.anthropocenemagazine.org/2017/07/the-carnery/, https://www.anthropocenemagazine.org/2017/07/the-carnery/

Datar, I., Kim, E., & d'Origny, G. (2016). New Harvest: Building the Cellular Agriculture Economy. In B. Donaldson & C. Carter (Eds.), *The Future of Meat without Animals* (pp. 121–131). Roman and Littlefield International.

Davis, H., & Todd, Z. (2017). On the Importance of a Date, or Decolonizing the Anthropocene. *ACME: An International Journal for Critical Geographies, 16*(4), 761–780.

Deleuze, G., & Guattari, F. (1987). *A Thousand Plateaus: Capitalism and Schizophrenia*. University of Minnesota Press.

Denis, J., Mongili, A., & Pontille, D. (2016). Maintenance & Repair in Science and Technology Studies. *Tecnoscienza: Italian Journal of Science & Technology Studies, 6*, 5–16.

Devyver, A. (2022). Freedom and Dialectics: On the Critical Theory of Moishe Postone and Theodor Adorno. *Philosophy & Social Criticism*, 01914537221137842. https://doi.org/10.1177/01914537221137842

Dodson, M. V., Martin, E. L., Brannon, M. A., Mathison, B. A., & McFarland, D. C. (1987). Optimization of Bovine Satellite Cell-derived Myotube Formation *In Vitro*. *Tissue and Cell, 19*(2), 159–166. https://doi.org/10.1016/0040-8166(87)90001-2

Douglas, M. (2001). *Purity and Danger: An Analysis of the Concepts of Pollution and Taboo* (Repr). Routledge.

Dürbeck, G. (2020). Narratives of the Anthropocene: From the Perspective of Postcolonial Ecocriticism and Environmental Humanities. In *Post-colonialism Cross-examined: Multidirectional Perspectives on Imperial and Colonial Pasts and the Neocolonial Present, Monika Albrecht Ed* (pp. 271–288). Routledge, Taylor & Francis Group.

Edelman, P. D., McFarland, D. C., Mironov, V. A., & Matheny, J. G. (2005). Commentary: In vitro-cultured meat production. *Tissue Engineering, 11*(5–6), 659–662. https://doi.org/10.1089/ten.2005.11.659

Edible Planet Ventures, I. (2023, May 9). *Mosa Meat Opens the World's Largest Cultivated Meat Campus*. Edible Planet. Retrieved from https://www.edibleplanetventures.com/post/mosa-meat-opens-the-world-s-largest-cultivated-meat-campus

Einhorn, B. (2020, April 22). The World's Top Pork Processor Is Battling Two Epidemics at Once. *Bloomberg.Com*. Retrieved from https://www.bloomberg.com/news/articles/2020-04-22/coronavirus-swine-fever-threaten-world-s-biggest-pork-producer

Eschner, K. (2022, February 7). Could Lab-grown Meat ever be Indigenized? *Canada's National Observer*. Retrieved from https://www.nationalobserver.com/2022/08/02/news/could-lab-grown-meat-ever-be-indigenized

ETC Group. (2017). *Who Will Feed Us?* Retrieved from https://www.etcgroup.org/sites/www.etcgroup.org/files/files/etc-whowillfeedus-english-webshare.pdf

Evans, A. B., & Miele, M. (2012). Between Food and Flesh: How Animals are Made to Matter (and Not Matter) within Food Consumption Practices. *Environment and Planning D: Society and Space, 30*(2), 298–314. https://doi.org/10.1068/d12810

FAO, & WHO. (2023). *Food Safety Aspects of Cell-Based Food*. FAO; WHO. Retrieved from https://openknowledge.fao.org/handle/20.500.14283/cc4855en

FAO (Food and Agriculture Organization of the United Nations). (2006). *Livestock's long shadow: Environmental issues and options*. FAO. https://www.fao.org/3/a0701e/a0701e.pdf

FAO (Food and Agriculture Organization of the United Nations). (2009). *The state of food and agriculture: Livestock in the balance*. FAO. https://doi.org/10.18356/1c04f5eb-en

FAO (Food and Agriculture Organization of the United Nations). (2011). *World livestock 2011: Livestock in food security*. FAO. https://doi.org/10.18356/3d6e8c5d-en

Fassler, J. (2021, September 22). *Lab-grown Meat is Supposed to be Inevitable. The Science Tells a Different Story.* The Counter. Retrieved from https://thecounter.org/lab-grown-cultivated-meat-cost-at-scale/

Fassler, J. (2024, June 29). The Revolution That Died on Its Way to Dinner. *New York Times*.

Federau, A. (2017). *Pour une philosophie de l'Anthropocène* (1re éd.). Presses universitaires de France.

Fernandes, A. M., Teixeira, O. D. S., Fantinel, A. L., Revillion, J. P. P., & Souza, Â. R. L. D. (2022). Technological Prospecting: The Case of Cultured Meat. *Future Foods, 6*, 100156. https://doi.org/10.1016/j.fufo.2022.100156

Fish, K. (2013). *Living Factories: Biotechnology and the Unique Nature of Capitalism*. McGill-Queen's University Press.

Fleming, A. (2017, September 20). Could Lab-grown Fish and Meat Feed the World—Without Killing a Single Animal? *The Guardian*. Retrieved from https://www.theguardian.com/lifeandstyle/2017/sep/20/lab-grown-meat-fish-feed-the-world-frankenmeat-startups

Flichy, P. (2007). *The Internet Imaginaire*. The MIT Press.

Flood, E. (2023, July 31). *From Prediction to Production: A Timeline of Cultivated Meat.*

Food Dive. Retrieved from https://www.fooddive.com/news/from-prediction-production-timeline-cultivated-meat-industry/689441/

Fraeye, I., Kratka, M., Vandenburgh, H., & Thorrez, L. (2020). Sensorial and Nutritional Aspects of Cultured Meat in Comparison to Traditional Meat: Much to Be Inferred. *Frontiers in Nutrition, 7*, 35. https://doi.org/10.3389/fnut.2020.00035

Fremaux, A. (2019). The Return of Nature in the Capitalocene: A Critique of the Ecomodernist Version of the "Good Anthropocene". In *Rethinking the Environment for the Anthropocene: Political Theory and Socionatural Relations in the New Geological Epoch*. Routledge.

From Fauna. (2024). *What We Do*. Retrieved from https://fromfauna.org/work/

Galpin, K. (2024, February 15). *Tech Expert Shares Opinion after Trying Chicken-like Meat Grown in a Lab*. The Focus. Retrieved from https://www.thefocus.news/lifestyle/tech-expert-shares-opinion-after-trying-chicken-like-meat-grown-in-a-lab/

Galusky, W. (2010). Playing Chicken: Technologies of Domestication, Food, and Self. *Science as Culture, 19*(1), 15–35. https://doi.org/10.1080/09505430903557874

Galusky, W. (2014). Technology as Responsibility: Failure, Food Animals, and Lab-grown Meat. *Journal of Agricultural & Environmental Ethics, 27*(6), 931–948. https://doi.org/10.1007/s10806-014-9508-9

Galusky, W. (2022). *Protein Machines, Technology, and the Nature of he Future*. Springer International Publishing. https://doi.org/10.1007/978-3-031-08717-2

Gasteratos, K. (2017). *Nature and the Neomnivore*. Retrieved from http://nrs.harvard.edu/urn-3:HUL.InstRepos:33839952

Georgescu-Roegen, N. (2011). *From bioeconomics to degrowth: Georgescu-Roegen's 'new economics' in eight essays*. Routledge.

GFI. (2021, January 19). *Plant-Based Retail Market Overview | GFI*. Retrieved from https://gfi.org/marketresearch/

GFI, G. F. I. (2022). *2022 Cultivated Meat State of the Industry Report*. Retrieved from https://gfi.org/wp-content/uploads/2023/01/2022-Cultivated-Meat-State-of-the-Industry-Report-2-1.pdf

GFI, G. F. I. (2023). *New Research Supports "Cultivated Meat" Term*. Retrieved from https://gfi.org/wp-content/uploads/2023/01/COR23008_CM-nomenclature-fact-sheet_Final.pdf

Gillespie, K. (2018). *The Cow with Ear Tag #1389*. The University of Chicago Press.

Good Food Institute. (2024). *Alternative Protein Company Database* [Dataset]. Retrieved from https://gfi.org/resource/alternative-protein-company-database/

Goodman, M. K., & Sage, C. (Eds.). (2016). *Food Transgressions: Making Sense of Contemporary Food Politics* (1st ed.). Routledge. https://doi.org/10.4324/9781315582702

Goodman, D., Wilkinson, J., & Sorj, B. (1987). *From Farming to Biotechnology: A Theory of Agro-industrial Development*. Basil Blackwell. Retrieved from https://trove.nla.gov.au/version/45809245

Gordon, P. E. (2016). *Adorno and Existence*. Harvard University Press.

Groll, J., Boland, T., Blunk, T., Burdick, J. A., Cho, D.-W., Dalton, P. D., Derby, B., Forgacs, G., Li, Q., Mironov, V. A., Moroni, L., Nakamura, M., Shu, W., Takeuchi, S., Vozzi, G., Woodfield, T. B. F., Xu, T., Yoo, J. J., & Malda, J. (2016). Biofabrication: Reappraising the Definition of an Evolving Field. *Biofabrication, 8*(1), 013001. https://doi.org/10.1088/1758-5090/8/1/013001

Guo, D., Zhang, L., Wang, X., Zheng, J., & Lin, S. (2022). Establishment Methods and Research Progress of Livestock and Poultry Immortalized Cell Lines: A Review. *Frontiers in Veterinary Science, 9*, 956357. https://doi.org/10.3389/fvets.2022.956357

Guo, X., Wang, D., He, B., Hu, L., & Jiang, G. (2024). 3D Bioprinting of Cultured Meat: A Promising Avenue of Meat Production. *Food and Bioprocess Technology, 17*(7), 1659–1680. https://doi.org/10.1007/s11947-023-03195-x

Guthman, J. (2009). Teaching the Politics of Obesity: Insights into Neoliberal Embodiment and Contemporary Biopolitics. *Antipode, 41*(5), 1110–1133. https://doi.org/10.1111/j.1467-8330.2009.00707.x

Guthman, J. (2022). The CAFO in the Bioreactor. *Environmental Humanities, 14*(1), 71–88. https://doi.org/10.1215/22011919-9481440

Guthman, J., & Biltekoff, C. (2023). Agri-food Tech's Building Block: Narrating Protein, Agnostic of Source, in the Face of Crisis. *BioSocieties, 18*(3), 656–678. https://doi.org/10.1057/s41292-022-00287-3

Hahn, J. (2020, November 13). *Grow-your-own Human Meat Kit is "Technically" not Cannibalism.* Dezeen. Retrieved from https://www.dezeen.com/2020/11/13/ouroboros-steak-meal-kit-andrew-pelling-grace-knight-orkan-telhan/

Haldane, J. B. S., & Price, C. A. (2001). The Future of Biology. In *Possible Worlds.* Routledge.

Handral, K. H., Hua Tay, S., Wan Chan, W., & Choudhury, D. (2022). 3D Printing of Cultured Meat Products. *Critical Reviews in Food Science and Nutrition, 62*(1), 272–281. https://doi.org/10.1080/10408398.2020.1815172

Hannan, J. (2020). *Meatsplaining: The Animal Agriculture Industry and the Rhetoric of Denial.* Sydney University Press.

Haraway, D. (2003). *The Companion Species Manifesto: Dogs, People, and Significant Otherness.* Prickly Paradigm Press.

Haraway, D. (2015). Anthropocene, Capitalocene, Plantationocene, Chthulucene: Making Kin. *Environmental Humanities, 6*(1), 159–165. https://doi.org/10.1215/22011919-3615934

Haraway, D. (2016). *Staying with the Trouble: Making Kin in the Chthulucene.* Duke University Press.

Harper, E. T., & Specht, D. (Eds.). (2022). *Imagining Apocalyptic Politics in the Anthropocene*. Routledge.

Hart, J. (2016, October 24). *Cellular Agriculture—Can it Help Feed the World?* Farm Progress. Retrieved from https://www.farmprogress.com/soybeans/cellular-agriculture-can-it-help-feed-world

Harvey, D. (2005). *A Brief History of Neoliberalism*. Oxford University Press.

Haug, T. (2018). The Capitalist Metabolism: An Unachieved Subsumption of Life under the Value-form. *Journal for Cultural Research, 22*(2), 191–203. https://doi.org/10.1080/14797585.2018.1461352

Hauser, J. (with Le Lieu unique). (2003). *L'art biotech': Exposition, Nantes, Lieu unique, 14 mars-4 mai 2003*. Filigranes éd. le Lieu unique.

Hayek, M. N., & Miller, S. M. (2021). Underestimates of Methane from Intensively Raised Animals Could Undermine Goals of Sustainable Development. *Environmental Research Letters, 16*(6), 063006. https://doi.org/10.1088/1748-9326/ac02ef

Hebrew University of Jerusalem. (2024, August 23). Unprecedented Efficiency: New Manufacturing Method Could Make Cultivated Meat Cheaper Than Organic Chicken. *SciTechDaily*. Retrieved from https://scitechdaily.com/unprecedented-efficiency-new-manufacturing-method-could-make-cultivated-meat-cheaper-than-organic-chicken/

Heine, S., Ahlfeld, T., Albrecht, F. B., Gelinsky, M., & Kluger, P. J. (2024). How Biofabrication can Accelerate Cultured Meat's Path to Market. *Nature Reviews Materials, 9*(2), 83–85. https://doi.org/10.1038/s41578-024-00650-9

Hennion, A. (2007). Those Things That Hold Us Together: Taste and Sociology. *Cultural Sociology, 1*(1), 97–114. https://doi.org/10.1177/1749975507073923

Henrekson, M., Sandström, C., & Stenkula, M. (Eds.). (2024). *Moonshots and the New Industrial Policy: Questioning the Mission Economy* (Vol. 56). Springer Nature Switzerland. https://doi.org/10.1007/978-3-031-49196-2

Hickman, L. (2009, December 1). Why I'd Happily Eat Lab-grown Meat. *The Guardian*. Retrieved from https://www.theguardian.com/science/2009/dec/01/lab-grown-meat

Hinchliffe, S., Allen, J., Lavau, S., Bingham, N., & Carter, S. (2013). Biosecurity and the Topologies of Infected Life: From Borderlines to Borderlands. *Transactions of the Institute of British Geographers, 38*(4), 531–543. https://doi.org/10.1111/j.1475-5661.2012.00538.x

Hocquette, A., Lambert, C., Sinquin, C., Peterolff, L., Wagner, Z., Bonny, S. P. F., Lebert, A., & Hocquette, J.-F. (2015). Educated Consumers Don't Believe Artificial Meat is the Solution to the Problems with the Meat Industry.

Journal of Integrative Agriculture, 14(2), 273–284. https://doi.org/10.1016/S2095-3119(14)60886-8

Holmes, R. I. (2017). Technology and/or Nature: Denatured/Renatured/Engineered/Artifacted Life? *Ethics and the Environment, 22*(1), 41. https://doi.org/10.2979/ethicsenviro.22.1.03

Holt-Giménez, E. (2017). *A Foodie's Guide to Capitalism: Understanding the Political Economy of What We Eat*. Monthly Review Press: Food FirstBooks.

Hornborg, A. (2014). Technology as Fetish: Marx, Latour, and the Cultural Foundations of Capitalism. *Theory, Culture & Society, 31*(4), 119–140. https://doi.org/10.1177/0263276413488960

Hornborg, A. (2021). Objects Don't Have Desires: Toward an Anthropology of Technology beyond Anthropomorphism. *American Anthropologist, 123*(4), 753–766. https://doi.org/10.1111/aman.13628

Hribal, J. (2003). "Animals are Part of the Working Class": A Challenge to Labor History. *Labor History, 44*(4), 435–453.

Hughes-Castleberry, K. (2023, April 21). The Woolly Mammoth Meatball Could Kick Off a Trend of Eating Extinct Meats. *Discover Magazine*. Retrieved from https://www.discovermagazine.com/planet-earth/the-woolly-mammoth-meatball-could-kick-off-a-trend-of-eating-extinct-meats

Humbird, D. (2020). *Scale-Up Economics for Cultured Meat: Techno-Economic Analysis and Due Diligence*. Engineering Archive. https://doi.org/10.31224/osf.io/795su

Humbird, D. (2021). Scale-up Economics for Cultured Meat. *Biotechnology and Bioengineering, 118*(8), 3239–3250. https://doi.org/10.1002/bit.27848

Inskeep, S., & Green, A. (2024, May 17). *Florida Gov. DeSantis Signs Bill that Deletes Climate Change from State Law*. NPR. Retrieved from https://www.npr.org/2024/05/17/1252012825/florida-gov-desantis-signs-bill-that-deletes-climate-change-from-state-law

Jack, J. H., Simon. (2020, April 23). "Fake" Meat on the Menu as China Reopens Restaurants. *BBC News*. Retrieved from https://www.bbc.com/news/business-52391517

Jasanoff, S., & Kim, S. H. (2015). *Dreamscapes of Modernity: Sociotechnical Imaginaries and the Fabrication of Power*. University of Chicago Press. Retrieved from https://www.press.uchicago.edu/ucp/books/book/chicago/D/bo20836025.html

Jha, A. (2013, August 6). First Lab-grown Hamburger Gets Full Marks for "Mouth Feel." *The Guardian*. Retrieved from https://www.theguardian.com/science/2013/aug/05/world-first-synthetic-hamburger-mouth-feel

Johnson, A. (2024, February 23). *What is Lab-grown Meat, and Why does Ron DeSantis want to Ban it in Florida?* Miami Herald. Retrieved from https://www.miamiherald.com/news/politics-government/state-politics/article285829481.html

Joly, P.-B. (2015). Le régime des promesses technoscientifiques. In M. Audétat (Ed.), *Pourquoi tant de promesses?* (pp. 31–48). Harmann Éditeurs.

Jones, N. (2023a). Lab-grown Meat: The Science of Turning Cells into Steaks and Nuggets. *Nature, 619*(7968), 22–24. https://doi.org/10.1038/d41586-023-02095-6

Jones, N. (2023b). Will Cultured Meat Replace The Real Thing? *Nature, 619*, 22–24.

Jones, A. (2024, February 16). *Our Founder Ate Chicken for the First Time in 40 Years (And No Animals Died).* Species Unite. Retrieved from https://www.speciesunite.com/news-stories/our-founder-tried-cultivated-meat

Jönsson, E. (2016). Benevolent Technotopias and Hitherto Unimaginable Meats: Tracing the Promises of In Vitro Meat. *Social Studies of Science, 46*(5), 725–748. https://doi.org/10.1177/0306312716658561

Jørgensen, D. (2015). Rethinking Rewilding. *Geoforum, 65*, 482–488. https://doi.org/10.1016/j.geoforum.2014.11.016

Joshi, V., & Kumar, S. (2015). Meat Analogues: Plant based Alternatives to Meat Products—A Review. *International Journal of Food and Fermentation Technology, 5*(2), 107. https://doi.org/10.5958/2277-9396.2016.00001.5

Just Egg (Producer). (2018, May 21). *Cultured Meat: A Vision of the Future* [Video recording]. Retrieved from https://www.youtube.com/watch?v=f8Ii3DB6ejE

Kang, D.-H., Louis, F., Liu, H., Shimoda, H., Nishiyama, Y., Nozawa, H., Kakitani, M., Takagi, D., Kasa, D., Nagamori, E., Irie, S., Kitano, S., & Matsusaki, M. (2021). Engineered Whole Cut Meat-like Tissue by the Assembly of Cell Fibers using Tendon-Gel Integrated Bioprinting. *Nature Communications, 12*(1), 5059. https://doi.org/10.1038/s41467-021-25236-9

Kent, G. (2020). Why make Fake Breastmilk? *World Nutrition, 11*(3), 2–11. https://doi.org/10.26596/wn.20201132-11

Kibbe, B. (2014). *Mindful Mending: The Repair of Thought and Action Amidst Technologies* [Doctoral Dissertation, Loyola University Chicago]. Retrieved from https://ecommons.luc.edu/luc_diss/1093/?utm_source=ecommons.luc.edu%2Fluc_diss%2F1093&utm_medium=PDF&utm_campaign=PDFCoverPages

Klein, E. (2019, January 29). *Ending the Age of Animal Cruelty*. Vox. Retrieved from https://www.vox.com/future-perfect/2019/1/29/18197907/clean-meat-cell-plant-impossible-beyond-animal-cruelty

Klein, E. (2021, April 2024). Let's Launch a Moonshot for Meatless Meat. *The New York Times*. https://nytimes.com/2021/04/24/opinion/climate-change-meatless-meat.html

Knorr-Cetina, K. (1999). *Epistemic Cultures: How the Sciences make Knowledge*. Harvard University Press.

Krafts, K. P. (2010). Tissue Repair. *Organogenesis, 6*(4), 225–233. https://doi.org/10.4161/org.6.4.12555

Lafontaine, C., Wolfe, M., Gagné, J., & Abergel, E. (2021). Bioprinting as a Sociotechnical Project: Imaginaries, Promises and Futures. *Science as Culture, 30*(4), 556–580. https://doi.org/10.1080/09505431.2021.1977264

Lagally, C., Clayton, E. R., & Specht, L. (2017). Plant-based meat mind maps. *The Good Food Institute*. https://gfi.org/wp-content/uploads/2021/02/PBMap.pdf

Lamas, M. (2021). *How Scientists make Plant-Based Foods Taste and Look More Like Meat*. The Conversation. Retrieved from http://theconversation.com/how-scientists-make-plant-based-foods-taste-and-look-more-like-meat-156839

Landecker, H. (2002). New Times for Biology: Nerve Cultures and the Advent of Cellular Life In Vitro. *Studies in History and Philosophy of Science Part C: Studies in History and Philosophy of Biological and Biomedical Sciences, 33*(4), 667–694. https://doi.org/10.1016/S1369-8486(02)00026-2

Landecker, H. (2004). Living Differently in Time: Plasticity, Temporality and Cellular Biotechnologies. *Culture Machine, 7*(0). Retrieved from http://svr91.edns1.com/~culturem/index.php/cm/article/view/26

Landecker, H. (2007). *Culturing Life: How Cells Became Technologies* (First paperback ed). Harvard Univ. Press.

Laskawy, T. (2013, August 6). *Don't have a Cow: Lab-grown Meat Inches Closer to your Plate*. Grist. Retrieved from https://grist.org/food/dont-have-a-cow-lab-grown-meat-inches-closer-to-your-plate/

Latour, B. (2010). *Nous n'avons jamais été modernes: Essai d'anthropologie symétrique* (Nachdr.). Editions La Découverte [u.a.].

Latour, B. (2017). *Facing Gaia: Eight Lectures on the New Climactic Regime*. Polity Press.

Law, J. (2008). Material Semiotics. Heterogeneities.Net: John Law's STS Web Page. Retrieved from http://heterogeneities.net/publications/Law2019MaterialSemiotics.pdf

LeBesco, K., & Naccarato, P. (Eds.). (2008). *Edible Ideologies: Representing Food and Meaning*. State Univ. of New York Press.

Lee, L. Y. (2005). *Dialectics of the Body: Corporeality in the Philosophy of T.W. Adorno*. Routledge.

Lemkow, L. (2015). Science Fiction. *Food Phreaking, 2*, 42–47.

Lepore, J. (2014, June 16). *What the Gospel of Innovation Gets Wrong*. Retrieved from https://www.newyorker.com/magazine/2014/06/23/the-disruption-machine

Lewis, S. L., & Maslin, M. A. (2015). Defining the Anthropocene. *Nature, 519*(7542), 171–180. https://doi.org/10.1038/nature14258

Liboiron, M. (2016). Redefining Pollution and Action: The Matter of Plastics. *Journal of Material Culture, 21*(1), 87–110. https://doi.org/10.1177/1359183515622966

Lin, L. (2017). *Freud's Jaw and Other Lost Objects: Fractured Subjectivity in the Face of Cancer*. Fordham University Press.

Listek, V. (2019, May 6). *Organovo: Bioprinting Could be the New Solution to Organ Transplantation*. 3DPrint.Com | The Voice of 3D Printing / Additive Manufacturing. Retrieved from https://3dprint.com/243160/organovo-bioprinting-could-be-the-new-solution-to-organ-transplantation/

Lorimer, A. (2022). Chapter 4: Embodied Brains: Why Science Studies Needs the Anthropology of Museums. In J. Edwards, P. Harvey, & P. Wade (Eds.), *Technologized Images, Technologized Bodies* (pp. 85–116). Berghahn Books. https://doi.org/10.1515/9781845458300-006

Lorimer, J., & Driessen, C. (2013). Bovine Biopolitics and the Promise of Monsters in the Rewilding of Heck Cattle. *Geoforum, 48*, 249–259. https://doi.org/10.1016/j.geoforum.2011.09.002

Lynch, J., & Pierrehumbert, R. (2019). Climate Impacts of Cultured Meat and Beef Cattle. *Frontiers in Sustainable Food Systems, 3*. https://doi.org/10.3389/fsufs.2019.00005

Malabou, C. (2006). Les régénérés: Cellules souches, thérapie génique, clonage. *Critique, 709–710*(6), 529. https://doi.org/10.3917/criti.709.0529

Malabou, C. (2012). *Ontology of the Accident: An Essay on Destructive Plasticity*. Polity Press.

Malm, A. (2015). The Anthropocene Myth: Blaming All of Humanity for Climate Change lets Capitalism off the Hook. *Jacobin Magazine*. Retrieved from https://jacobin.com/2015/03/anthropocene-capitalism-climate-change/

Malm, A., & Hornborg, A. (2014). The Geology of Mankind? A Critique of the Anthropocene Narrative. *The Anthropocene Review, 1*(1), 62–69. https://doi.org/10.1177/2053019613516291

Marcuse, H. (1991). *One-Dimensional Man: Studies in the Ideology of Advanced Industrial Society* (1st ed.). Beacon Press.

Marcus, G. E. (Ed.). (1995). *Technoscientific Imaginaries: Conversations, Profiles, and Memoirs* (Vol. 2). University of Chicago Press.

Martin, A., Fischer, A., McMorran, R., & Smith, M. (2021). Taming Rewilding—From the Ecological to the Social: How Rewilding Discourse in Scotland has Come to Include People. *Land Use Policy, 111*, 105677. https://doi.org/10.1016/j.landusepol.2021.105677

Martins, B., Bister, A., Dohmen, R. G. J., Gouveia, M. A., Hueber, R., Melzener, L., Messmer, T., Papadopoulos, J., Pimenta, J., Raina, D., Schaeken, L., Shirley, S., Bouchet, B. P., & Flack, J. E. (2024). Advances and Challenges in Cell Biology for Cultured Meat. *Annual Review of Animal Biosciences, 12*(1), annurev-animal-021022-055132. https://doi.org/10.1146/annurev-animal-021022-055132

Masterson, V. (2022, October 10). *What is Rewilding?* World Economic Forum. Retrieved from https://www.weforum.org/agenda/2022/10/what-is-rewilding-nature/

Mattick, C. S. (2018). Cellular Agriculture: The Coming Revolution in Food Production. *Bulletin of the Atomic Scientists, 74*(1), 32–35. https://doi.org/10.1080/00963402.2017.1413059

McFarland, D. C., Doumit, M. E., & Minshall, R. D. (1988). The Turkey Myogenic Satellite Cell: Optimization of *In Vitro* Proliferation and Differentiation. *Tissue and Cell, 20*(6), 899–908. https://doi.org/10.1016/0040-8166(88)90031-6

McHugh, S. (2010). Real Artificial: Tissue-cultured Meat, Genetically Modified Farm Animals, and Fictions. *Configurations, 18*(1–2), 181–197. https://doi.org/10.1353/con.2010.0006

McHugh, T. (2019, October 1). How Plant-Based Meat and Seafood Are Processed. *Food Technology Magazine, 73*(10). Retrieved from https://www.ift.org/news-and-publications/food-technology-magazine/issues/2019/october/columns/processing-how-plant-based-meat-and-seafood-are-processed

McKenna, M. (2012, December 28). The Father of the Chicken Nugget. *Slate*. Retrieved from https://slate.com/human-interest/2012/12/robert-c-baker-the-man-who-invented-chicken-nuggets.html

McMichael, P. (2013). *Food Regimes and Agrarian Questions*. Fernwood Publishing.

Meilgaard, M. C., Carr, B. T., & Carr, B. T. (2014). *Sensory Evaluation Techniques* (4th ed.). CRC Press. https://doi.org/10.1201/b16452

Meloni, M. (2019). *Impressionable Biologies: From the Archaeology of Plasticity to the Sociology of Epigenetics*. Routledge. https://doi.org/10.4324/9781315169583

Meyer, G., & Terazono, E. (2020, April 26). *Pandemic Accelerates Shift to Meat Substitutes*. Retrieved from https://www.ft.com/content/0127984a-6def-4040-9bca-002b6ffd4e0a

Miller, J. (2012). In Vitro Meat: Power, Authenticity and Vegetarianism. *Journal of Critical Animal Studies, 10*(4), 41–63.

Mirzoeff, N. (2016). *How to See the World: An Introduction to Images, from Self-portraits to Selfies, Maps to Movies, and More*. Basic Books, a member of the Perseus Books Group.

Mol, A. (2007). *The Body Multiple: Ontology in Medical Practice* (3rd print). Duke University Press.

Mol, A. (2008). I Eat an Apple. On Theorizing Subjectivities. *Subjectivity, 22*(1), 28–37. https://doi.org/10.1057/sub.2008.2

Monger, C. (2024, April 27). *House Approves Senate Bill to Ban the Sale of Lab-grown Meat* |.... 1819 News. Retrieved from https://1819news.com/news/item/house-approves-senate-bill-to-ban-the-sale-of-lab-grown-meat

Montgomery, R. A., Stern, J. M., Lonze, B. E., Tatapudi, V. S., Mangiola, M., Wu, M., Weldon, E., Lawson, N., Deterville, C., Dieter, R. A., Sullivan, B., Boulton, G., Parent, B., Piper, G., Sommer, P., Cawthon, S., Duggan, E., Ayares, D., Dandro, A., et al. (2022). Results of Two Cases of Pig-to-Human Kidney Xenotransplantation. *New England Journal of Medicine, 386*(20), 1889–1898. https://doi.org/10.1056/NEJMoa2120238

Moore, A. (2016). Anthropocene Anthropology: Reconceptualizing Contemporary Global Change. *Journal of the Royal Anthropological Institute, 22*(1), 1. https://doi.org/10.1111/1467-9655.12332

Morton, T. (2013). *Hyperobjects: Philosophy and Ecology after the End of the World*. University of Minnesota Press.

Moskovitch, K. (2013, January 10). Modern Meadow Aims to Print Raw Meat using Bioprinter. *BBC News*. Retrieved from https://www.bbc.com/news/technology-20972018

Mulvaney, K. (2023, July 19). De la viande cultivée en laboratoire commence à être servie aux États-Unis. *National Geographic*. Retrieved from https://www.

nationalgeographic.fr/sciences/alimentation-revolution-ecosysteme-de-la-viande-cultivee-en-laboratoire-commence-a-etre-servie-aux-etats-unis

Myelnikov, D. (2020). Tinkering with Genes and Embryos: The Multiple Invention of Transgenic Mice c. 1980. *History and Technology, 35*(4), 425–452. https://doi.org/10.1080/07341512.2019.1694126

New Harvest. (2024). *What is Cellular Agriculture.*

New Havest. (2022). *What We Do.* New Harvest. Retrieved from https://new-harvest.org/what-we-do/

Niamh, M. (2018). *Aleph Farms CEO on its 3D Cultured Beef: "Unlike Other Companies, Our Meat Grows Together Like Real Meat."* Foodnavigator.Com. Retrieved from https://www.foodnavigator.com/Article/2018/05/02/Aleph-Farms-CEO-on-its-3D-cultured-beef-Unlike-other-companies-our-meat-grows-together-like-real-meat

Nicholas, I., & Silver, M. (2021, October 15). *Tufts Receives $10 Million Grant to Help Develop Cultivated Meat | Tufts Now.* Retrieved from https://now.tufts.edu/2021/10/15/tufts-receives-10-million-grant-help-develop-cultivated-meat

Nicholsen, S. W. (2003). *The Love of Nature and the End of the World: The Unspoken Dimensions of Environmental Concern.* MIT Press.

Nicholson, D. J. (2019). Is the Cell Really a Machine? *Journal of Theoretical Biology, 477*, 108–126. https://doi.org/10.1016/j.jtbi.2019.06.002

Nie, X., Liang, Z., Li, K., Yu, H., Huang, Y., Ye, L., & Yang, Y. (2021). Novel Organoid Model in Drug Screening: Past, Present, and Future. *Liver Research, 5*(2), 72–78. https://doi.org/10.1016/j.livres.2021.05.003

Niman, N. H. (2021). *Defending Beef: The Case for Sustainable Meat Production.* Chelsea Green Publishing.

Nimmo, R. (2015). Apiculture in the Anthropocene: Between Posthumanism and Critical Animal Studies. In *Animals in the Anthropocene: Critical Perspectives on Non-human Futures.* Sydney University Press. Retrieved from https://www.academia.edu/8976378/Apiculture_in_the_Anthropocene_Between_Posthumanism_and_Critical_Animal_Studies

Nogués-Bravo, D., Simberloff, D., Rahbek, C., & Sanders, N. J. (2016). Rewilding is the New Pandora's Box in Conservation. *Current Biology, 26*(3), R87–R91. https://doi.org/10.1016/j.cub.2015.12.044

Novak, B. J. (2018). De-Extinction. *Genes, 9*(11), 548. https://doi.org/10.3390/genes9110548

Nowell, C. (2024, April 9). 'Political Efforts': The Republican States Trying to Ban Lab-grown Meat. *The Guardian.* Retrieved from https://www.theguardian.com/environment/2024/apr/09/us-states-republicans-banning-lab-grown-meat

Nyyssölä, A., Suhonen, A., Ritala, A., & Oksman-Caldentey, K.-M. (2022). The Role of Single Cell Protein in Cellular Agriculture. *Current Opinion in Biotechnology, 75*, 102686. https://doi.org/10.1016/j.copbio.2022.102686

O'Riordan, K., Fotopoulo, A., & Stephens, N. (2017). *The First Bite: Imaginaries, Promotional Publics and the Laboratory Grown Burger.* https://doi.org/10.1177/0963662516639001

OECD. (2021). *OECD Science, Technology and Innovation Outlook 2020: Science and Innovation in Times of Crisis.* OECD. https://doi.org/10.1787/75f79015-en

Ogles, J. (2024, March 4). Meat Institute Argues Cellular Agriculture Ban will Hurt Florida's Food Economy. *Florida Politics—Campaigns & Elections. Lobbying & Government.* Retrieved from https://floridapolitics.com/archives/662944-meat-institute-argues-cellular-agriculture-ban-will-hurt-floridas-food-economy/

Ong Guang-Xi, J. (2024, May 17). *After a Promising Beginning, is Singapore's Cultivated Meat Industry on the Way to the Slaughterhouse?* CNA. Retrieved from https://www.channelnewsasia.com/singapore/labgrown-cultivated-meat-technology-eat-just-acceptance-4332656

Oven, A., Yoxon, B., & Milburn, J. (2022). Investigating the Market for Cultivated Meat as Pet Food: A Survey Analysis. *PLoS One, 17*(12), e0275009. https://doi.org/10.1371/journal.pone.0275009

Palsson, G., & Swanson, H. A. (2016). Down to Earth: Geosocialities and Geopolitics. *Environmental Humanities, 8*(2), 149–171. Retrieved from http://dx.doi.org.proxy.bibliotheques.uqam.ca/10.1215/22011919-3664202

Panetta, A. (2024, May 8). In Budding Battle over Lab-Grown Meat, Florida takes Opening Stab. *CBC News.* Retrieved from https://www.cbc.ca/news/world/florida-ban-analysis-canada-1.7197136

Park, S., Lee, M., Jung, S., Lee, H., Choi, B., Choi, M., Lee, J. M., Yoo, K. H., Han, D., Lee, S. T., Koh, W.-G., Bang, G., Hwang, H., Lee, S., & Hong, J. (2024). Rice Grains Integrated with Animal Cells: A Shortcut to a Sustainable Food System. *Matter, 7*(4), 1677. https://doi.org/10.1016/j.matt.2024.03.007

Pavone, V., & Goven, J. (2017). *Bioeconomies—Life, Technology, and Capital in the 21st Century.* Palgrave Macmillan. Retrieved from https://www.palgrave.com/gp/book/9783319556505

Pell, R., & Allen, L. (2015). Preface to the Genealogy of the Postnatural. In A.-S. Springer & E. Turpin (Eds.), *Intercalations 2: Land & Animal & Non-Animal* (pp. 29–55). K. Verlag/Haus der Kulturen der Welt.

Pellizzoni, L. (2022). Anthropocene. In P. Rebughini & E. Colombo (Eds.), *Framing Social Theory* (pp. 40–54). Routledge.

Pellizzoni, L., & Ylönen, M. (2012). *Neoliberalism and Technoscience: Critical Assessment*. Ashgate.

Perle, D. (2022, November 16). *PETA Statement re Culture Meat's FDA Milestone*. Retrieved from https://www.peta.org/media/news-releases/peta-statement-re-cultured-meats-fda-milestone/

PETA, P. for the E. T. of A. (2008, October 6). *PETA's "In Vitro" Chicken Contest*. PETA. Retrieved from https://www.peta.org/features/vitro-meat-contest/

Pfotenhauer, S., Laurent, B., Papgeorgiou, K., & Stilgoe, J. (2022). The Politics of Scaling. *Social Studies of Science, 52*(1), 3–34. https://doi.org/10.1177/03063127211048945

Philpott, T. (2021, February 8). Is Lab Meat About to Hit Your Dinner Plate? *Mother Jones*. Retrieved from https://www.motherjones.com/food/2021/08/is-lab-meat-about-to-hit-your-dinner-plate/

Picot, P., & Guillaume, B. (2024). The Controllability of the Technosphere, an Impossible Question. *The Anthropocene Review, 11*(1), 91–109. https://doi.org/10.1177/20530196231187242

Pilgrim, K. (2013). 'Happy Cows,' 'Happy Beef': A Critique of the Rationales for Ethical Meat. *Environmental Humanities, 3*(1), 111–127. https://doi.org/10.1215/22011919-3611257

Poinski, M. (2022). How can Plant-Based Meat Get Back to Growth? *Food Dive*. Retrieved from https://www.fooddive.com/news/plant-based-meat-back-to-growth/636834/

Poinski, M. (2024, June 8). *Why Mission Barns Believes Cultivated Fat is the Key to Better Plant-Based Meat*. Food Dive. Retrieved from https://www.fooddive.com/news/mission-barns-cultivated-fat-plant-based-meat/624960/

Polanyi, K. (1944). *The Great Transformation: The Political and Economic Origins of our Time*. Beacon Press.

Pollan, M. (2016). *The Omnivore's Dilemma: A Natural History of Four Meals* (Edition with a new afterword). Penguin Books.

Pomiès, A., Arsel, Z., & Bean, J. (2021). Taste. *Consumption Markets & Culture, 24*(1), 1–5. https://doi.org/10.1080/10253866.2020.1803846

Porcher, J. (2010). La production de viandes in vitro, stade ultime? *Revue Politique et Parlementaire, 1057*, 97–104.

Porcher, J. (2016). Demain, une agriculture sans élevage? *Pour, 231*(3), 255. https://doi.org/10.3917/pour.231.0255

Porcher, J. (2017). *The Ethics of Animal Labor*. Springer International Publishing. https://doi.org/10.1007/978-3-319-49070-0

Porcher, J. (2020). Incarnation ou incubation?: Les animal studies et la "clean meat". *Zilsel, N° 7*(2), 292–306. https://doi.org/10.3917/zil.007.0292

Post, M. J. (2012). Cultured Meat from Stem Cells: Challenges and Prospects. *Meat Science, 92*(3), 297–301.

Post, M. J., & Hocquette, J.-F. (2017). Chapter 16—New Sources of Animal Proteins: Cultured Meat. In P. P. Purslow (Ed.), *New Aspects of Meat Quality* (pp. 425–441). Woodhead Publishing. https://doi.org/10.1016/B978-0-08-100593-4.00017-5

Postone, M. (2020). Capital and Temporality. In *The Marx Revival: Key Concepts and New Critical Interpretations* (p. 157). Cambridge University Press.

Powell, R. L., Dodson, M. V., & Cloud, J. G. (1989). Cultivation and Differentiation of Satellite Cells from Skeletal Muscle of the Rainbow Trout Salmo gairdneri. *Journal of Experimental Zoology, 250*(3), 333–338. https://doi.org/10.1002/jez.1402500314

Purdy, C. (2020). *Billion Dollar Burger: Inside Big Tech's Race for the Future of Food*. Portfolio/Penguin.

Rabinow, P. (1996). *Essays on the Anthropology of Reason*. Princeton University Press.

Rajan, K. S. (2006). *Biocapital: The Constitution of Postgenomic Life*. Duke University Press.

Ramirez, V. B. (2022, June 1). The World's Biggest Cultured Meat Factory Will Soon Be Built in the US. *Singularity Hub*. Retrieved from https://singularityhub.com/2022/06/01/the-worlds-biggest-cultured-meat-factory-will-soon-be-built-in-the-us/

Rao, D. (2024, March 25). *Lab-grown Meat Might be About to Meet its Maker*. Theweek. Retrieved from https://theweek.com/science/lab-grown-meat-controversy

Rasko, J., & Power, C. (2015, February 18). What Pushes Scientists to Lie? The Disturbing but Familiar Story of Haruko Obokata. *The Guardian*. Retrieved fromhttps://www.theguardian.com/science/2015/feb/18/haruko-obokata-stap-cells-controversy-scientists-lie

Ravikumar, M., & Powell, D. (2023). *Cell Line Development and Utilisation Trends in the Cultivated Meat Industry*.

Reinert, H. (2007). The Pertinence of Sacrifice—Some Notes on Larry the Luckiest Lamb. *Borderlandes E-Journal, 6*(3). Retrieved from http://www.borderlands.net.au/vol6no3_2007/reinert_larry.htm

Reisman, E., & Fairbairn, M. (2021). Agri-Food Systems and the Anthropocene. *Annals of the American Association of Geographers, 111*(3), 687–697. https://doi.org/10.1080/24694452.2020.1828025

Rezapour Sarabi, M., Yetisen, A. K., & Tasoglu, S. (2023). Bioprinting in Microgravity. *ACS Biomaterials Science & Engineering, 9*(6), 3074–3083. https://doi.org/10.1021/acsbiomaterials.3c00195

Rheinberger, H.-J. (1993). Experiment and Orientation: Early Systems of In Vitro Protein Synthesis. *Journal of the History of Biology, 26*(3), 443–471. https://doi.org/10.1007/BF01062057

Rheinberger, H.-J. (2015). Difference Machines: Time in Experimental Systems. *Configurations, 23*(2), 165–176. https://doi.org/10.1353/con.2015.0013

Ricœur, P. (1984). L'idéologie et l'utopie: Deux expressions de l'imaginaire social. *Autres Temps, 2*(1), 53–64. https://doi.org/10.3406/chris.1984.940

Risner, D., Kim, Y., Nguyen, C., Siegel, J. B., & Spang, E. S. (2023). *Environmental Impacts of Cultured Meat: A Cradle-to-gate Life Cycle Assessment* [Preprint]. Ecology. https://doi.org/10.1101/2023.04.21.537778

Rodríguez Fernández, C. (2019, February 19). *From Saving Endangered Rhinos to Enabling Lab-Grown Meat.* Labiotech.Eu. https://www.labiotech.eu/food/lab-grown-meat-biobank/

Rodríguez Escobar, M. I., Cadena, E., Nhu, T. T., Cooreman-Algoed, M., De Smet, S., & Dewulf, J. (2021). Analysis of the Cultured Meat Production System in Function of Its Environmental Footprint: Current Status, Gaps and Recommendations. *Foods, 10*(12), 12. https://doi.org/10.3390/foods10122941

Roe, E. J. (2006). Material Connectivity, the Immaterial and the Aesthetic of Eating Practices: An Argument for How Genetically Modified Foodstuff Becomes Inedible. *Environment and Planning A: Economy and Space, 38*(3), 465–481. https://doi.org/10.1068/a3835

Rose, N. S. (2007). *Politics of Life Itself: Biomedicine, Power, and Subjectivity in the Twenty-First Century.* Princeton University Press.

Rothe, D. (2020). Governing the End Times? Planet Politics and the Secular Eschatology of the Anthropocene. *Millennium, 48*(2), 143–164. https://doi.org/10.1177/0305829819889138

Russell, S. A. (2006). *Hunger: An Unnatural History* (Pbk. ed.). Basic Books.

Sargent, L. T. (2017). Ideology and Utopia: Karl Mannheim and Paul Ricoeur. In *Utopian Horizons: Ideology, Politics, Literature in Z. Czigànyik (Ed.)* (pp. 19–40). Central European University Press.

Schonwald, J. (2013, August 9). *Biting Into the First In Vitro Burger* [Interview]. Retrieved from https://www.npr.org/2013/08/09/210508782/biting-into-the-first-in-vitro-burger

Schwab, K. (2016). *The Fourth Industrial Revolution*. World Economic Forum.

Seamons, K. (2017, September 30). *A Hungry WWII POW Had Ideas. Some 60 Years Later, Silicon Valley Wanted Them*. Newser. Retrieved from http://www.newser.com/story/249128/hungry-wwii-pow-had-ideas-silicon-valley-now-wants.html

Selin, C. (2008). The Sociology of the Future: Tracing Stories of Technology and Time. *Sociology Compass, 2*(6), 1878–1895. https://doi.org/10.1111/j.1751-9020.2008.00147.x

Sentient Media. (2019, December 24). *Good Food Institute: Interview with Bruce Friedrich*. Retrieved from https://sentientmedia.org/good-food-institute-interview-with-bruce-friedrich/

Serreau, R., Terbeche, Y., & Rigourd, V. (2024). Pollutants in Breast Milk: A Scoping Review of the Most Recent Data in 2024. *Healthcare, 12*(6), 6. https://doi.org/10.3390/healthcare12060680

Sexton, A. E. (2018). Eating for the Post-Anthropocene: Alternative Proteins and the Biopolitics of Edibility. *Transactions of the Institute of British Geographers, 43*(4), 586–600. https://doi.org/10.1111/tran.12253

Sharma, S. (2018). *Mighty Giants: Leaders of the Global Meat Complex*. Retrieved from https://www.iatp.org/blog/leaders-global-meat-complex

Shaw, R. (2008). The Notion of the Gift in the Donation of Body Tissues. *Sociological Research Online, 13*(6), 41–50. https://doi.org/10.5153/sro.1832

Shay, J. W., & Wright, W. E. (2000). Hayflick, his Limit, and Cellular Ageing. *Nature Reviews Molecular Cell Biology, 1*(1), 72–76. https://doi.org/10.1038/35036093

Shilton, A. C. (2019, March 18). *Protein of the Future*. The Hub. Retrieved from https://hub.jhu.edu/magazine/2019/spring/lab-grown-alternative-meat-protein-ethics/

Shiva, V. (2019, January 20). *A New Report Sustains Unsustainable Food Systems*. Seed Freedom. Retrieved from https://seedfreedom.info/poison-cartel-toxic-food-eat-report/

Shruti, V. C., Kutralam-Muniasamy, G., & Pérez-Guevara, F. (2023). New Forms of Particulate Plastics in the Anthropocene. *Earth-Science Reviews, 246*, 104601. https://doi.org/10.1016/j.earscirev.2023.104601

Shukin, N. (2009). *Animal Capital: Rendering Life in Biopolitical Times*. University of Minnesota Press.

Siegner, C. (2019). *Plant-Based Meat Market Forecast to Reach $85B by 2030, Report Says*. Food Dive. Retrieved from https://www.fooddive.com/news/plant-based-meat-market-forecast-to-reach-85b-by-2030-report-says/559170/

Simondon, G. (2016). *On the Mode of Existence of Technical Objects.* Univocal Pub..

Sinke, P., Swartz, E., Sanctorum, H., van der Giesen, C., & Odegard, I. (2023). Ex-ante Life Cycle Assessment of Commercial-Scale Cultivated Meat Production in 2030. *The International Journal of Life Cycle Assessment, 28*(3), 234–254. https://doi.org/10.1007/s11367-022-02128-8

Sklar, J. (2019, September 9). *With Lab-Grown Chicken Nuggets, Dumplings, and Burgers, Futuristic Foods Aim to Seem Familiar.* Upworthyscience.Com. Retrieved from https://upworthyscience.com/with-lab-grown-chicken-nuggets-dumplings-and-burgers-futuristic-foods-aim-to-seem-familiar/

Smaje, C. (2020). *A Small Farm Future: Making the Case for a Society Built Around Local Economies, Self-provisioning, Agricultural Diversity, and a Shared Earth.* Chelsea Green Publishing.

Smaje, C. (2023). *Saying No to a Farm-free Future: The Case for an Ecological Food System and Against Manufactured Foods.* Chelsea Green Publishing.

Soice, E., & Johnston, J. (2021). Immortalizing Cells for Human Consumption. *International Journal of Molecular Sciences, 22*(21), 11660. https://doi.org/10.3390/ijms222111660

Southey, F. (2024, March 6). *From Meat to 3D Printed Plants: Why Cargill is Diversifying its Protein Investments.* Foodnavigator.Com. Retrieved from https://www.foodnavigator.com/Article/2024/03/06/Why-Cargill-is-diversifying-its-protein-investments

Specht, L. (2018). Meat by the Molecule: Making meat with plants and cells. *The Biochemist, 40*(4), 18–21.

Specht, L. (2020). *An analysis of culture medium costs and production volumes for cultivated meat.* Good Food Institute. https://gfi.org/wp-content/uploads/2021/01/clean-meat-production-volume-and-medium-cost.pdf.

Specht, E. A., Welch, D. R., Rees Clayton, E. M., & Lagally, C. D. (2018). Opportunities for Applying Biomedical Production and Manufacturing Methods to the Development of the Clean Meat Industry. *Biochemical Engineering Journal, 132,* 161–168. https://doi.org/10.1016/j.bej.2018.01.015

Speciale, A. (2023, March 28). Italy Wants to Ban Cultivated Food to Protect Culinary Heritage. *Bloomberg.Com.* Retrieved from https://www.bloomberg.com/news/articles/2023-03-28/italy-wants-to-ban-cultivated-food-to-protect-culinary-heritage

Specter, M. (2011, May 16). *Test-Tube Burgers.* Retrieved from https://www.newyorker.com/magazine/2011/05/23/test-tube-burgers

Spelman, E. V. (2002). *Repair: The Impulse to Restore in a Fragile World.*

Stafford, B. M. (2007). From Genetic Perspective to Biohistory: The Ambiguities of Looking Down, Across, and Beyond. In E. Kac (Ed.), *Signs of Life: Bio Art and Beyond* (pp. 373–386). The MIT Press.

Stahl, G. (2017). Making a Mockery of Meat: Troubling Texture and the Failings of the "Flesh". *Journal of Asia-Pacific Pop Culture.* https://doi.org/10.5325/jasiapacipopcult.2.2.0184

Stanescu, V. (2011). "Green" Eggs and Ham? The Myth of Sustainable Meat and the Danger of the Local. In J. Sanbonmatsu (Ed.), *Critical Theory and Animal Liberation* (pp. 239–255). Rowman & Littlefield Publishers Inc..

Steffen, W., Broadgate, W., Deutsch, L., Gaffney, O., & Ludwig, C. (2015). The Trajectory of the Anthropocene: The Great Acceleration. *The Anthropocene Review, 2*(1), 81–98. https://doi.org/10.1177/2053019614564785

Steinfeld, H. (2006). *Livestock's Long Shadow: Environmental Issues and Options.* Food and Agriculture Organization of the United Nations.

Stephens, N., & Ruivenkamp, M. (2016). Promise and Ontological Ambiguity in the *In vitro* Meat Imagescape: From Laboratory Myotubes to the Cultured Burger. *Science as Culture, 25*(3), 327–355. https://doi.org/10.1080/09505431.2016.1171836

Stephens, N., King, E., & Lyall, C. (2018). Blood, Meat, and Upscaling Tissue Engineering: Promises, Anticipated Markets, and Performativity in the Biomedical and Agri-food Sectors. *BioSocieties, 13*(2), 368–388. https://doi.org/10.1057/s41292-017-0072-1

Stephens, N., Sexton, A. E., & Driessen, C. (2019). Making Sense of Making Meat: Key Moments in the First 20 Years of Tissue Engineering Muscle to Make Food. *Frontiers in Sustainable Food Systems, 3*. https://doi.org/10.3389/fsufs.2019.00045

Stiegler, B. (2019). *The Age of Disruption: Technology and Madness in Computational Capitalism* (English ed.). Polity Press.

Stiffman, E. (2023, October 16). *Does Alternative Meat Need Philanthropy to Take It Beyond?* The Chronicle of Philanthropy. Retrieved from https://www.philanthropy.com/article/does-alternative-meat-need-philanthropy-to-take-it-beyond

Stout, A. J., Zhang, X., Letcher, S. M., Rittenberg, M. L., Shub, M., Chai, K. M., Kaul, M., & Kaplan, D. L. (2024). Engineered Autocrine Signaling Eliminates Muscle Cell FGF2 Requirements for Cultured Meat Production. *Cell Reports Sustainability, 1*(1), 100009. https://doi.org/10.1016/j.crsus.2023.100009

Su, L., Jing, L., Zeng, X., Chen, T., Liu, H., Kong, Y., Wang, X., Yang, X., Fu, C., Sun, J., & Huang, D. (2023). 3D-Printed Prolamin Scaffolds for Cell-Based Meat Culture. *Advanced Materials, 35*(2), 2207397. https://doi.org/10.1002/adma.202207397

Subramaniam, R. (2021). The Ethical Impurative: Elemental Frontiers of Technologized Meat. In S. Chatterjee & R. Subramaniam (Eds.), *Meat! A Transnational Analysis* (pp. 254–278). Duke University Press.

Sunder Rajan, K. (2012). *Lively Capital: Biotechnologies, Ethics, and Governance in Global Markets*. Duke University Press.

SuperMeat. (2024). *Manufacturing—SuperMeat*. Retrieved from https://supermeat.com/manufacturing/

Swyngedouw, E., & Ernstson, H. (2018). Interrupting the Anthropo-obScene: Immuno-biopolitics and Depoliticizing Ontologies in the Anthropocene. *Theory, Culture & Society, 35*(6), 3–30. https://doi.org/10.1177/0263276418757314

Szondy, D. (2013, August 6). *First Public Tasting of US$330,000 Lab-grown Burger*. New Atlas. Retrieved from https://newatlas.com/cultured-beef/28584/

Tamminen, S., & Vermeulen, N. (2012). Bio-objects and Generative Relations. *Croatian Medical Journal, 53*(2), 198–200. https://doi.org/10.3325/cmj.2012.53.198

Tayag, Y. (2024, March 8). Everything Can Be Meat. *The Atlantic*. Retrieved from https://www.theatlantic.com/health/archive/2024/03/lab-grown-meat-beef-rice/677689/

TC&A Project, T. T. C., & A. P. (2000). Semi-living Steak. *The Tissue Culture & Art Project*. Retrieved from https://tcaproject.net/portfolio/semi-living-steak/

Thacker, Eugene. (2010). *After Life*. The University of Chicago.

The White House. (2022, September 12). *Background Press Call on President Biden's Executive Order to Launch a National Biotechnology and Biomanufacturing Initiative*. The White House. Retrieved from https://www.whitehouse.gov/briefing-room/press-briefings/2022/09/12/background-press-call-on-president-bidens-executive-order-to-launch-a-national-biotechnology-and-biomanufacturing-initiative/

The World Staff. (2019, October 31). *Pork Prices Soar as Millions of Pigs Die due to African Swine Fever*. Public Radio International. Retrieved from https://www.pri.org/stories/2019-10-31/pork-prices-soar-millions-pigs-die-due-african-swine-fever

Tønnesen, M., Armstong Oma, K., & Rattasepp, S. (Eds.). (2016). *Thinking about Animals in the Age of the Anthropocene*. Lexington Books.

Travis, J. (2014, March 3). *PETA Abandons $1 Million Prize for Artificial Chicken*. Retrieved from https://www.science.org/content/article/peta-abandons-1-million-prize-artificial-chicken

Tree, I. (2018, August 15). If you want to save the world, veganism isn't the answer. *The Guardian*. https://theguardian.com/commentisfree/2018/aug/25/veganism-intensively-farmed-meat-dairy-soya-maize

Tsing, A. L. (2017). *The Mushroom at the End of the World: On the Possibility of Life in Capitalist Ruins* (First paperback printing). Princeton University Press.

Tubb, C., & Seba, T. (2020). *RethinkX: Disruption, Implications and Choices. Rethinking Food and Agriculture 2020–2050*. RethinkX.

Tuomisto, H. L. (2022). Challenges of Assessing the Environmental Sustainability of Cellular Agriculture. *Nature Food, 3*(10), 10. https://doi.org/10.1038/s43016-022-00616-6

Tuomisto, H. L., & de Mattos, M. J. T. (2011). Environmental Impacts of Cultured Meat Production. *Environmental Science & Technology, 45*(14), 6117–6123. https://doi.org/10.1021/es200130u

Tutton, R. (2018). Multiplanetary Imaginaries and Utopia: The Case of Mars One. *Science, Technology, & Human Values, 43*(3), 518–539. https://doi.org/10.1177/0162243917737366

Twine, R. (2007). Animal Genomics and Ambivalence: A Sociology of Animal Bodies in Agricultural Biotechnology. *Genomics, Society and Policy, 3*(2), 99. https://doi.org/10.1186/1746-5354-3-2-99

Ung-Lanki, S. (2014). Constructing the Biotech Nonhuman Animal: Instrumentalism and Ambivalence. *Society & Animals, 22*(5), 439–458. https://doi.org/10.1163/15685306-12341349

Van Der Weele, C., & Driessen, C. (2013). Emerging Profiles for Cultured Meat; Ethics through and as Design. *Animals, 3*(3), 647–662. https://doi.org/10.3390/ani3030647

Van Der Weele, C., & Tramper, J. (2014). Cultured Meat. Every Village its Own Factory? *Trends in Biotechnology, 32*(6), 294–296. https://doi.org/10.1016/j.tibtech.2014.04.009

van Mensvoort, K. (2015, March 3). *In Vitro Meat Godfather Dies at Age of 91*. Next Nature Network. Retrieved from https://nextnature.net/magazine/story/2015/in-vitro-meat-godfather-dies-at-age-of-91

Verbeke, W., Marcu, A., Rutsaert, P., Prada, M., Gaspar, R., Seibt, B., Fletcher, D., & Barnett, J. (2015). "Would You Eat Cultured Meat?": Consumers'

Reactions and Attitude Formation in Belgium, Portugal and the United Kingdom. *Meat Science*. https://doi.org/10.1016/j.meatsci.2014.11.013

Vermeulen, N., Tamminen, S., & Webster, A. (2017). *Bio-objects Life in the 21st Century*. Routledge, Taylor & Francis Group.

Vialles, N. (1988). La viande ou la bête. *Terrain. Anthropologie & Sciences Humaines, 10*, 86–96. https://doi.org/10.4000/terrain.2932

Vialles, N. (1995). De l'animal à la viande: Une mort sans cadavre. *French Cultural Studies, 6*(18), 335–350. https://doi.org/10.1177/095715589500601805

Vogelpohl, T., & Töller, A. E. (2021). Perspectives on the Bioeconomy as an Emerging Policy Field. *Journal of Environmental Policy & Planning, 23*(2), 143–151. https://doi.org/10.1080/1523908X.2021.1901394

Vogt, M. A. B. (2021). Agricultural Wilding: Rewilding for Agricultural Landscapes through an Increase in Wild Productive Systems. *Journal of Environmental Management, 284*, 112050. https://doi.org/10.1016/j.jenvman.2021.112050

Vow Food (Director). (2023). *Forged by Vow—Introducing the Mammoth Meatball* [Video recording]. Retrieved from https://www.forgedbyvow.com/the-feed/the-mammoth-meatball/

Waldby, C. (2002a). Biomedicine, Tissue Transfer and Intercorporeality. *Feminist Theory, 3*(3), 239–254. https://doi.org/10.1177/146470002762491980

Waldby, C. (2002b). Stem Cells, Tissue Cultures and the Production of Biovalue. *Health: An Interdisciplinary Journal for the Social Study of Health, Illness and Medicine, 6*(3), 305–323. https://doi.org/10.1177/136345930200600304

Waldby, C., & Cooper, M. (2014). *Clinical Labor: Tissue Donors and Research Subjects in the Global Bioeconomy*. Duke University Press.

Waldby, C., & Mitchell, R. (2006). *Tissue Economies: Blood, Organs, and Cell Lines in Late Capitalism*. Duke University Press.

Wall, T. (2024, March 19). *Cell-cultured Meat for Pet Food Nearing Market Appearance*. PetfoodIndustry. Retrieved from https://www.petfoodindustry.com/blogs-columns/news-hound/article/15666679/cellcultured-meat-for-pet-food-nearing-market-appearance

Wallace, R. G. (2016). *Big Farms make Big Flu: Dispatches on Infectious Disease, Agribusiness, and the Nature of Science*. Monthly Review Press.

Wallace, R., Liebman, A., Chaves, L. F., & Wallace, R. (2020, April 1). Monthly Review | COVID-19 and Circuits of Capital. *Monthly Review*. Retrieved from https://monthlyreview.org/2020/04/01/covid-19-and-circuits-of-capital/

Walsham, A. (2010). Introduction: Relics and Remains. *Past & Present, 206*(Suppl_5), 9–36. https://doi.org/10.1093/pastj/gtq026

Watson. (2022, November 16). *'A Watershed Moment in the History of Food…' Cultivated Meat Edges Closer to Commercialization in the US as UPSIDE Foods gets FDA Green Light*. Foodnavigator-Usa.Com. Retrieved from https://www.foodnavigator-usa.com/Article/2022/11/17/A-watershed-moment-in-the-history-of-food-Cultivated-meat-edges-closer-to-commercialization-in-the-US-as-UPSIDE-Foods-gets-FDA-green-light

Watson, E. (2023, November 20). *Crunch Time for Cultivated Meat: 'Probably 70–90% of Players will Fail in the Next Year'*. AgFunderNews. Retrieved from https://agfundernews.com/crunch-time-for-cultivated-meat-probably-70-90-of-players-will-fail-in-the-next-year

Watts, J. (2023, May 4). Grain Trader Cargill Faces Legal Challenge in US over Brazilian Soya Supply Chain. *The Guardian*. Retrieved from https://www.theguardian.com/environment/2023/may/04/cargill-faces-legal-challenge-us-over-brazilian-soya-supply-chain-deforestation

WEF, W. E. F. (2018). *Meat the Future: Time for a Protein Portfolio to Meet Tomorrow's Demand*. World Economic Forum. Retrieved from https://www.weforum.org/publications/meat-the-future-time-for-a-protein-portfolio-to-meet-tomorrow-s-demand/

WEF, W. E. F. (2024). *Fourth Industrial Revolution*. World Economic Forum. Retrieved from https://www.weforum.org/focus/fourth-industrial-revolution/

Weis, T. (2018). Ghosts and Things: Agriculture and Animal Life. *Global Environmental Politics, 18*(2), 134–142. https://doi.org/10.1162/glep_a_00457

Weisberg, Z. (2015). Biotechnology as End Game: Ontological and Ethical Collapse in the "Biotech Century". *NanoEthics, 9*(1), 39–54. https://doi.org/10.1007/s11569-014-0219-5

Welin, S., & van der Weele, C. (2012). Cultured Meat; Will it Separate us from Nature? In *Climate Change and Sustainable Development. Ethical Perspectives on Land Use and Food Production* (pp. 348–351). Wageningen Academic Publishers. Retrieved from http://library.wur.nl/WebQuery/wurpubs/479212

Whittall, E. (2017). Taste, Performed: M. Clavel and the Live Feed. *Performance Research, 22*(7), 44–47. https://doi.org/10.1080/13528165.2017.1353191

Whyte, K. P. (2018). Indigenous Science (Fiction) for the Anthropocene: Ancestral Dystopias and Fantasies of Climate Change Crises. *Environment and Planning E: Nature and Space, 1*(1–2), 224–242. https://doi.org/10.1177/2514848618777621

Willett, W., Rockström, J., Loken, B., Springmann, M., Lang, T., Vermeulen, S., Garnett, T., Tilman, D., DeClerck, F., Wood, A., Jonell, M., Clark, M., Gordon, L. J., Fanzo, J., Hawkes, C., Zurayk, R., Rivera, J. A., De Vries, W., Majele Sibanda, L., et al. (2019). Food in the Anthropocene: The EAT–Lancet Commission on Healthy Diets from Sustainable Food Systems. *The Lancet, 393*(10170), 447–492. https://doi.org/10.1016/S0140-6736(18)31788-4

Wilschut, K. J., Jaksani, S., Van Den Dolder, J., Haagsman, H. P., & Roelen, B. A. J. (2008). Isolation and Characterization of Porcine Adult Muscle-Derived Progenitor Cells. *Journal of Cellular Biochemistry, 105*(5), 1228–1239. https://doi.org/10.1002/jcb.21921

Winer, S. (2021, February 10). Higher Steaks as Israeli Company Makes First Lab-Cultivated Rib-Eye Cut. *The Times of Israel.* Retrieved from https://www.timesofisrael.com/higher-steaks-as-israeli-company-makes-first-lab-cultivated-rib-eye-cut/

Wolpa, A. (2016). Seeing Meat without Animals: Attitudes for the Future. In B. Donaldson & C. Carter (Eds.), *The Future of Meat without Animals* (pp. 87–96). Roman and Littlefield International.

Wood, P., Thorrez, L., Hocquette, J.-F., Troy, D., & Gagaoua, M. (2023). "Cellular Agriculture": Current Gaps Between Facts and Claims Regarding "Cell-Based Meat". *Animal Frontiers, 13*(2), 68–74. https://doi.org/10.1093/af/vfac092

World Economic Forum. (2018). *Meat: The Future. Time for a Protein Portfolio to Meet Tomorrow's Demand* [White Paper]. Retrieved from https://www3.weforum.org/docs/White_Paper_Meat_the_Future_Time_Protein_Portfolio_Meet_Tomorrow_Demand_report_2018.pdf

World Economic Forum. (2019). *Meat: The Future Series. Alternative Proteins* [White Paper]. World Economic Forum. Retrieved from https://www3.weforum.org/docs/WEF_White_Paper_Alternative_Proteins.pdf

World's First Lab-Grown Burger is Eaten in London. (2013, August 5). *BBC News.* Retrieved from https://www.bbc.com/news/science-environment-23576143

Worster, D. (1990). Transformations of the Earth: Toward an Agroecological Perspective in History. *Journal of American History, 76*(4), 1087–1106. https://doi.org/10.2307/2936586

Wurgaft, B. A. (2019). *Meat Planet: Artificial Flesh and the Future of Food.* University of California Press.

Wynne-Jones, S. (2022). Rewilding: An Emotional Nature. *Area, n/a*(n/a). https://doi.org/10.1111/area.12810

Yablonka-Reuveni, Z., Quinn, L. S., & Nameroff, M. (1987). Isolation and Clonal Analysis of Satellite Cells from Chicken Pectoralis Muscle. *Developmental Biology, 119*(1), 252–259. https://doi.org/10.1016/0012-1606(87)90226-0

Yaman, R. (2023). *Is Cultivated Meat For Real?—Asterisk*. Asterisk Mag. Retrieved from https://asteriskmag.com/issues/02/is-cultivated-meat-for-real

Yang, B., Wang, J., Tang, B., Liu, Y., Guo, C., Yang, P., Yu, T., Li, R., Zhao, J., Zhang, L., Dai, Y., & Li, N. (2011). Characterization of Bioactive Recombinant Human Lysozyme Expressed in Milk of Cloned Transgenic Cattle. *PLoS One, 6*(3), e17593. https://doi.org/10.1371/journal.pone.0017593

Yao, T., & Asayama, Y. (2017). Animal-Cell Culture Media: History, Characteristics, and Current Issues. *Reproductive Medicine and Biology, 16*(2), 99–117. https://doi.org/10.1002/rmb2.12024

Yeung, J. (2024, February 15). *Move over, Plant-Based Beef: Hello, Beef-infused Rice*. CNN. Retrieved from https://www.cnn.com/2024/02/15/asia/south-korea-beef-rice-study-intl-hnk-scn/index.html

Yusoff, K. (2018). *A Billion Black Anthropocenes or None*. University of Minnesota Press.

Zalasiewicz, J., Waters, C. N., Ellis, E. C., Head, M. J., Vidas, D., Steffen, W., Thomas, J. A., Horn, E., Summerhayes, C. P., Leinfelder, R., McNeill, J. R., Gałuszka, A., Williams, M., Barnosky, A. D., de B. Richter, D., Gibbard, P. L., Syvitski, J., Jeandel, C., Cearreta, A., et al. (2021). The Anthropocene: Comparing Its Meaning in Geology (Chronostratigraphy) with Conceptual Approaches Arising in Other Disciplines. *Earth's Future, 9*(3), e2020EF001896. https://doi.org/10.1029/2020EF001896

Zaleski. (2024, March 14). *Meet the DC Activist Behind the Alt-Meat Revolution*. Retrieved from https://www.washingtonian.com/2024/03/14/meet-the-dc-activist-behind-the-alt-meat-revolution/

Zassenhaus, M. (2021, March 27). *How the New Harvest Community Coined the Term Cellular Agriculture*. New Harvest. Retrieved from https://new-harvest.org/community-coined-cellular-agriculture/

Zioga, E., Tøstesen, M., Kjærulf Madsen, S., Shetty, R., & Bang-Berthelsen, C. H. (2022). Bringing Plant-Based Cli-meat Closer to Original Meat Experience: Insights in Flavor. *Future Foods, 5*, 100138. https://doi.org/10.1016/j.fufo.2022.100138

Zurr, I. (2008). *Growing Semi-Living Art* [PhD, University of Western Australia]. Retrieved from https://api.research-repository.uwa.edu.au/ws/portalfiles/portal/3225984/Zurr_Ionat_2008.pdfb

Zurr, I., & Catts, O. (2023). *Tissues, Cultures, Art.* Palgrave Macmillan.

Zylinska, J. (2009). *Bioethics in the Age of New Media.* MIT Press.

Index[1]

[1] Note: Page numbers followed by 'n' refer to notes.

E. Abergel, *Dead Meat*, https://doi.org/10.1007/978-981-97-9049-4

SPRINGER NATURE

GPSR Compliance

The European Union's (EU) General Product Safety Regulation (GPSR) is a set of rules that requires consumer products to be safe and our obligations to ensure this.

If you have any concerns about our products, you can contact us on ProductSafety@springernature.com

In case Publisher is established outside the EU, the EU authorized representative is:

Springer Nature Customer Service Center GmbH
Europaplatz 3
69115 Heidelberg, Germany

The manufacturer's authorised representative in the EU is Springer Nature Customer Service Centre GmbH, Europaplatz 3, 69115 Heidelberg, Germany. If you have any concerns regarding our products, please contact ProductSafety@springernature.com

Printed and bound by CPI Group (UK) Ltd, Croydon, CR0 4YY

01/10/2026

02225924-0010